Examples in Markov Decision Processes

Imperial College Press Optimization Series

ISSN 2041-1677

Series Editor: Jean Bernard Lasserre *(LAAS-CNRS and Institute of Mathematics, University of Toulouse, France)*

Vol. 1: Moments, Positive Polynomials and Their Applications
by Jean Bernard Lasserre

Vol. 2: Examples in Markov Decision Processes
by A. B. Piunovskiy

Imperial College Press Optimization Series — Vol. 2

Examples in Markov Decision Processes

A. B. Piunovskiy
The University of Liverpool, UK

Imperial College Press

Published by

Imperial College Press
57 Shelton Street
Covent Garden
London WC2H 9HE

Distributed by

World Scientific Publishing Co. Pte. Ltd.
5 Toh Tuck Link, Singapore 596224
USA office: 27 Warren Street, Suite 401-402, Hackensack, NJ 07601
UK office: 57 Shelton Street, Covent Garden, London WC2H 9HE

British Library Cataloguing-in-Publication Data
A catalogue record for this book is available from the British Library.

Imperial College Press Optimization Series — Vol. 2
EXAMPLES IN MARKOV DECISION PROCESSES
Copyright © 2013 by Imperial College Press

All rights reserved. This book, or parts thereof, may not be reproduced in any form or by any means, electronic or mechanical, including photocopying, recording or any information storage and retrieval system now known or to be invented, without written permission from the Publisher.

For photocopying of material in this volume, please pay a copying fee through the Copyright Clearance Center, Inc., 222 Rosewood Drive, Danvers, MA 01923, USA. In this case permission to photocopy is not required from the publisher.

ISBN 978-1-84816-793-3

Printed in Singapore by Mainland Press Pte Ltd.

Preface

Markov Decision Processes (MDP) is a branch of mathematics based on probability theory, optimal control, and mathematical analysis. Several books with counterexamples/paradoxes in probability [Stoyanov(1997); Szekely(1986)] and in analysis [Gelbaum and Olmsted(1964)] are in existence; it is therefore not surprising that MDP is also replete with unexpected counter-intuitive examples. The main goal of the current book is to collect together such examples. Most of them are based on earlier publications; the remainder are new. This book should be considered as a complement to scientific monographs on MDP [Altman(1999); Bertsekas and Shreve(1978); Hernandez-Lerma and Lasserre(1996a); Hernandez-Lerma and Lasserre(1999); Piunovskiy(1997); Puterman(1994)]. It can also serve as a reference book to which one can turn for answers to curiosities that arise while studying or teaching MDP. All the examples are self-contained and can be read independently of each other. Concerning uncontrolled Markov chains, we mention the illuminating collection of examples in [Suhov and Kelbert(2008)].

A survey of meaningful applications is beyond the scope of the current book. The examples presented either lead to counter-intuitive solutions, or illustrate the importance of conditions in the known theorems. Not all examples are equally simple or complicated. Several examples are aimed at undergraduate students, whilst others will be of interest to professional researchers.

The book has four chapters in line with the four main different types of MDP: the finite-horizon case, infinite horizon with total or discounted loss, and average loss over an infinite time interval. Some basic theoretical statements and proofs of auxiliary assertions are included in the Appendix.

The following notations and conventions will often be used without explanation.

$\stackrel{\triangle}{=}$ means 'equals by definition';
$\mathbf{C}^\infty$ is the space of infinitely differentiable functions;
$\mathbf{C}(\mathbf{X})$ is the space of continuous bounded functions on a (topological) space $\mathbf{X}$;
$\mathbf{B}(\mathbf{X})$ is the space of bounded measurable functions on a (Borel) space $\mathbf{X}$; in discrete (finite or countable) spaces, the discrete topology is usually supposed to be fixed;
$\mathbf{P}(\mathbf{X})$ is the space of probability measures on the (metrizable) space $\mathbf{X}$, equipped with the weak topology;
If Γ is a subset of space $\mathbf{X}$ then Γ^c is the complement;
$\mathbb{N} = \{1, 2, \ldots\}$ is the set of natural numbers; $\mathbb{N}_0 = \mathbb{N} \cup \{0\}$;
$\mathbb{R}^N$ is the N-dimensional Euclidean space; $\mathbb{R} = \mathbb{R}^1$ is the straight line;
$\mathbb{R}^* = [-\infty, +\infty]$ is the extended straight line;
$\mathbb{R}^+ = \{y > 0\}$ is the set of strictly positive real numbers;
$I\{statement\} = \begin{cases} 1, & \text{if the statement is correct;} \\ 0, & \text{if the statement is false;} \end{cases}$ is the indicator function;
$\delta_a(dy)$ is the Dirac measure concentrated at point a: $\delta_a(\Gamma) = I\{\Gamma \ni a\}$;
If $r \in \mathbb{R}^*$ then $r^+ \stackrel{\triangle}{=} \max\{0, r\}$, $r^- \stackrel{\triangle}{=} \min\{0, r\}$;
$\sum_{i=n}^{m} f_i \stackrel{\triangle}{=} 0$ and $\prod_{i=n}^{m} f_i \stackrel{\triangle}{=} 1$ if $m < n$;
$\lfloor r \rfloor$ is the integer part, the maximal integer i such that $i \leq r$.

Throughout the current book $\mathbf{X}$ is the state space, $\mathbf{A}$ is the action space, $p_t(dy|x, a)$ is the transition probability, $c_t(x, a)$ and $C(x)$ are the loss functions.

Normally, we denote random variables with capital letters (X), small letters (x) being used just for variables, arguments of functions, etc. Bold case $(\mathbf{X})$ is for spaces. All functions, mappings, and stochastic kernels are assumed to be Borel-measurable unless their properties are explicitly specified.

We say that a function on $\mathbb{R}^1$ with the values in a Borel space $\mathbf{A}$ is *piece-wise continuous* if there exists a sequence y_i such that $\lim_{i \to \infty} y_i = \infty$; $\lim_{i \to -\infty} y_i = -\infty$, this function is continuous on each open interval

(y_i, y_{i+1}) and there exists a right (left) limit as $y \to y_i + 0$ ($y \to y_{i+1} - 0$), $i = 0, \pm 1, \pm 2 \ldots$. A similar definition is accepted for real-valued piece-wise Lipschitz, continuously differentiable functions.

If **X** is a measurable space and ν is a measure on it, then both formulae

$$\int_{\mathbf{X}} f(x) d\nu(x) \quad \text{and} \quad \int_{\mathbf{X}} f(x) \nu(dx)$$

denote the same integral of a real-valued function f with respect to ν.

w.r.t. is the abbreviation for 'with respect to', a.s. means 'almost surely', and CDF means 'cumulative distribution function'.

We consider only minimization problems. When formulating theorems and examples published in books (articles) devoted to maximization, we always adjust the statements for our case without any special remarks.

It should be emphasized that the terminology in MDP is not entirely fixed. For example, very often strategies are called policies. There exist several slightly different definitions of a semi-continuous model, and so on.

The author is thankful to Dr.R. Sheen and to Dr.M. Ruck for the proof reading of all the text.

A.B. Piunovskiy

Contents

Preface v

1. Finite-Horizon Models 1
   - 1.1 Preliminaries 1
   - 1.2 Model Description 3
   - 1.3 Dynamic Programming Approach 5
   - 1.4 Examples 8
     - 1.4.1 Non-transitivity of the correlation 8
     - 1.4.2 The more frequently used control is not better . . 9
     - 1.4.3 Voting 11
     - 1.4.4 The secretary problem 13
     - 1.4.5 Constrained optimization 14
     - 1.4.6 Equivalent Markov selectors in non-atomic MDPs 17
     - 1.4.7 Strongly equivalent Markov selectors in non-atomic MDPs 20
     - 1.4.8 Stock exchange 25
     - 1.4.9 Markov or non-Markov strategy? Randomized or not? When is the Bellman principle violated? . . 27
     - 1.4.10 Uniformly optimal, but not optimal strategy . . . 31
     - 1.4.11 Martingales and the Bellman principle 32
     - 1.4.12 Conventions on expectation and infinities 34
     - 1.4.13 Nowhere-differentiable function $v_t(x)$; discontinuous function $v_t(x)$ 38
     - 1.4.14 The non-measurable Bellman function 43
     - 1.4.15 No one strategy is uniformly ε-optimal 44
     - 1.4.16 Semi-continuous model 46

2. **Homogeneous Infinite-Horizon Models: Expected Total Loss**   51

    2.1   Homogeneous Non-discounted Model   51
    2.2   Examples   54
           2.2.1   Mixed Strategies   54
           2.2.2   Multiple solutions to the optimality equation . . .   56
           2.2.3   Finite model: multiple solutions to the optimality equation; conserving but not equalizing strategy .   58
           2.2.4   The single conserving strategy is not equalizing and not optimal   58
           2.2.5   When strategy iteration is not successful   61
           2.2.6   When value iteration is not successful   63
           2.2.7   When value iteration is not successful: positive model I   67
           2.2.8   When value iteration is not successful: positive model II   69
           2.2.9   Value iteration and stability in optimal stopping problems   71
           2.2.10   A non-equalizing strategy is uniformly optimal . .   73
           2.2.11   A stationary uniformly ε-optimal selector does not exist (positive model)   75
           2.2.12   A stationary uniformly ε-optimal selector does not exist (negative model)   77
           2.2.13   Finite-action negative model where a stationary uniformly ε-optimal selector does not exist   80
           2.2.14   Nearly uniformly optimal selectors in negative models   83
           2.2.15   Semi-continuous models and the blackmailer's dilemma   85
           2.2.16   Not a semi-continuous model   88
           2.2.17   The Bellman function is non-measurable and no one strategy is uniformly ε-optimal   91
           2.2.18   A randomized strategy is better than any selector (finite action space)   92
           2.2.19   The fluid approximation does not work   94
           2.2.20   The fluid approximation: refined model   97
           2.2.21   Occupation measures: phantom solutions   101
           2.2.22   Occupation measures in transient models   104
           2.2.23   Occupation measures and duality   107

		2.2.24	Occupation measures: compactness	109
		2.2.25	The bold strategy in gambling is not optimal (house limit)	112
		2.2.26	The bold strategy in gambling is not optimal (inflation)	115
		2.2.27	Search strategy for a moving target	119
		2.2.28	The three-way duel ("Truel")	122
3.	Homogeneous Infinite-Horizon Models: Discounted Loss			127
	3.1	Preliminaries		127
	3.2	Examples		128
		3.2.1	Phantom solutions of the optimality equation . .	128
		3.2.2	When value iteration is not successful: positive model	130
		3.2.3	A non-optimal strategy $\hat{\pi}$ for which $v_x^{\hat{\pi}}$ solves the optimality equation	132
		3.2.4	The single conserving strategy is not equalizing and not optimal	134
		3.2.5	Value iteration and convergence of strategies . . .	135
		3.2.6	Value iteration in countable models	137
		3.2.7	The Bellman function is non-measurable and no one strategy is uniformly ε-optimal	140
		3.2.8	No one selector is uniformly ε-optimal	141
		3.2.9	Myopic strategies	141
		3.2.10	Stable and unstable controllers for linear systems	143
		3.2.11	Incorrect optimal actions in the model with partial information	146
		3.2.12	Occupation measures and stationary strategies . .	149
		3.2.13	Constrained optimization and the Bellman principle	152
		3.2.14	Constrained optimization and Lagrange multipliers	153
		3.2.15	Constrained optimization: multiple solutions . . .	157
		3.2.16	Weighted discounted loss and (N, ∞)-stationary selectors	158
		3.2.17	Non-constant discounting	160
		3.2.18	The nearly optimal strategy is not Blackwell optimal	163
		3.2.19	Blackwell optimal strategies and opportunity loss	164

    3.2.20 Blackwell optimal and n-discount optimal
           strategies 165
    3.2.21 No Blackwell (Maitra) optimal strategies 168
    3.2.22 Optimal strategies as $\beta \to 1-$ and MDPs with the
           average loss – I 171
    3.2.23 Optimal strategies as $\beta \to 1-$ and MDPs with the
           average loss – II 172

4. **Homogeneous Infinite-Horizon Models: Average Loss and Other Criteria**   **177**
    4.1 Preliminaries 177
    4.2 Examples 179
        4.2.1 Why lim sup? 179
        4.2.2 AC-optimal non-canonical strategies 181
        4.2.3 Canonical triplets and canonical equations 183
        4.2.4 Multiple solutions to the canonical equations in
               finite models 186
        4.2.5 No AC-optimal strategies 187
        4.2.6 Canonical equations have no solutions: the finite
               action space 188
        4.2.7 No AC-ε-optimal stationary strategies in a finite
               state model........................ 191
        4.2.8 No AC-optimal strategies in a finite-state semi-
               continuous model 192
        4.2.9 Semi-continuous models and the sufficiency of
               stationary selectors 194
        4.2.10 No AC-optimal stationary strategies in a unichain
               model with a finite action space 195
        4.2.11 No AC-ε-optimal stationary strategies in a finite
               action model 198
        4.2.12 No AC-ε-optimal Markov strategies 199
        4.2.13 Singular perturbation of an MDP 201
        4.2.14 Blackwell optimal strategies and AC-optimality . 203
        4.2.15 Strategy iteration in a unichain model 204
        4.2.16 Unichain strategy iteration in a finite
               communicating model 207
        4.2.17 Strategy iteration in semi-continuous models ... 208
        4.2.18 When value iteration is not successful 211
        4.2.19 The finite-horizon approximation does not work . 213

	4.2.20 The linear programming approach to finite models	215
	4.2.21 Linear programming for infinite models	219
	4.2.22 Linear programs and expected frequencies in finite models	223
	4.2.23 Constrained optimization	225
	4.2.24 AC-optimal, bias optimal, overtaking optimal and opportunity-cost optimal strategies: periodic model	229
	4.2.25 AC-optimal and average-overtaking optimal strategies	232
	4.2.26 Blackwell optimal, bias optimal, average-overtaking optimal and AC-optimal strategies . .	235
	4.2.27 Nearly optimal and average-overtaking optimal strategies	238
	4.2.28 Strong-overtaking/average optimal, overtaking optimal, AC-optimal strategies and minimal opportunity loss	239
	4.2.29 Strong-overtaking optimal and strong*-overtaking optimal strategies	242
	4.2.30 Parrondo's paradox	247
	4.2.31 An optimal service strategy in a queueing system	249

Afterword   253

Appendix A   Borel Spaces and Other Theoretical Issues   257

  A.1   Main Concepts   257
  A.2   Probability Measures on Borel Spaces   260
  A.3   Semi-continuous Functions and Measurable Selection . . .   263
  A.4   Abelian (Tauberian) Theorem   265

Appendix B   Proofs of Auxiliary Statements   267

Notation   281

List of the Main Statements   283

Bibliography   285

Index   291

Chapter 1

Finite-Horizon Models

1.1 Preliminaries

A decision maker is faced with the problem of influencing the behaviour of a probabilistic system as it evolves through time. Decisions are made at discrete points in time referred to as *decision epochs* and denoted as $t = 1, 2, \ldots, T < \infty$. At each time t, the system occupies a *state* $x \in \mathbf{X}$. The *state space* $\mathbf{X}$ can be either discrete (finite or countably infinite) or continuous (non-empty uncountable Borel subset of a complete, separable metric space, e.g. $\mathbb{R}^1$). If the state at time t is considered as a random variable, it is denoted by a capital letter X_t; small letters x_t are just for possible values of X_t. Therefore, the behaviour of the system is described by a *stochastic (controlled) process*

$$X_0, X_1, X_2, \ldots, X_T.$$

In case of uncontrolled systems, the theory of Markov processes is well developed: the initial probability distribution for X_0, $P_0(dx)$, is given, and the dynamics are defined by *transition probabilities* $p_t(dy|x)$. When $\mathbf{X}$ is finite and the process is time-homogeneous, those probabilities form a transition matrix with elements $p(j|i) = P(X_{t+1} = j | X_t = i)$.

In the case of controlled systems, we assume that the *action space* $\mathbf{A}$ is given, which again can be an arbitrary Borel space (including the case of finite or countable $\mathbf{A}$). As soon as the state X_{t-1} becomes known (equals x_{t-1}), the decision maker must choose an action/control $A_t \in \mathbf{A}$; in general this depends on all the realized values of $X_0, X_1, \ldots, X_{t-1}$ along with past actions $A_1, A_2, \ldots, A_{t-1}$. Moreover, that decision can be randomized. The rigorous definition of a control strategy is given in the next section.

As a result of choosing action a at decision epoch t in state x, the decision maker loses $c_t(x, a)$ units, and the system state at the next decision epoch is determined by the probability distribution $p_t(dy|x,a)$. The function $c_t(x,a)$ is called a *one-step loss*. The final/terminal loss equals $C(x)$ when the final state $X_T = x$ is realized.

We assume that the *initial distribution* $P_0(dx)$ for X_0 is given. Suppose a control strategy π is fixed (that is, the rule of choosing actions a_t; see the next section). Then the random sequence

$$X_0, A_1, X_1, A_2, X_2, \ldots, A_T, X_T$$

is well defined: there exists a single probability measure $P_{P_0}^\pi$ on the space of trajectories

$$(x_0, a_1, x_1, a_2, x_2, \ldots, a_T, x_T) \in \mathbf{X} \times (\mathbf{A} \times \mathbf{X})^T.$$

For example, if $\mathbf{X}$ is finite and the control strategy is defined by the map $a_t = \varphi_t(x_{t-1})$, then

$$P_{P_0}^\varphi \{X_0 = i, A_1 = a_1, X_1 = j, A_2 = a_2, X_2 = k, \ldots, X_{T-1} = l, A_T = a_T, X_T = m\}$$

$$= P_0(i) I\{a_1 = \varphi_1(i)\} p_1(j|i, a_1) I\{a_2 = \varphi_2(j)\} \ldots p_T(m|l, a_T).$$

Here and below, $I\{\cdot\}$ is the indicator function; if $\mathbf{X}$ is discrete then transition probabilities $p_t(\cdot|x,a)$ are defined by the values on singletons $p_t(y|x,a)$. The same is true for the initial distribution.

Therefore, for a fixed control strategy π, the *total expected loss* equals $v^\pi = E_{P_0}^\pi[W]$, where

$$W = \sum_{t=1}^{T} c_t(X_{t-1}, A_t) + C(X_T)$$

is the *total realized loss*. Here and below, $E_{P_0}^\pi$ is the mathematical expectation with respect to probability measure $P_{P_0}^\pi$.

The aim is to find an *optimal* control strategy π^* solving the problem

$$v^\pi = E_{P_0}^\pi \left[\sum_{t=1}^{T} c_t(X_{t-1}, A_t) + C(X_T) \right] \longrightarrow \inf_\pi. \quad (1.1)$$

Sometimes we call v^π the *performance functional*.

Using the dynamic programming approach, under some technical conditions, one can prove the following statement. Suppose function $v_t(x)$ on $\mathbf{X}$ satisfies the following equation

$$\begin{cases} v_T(x) = C(x); \\ v_{t-1}(x) = \inf_{a \in \mathbf{A}} \left\{ c_t(x,a) + \int_X v_t(y) p_t(dy|x,a) \right\} \\ \qquad = c_t(x, \varphi_t^*(x)) + \int_X v_t(y) p_t(dy|x, \varphi_t^*(x)); \quad t = T, T-1, \ldots, 1. \end{cases}$$
$$(1.2)$$

Then, the control strategy defined by the map $a_t = \varphi_t^*(x_{t-1})$ solves problem (1.1), i.e. it is optimal; $\inf_\pi v^\pi = \int_\mathbf{X} v_0(x) P_0(dx)$. Therefore, control strategies of the type presented are usually sufficient for solving standard problems. They are called *Markov selectors*.

1.2 Model Description

We now provide more rigorous definitions.

The *Markov Decision Process (MDP)* with a finite horizon is defined by the collection

$$\{\mathbf{X}, \mathbf{A}, T, p, c, C\},$$

where $\mathbf{X}$ and $\mathbf{A}$ are the state and action spaces (Borel); T is the *time horizon*; $p_t(dy|x,a)$, $t = 1, 2, \ldots, T$, are measurable stochastic kernels on $\mathbf{X}$ given $\mathbf{X} \times \mathbf{A}$; $c_t(x, a)$ are measurable functions on $\mathbf{X} \times \mathbf{A}$ with values on the extended straight-line $\mathbb{R}^* = [-\infty, +\infty]$; $C(x)$ is a measurable map $C : \mathbf{X} \to \mathbb{R}^*$. Necessary statements about Borel spaces are presented in Appendix A.

The *space of trajectories* (or *histories*) up to decision epoch t is

$$\mathbf{H}_{t-1} \overset{\triangle}{=} \mathbf{X} \times (\mathbf{A} \times \mathbf{X})^{t-1}, \quad t = 1, 2, \ldots, T: \quad \mathbf{H} \overset{\triangle}{=} \mathbf{X} \times (\mathbf{A} \times \mathbf{X})^T.$$

A *control strategy* $\pi = \{\pi_t\}_{t=1}^T$ is a sequence of measurable stochastic kernels

$$\pi_t(da|x_0, a_1, x_1, \ldots, a_{t-1}, x_{t-1}) = \pi_t(da|h_{t-1})$$

on $\mathbf{A}$, given $\mathbf{H}_{t-1}$. If a strategy π^m is defined by (measurable) stochastic kernels $\pi_t^m(da|x_{t-1})$ then it will be called a *Markov strategy*. It is called *semi-Markov* if it has the form $\pi_t(da|x_0, x_{t-1})$. A Markov strategy π^{ms} is called *stationary* if none of the kernels $\pi^{ms}(da|x_{t-1})$ depends on the time t. Very often, stationary strategies are denoted as π^s. If for any $t = 1, 2, \ldots, T$ there exists a measurable mapping $\varphi_t(h_{t-1}) : \mathbf{H}_{t-1} \to \mathbf{A}$ such that $\pi_t(\Gamma|h_{t-1}) = I\{\Gamma \ni \varphi_t(h_{t-1})\}$ for any $\Gamma \in \mathcal{B}(\mathbf{A})$, then the strategy is denoted by the symbol φ and is called a *selector* or *non-randomized* strategy. Selectors of the form $\varphi_t(x_{t-1})$ and $\varphi(x_{t-1})$ are called *Markov* and *stationary* respectively. Stationary semi-Markov strategies and semi-Markov (stationary) selectors are defined in the same way. In what follows, Δ^{All} is the collection of all strategies, Δ^{M} is the set of all Markov strategies, Δ^{MN} is the set of all Markov selectors. In this connection, letter N corresponds to non-randomized strategies. Further, Δ^S and Δ^{SN} are the sets of all stationary strategies and of all stationary selectors.

We assume that *initial* probability distribution $P_0(dx)$ is fixed. If a control strategy π is fixed too, then there exists a unique probability measure $P_{P_0}^\pi$ on $\mathbf{H}$ such that $P_{P_0}^\pi(\Gamma^X) = P_0(\Gamma^X)$ for $\Gamma \in \mathcal{B}(\mathbf{H}_0) = \mathcal{B}(\mathbf{X})$ and, for all $t = 1, 2, \ldots, T$, for $\Gamma^G \in \mathcal{B}(\mathbf{H}_{t-1} \times \mathbf{A})$, $\Gamma^X \in \mathcal{B}(\mathbf{X})$

$$P_{P_0}^\pi(\Gamma^G \times \Gamma^X) = \int_{\Gamma^G} p_t(\Gamma^X | x_{t-1}) P_{P_0}^\pi(dg_t)$$

and

$$P_{P_0}^\pi(\Gamma^H \times \Gamma^A) = \int_{\Gamma^H} \pi_t(\Gamma^A | h_{t-1}) P_{P_0}^\pi(dh_{t-1})$$

for $\Gamma^H \in \mathcal{B}(\mathbf{H}_{t-1})$, $\Gamma^A \in \mathcal{B}(\mathbf{A})$. Here, with some less-than-rigorous notation, we also denote $P_{P_0}^\pi(\cdot)$ the images of $P_{P_0}^\pi$ relative to projections of the types

$$\mathbf{H} \to \mathbf{H}_{t-1} \times \mathbf{A} \stackrel{\triangle}{=} \mathbf{G}_t, \quad t = 1, 2, \ldots, T, \text{ and } \mathbf{H} \to \mathbf{H}_t, \quad t = 0, 1, 2, \ldots, T. \tag{1.3}$$

$g_t = (x_0, a_1, x_1, \ldots, a_t)$ and $h_t = (x_0, a_1, x_1, \ldots, a_t, x_t)$ are the generic elements of $\mathbf{G}_t$ and $\mathbf{H}_t$. Where they are considered as random elements on $\mathbf{H}$, we use capital letters G_t and H_t, as usual.

Measures $P_{P_0}^\pi(\cdot)$ on $\mathbf{H}$ are called *strategic* measures; they form space $\mathcal{D}$.

One can introduce σ-algebras $\mathcal{G}_t$ and $\mathcal{F}_t$ in $\mathbf{H}$ as the pre-images of $\mathcal{B}(\mathbf{G}_t)$ and $\mathcal{B}(\mathbf{H}_t)$ with respect to (1.3). Now the trivial projections

$$(x_0, a_1, x_1, \ldots, a_T, x_T) \to x_t \text{ and } (x_0, a_1, x_1, \ldots, a_T, x_T) \to a_t$$

define $\mathcal{F}$-adapted and $\mathcal{G}$-adapted stochastic processes $\{X_t\}_{t=0}^T$ and $\{A_t\}_{t=1}^T$ on the *stochastic basis* $(\mathbf{H}, \mathcal{B}(\mathbf{H}), \{\mathcal{F}_0, \mathcal{G}_1, \mathcal{F}_1, \ldots, \mathcal{G}_T, \mathcal{F}_T\}, P_{P_0}^\pi)$, which is completed as usual. Note that the process A_t is $\mathcal{F}$-predictable, and that this property is natural. That is the main reason for considering sequences $(x_0, a_1, x_1, \ldots, a_T, x_T)$, not $(x_0, a_0, x_1, \ldots, a_{T-1}, x_T)$. The latter notation is also widely used by many authors.

For each $h \in \mathbf{H}$ the (realized) *total loss* equals

$$w(h) = \sum_{t=1}^T c_t(x_{t-1}, a_t) + C(x_T),$$

where we put "$+\infty$" + "$-\infty$" $\stackrel{\triangle}{=}$ "$+\infty$". The map $W: h \to w(h)$ defines the random total loss, and the performance of control strategy π is given by $v^\pi = E_{P_0}^\pi[W]$. Here and below,

$$E_{P_0}^\pi[W] \stackrel{\triangle}{=} E_{P_0}^\pi[W^+] + E_{P_0}^\pi[W^-]; \quad \text{"}+\infty\text{"} + \text{"}-\infty\text{"} \stackrel{\triangle}{=} \text{"}+\infty\text{"};$$

$$W^+ \triangleq \max\{0, W\}, \qquad W^- \triangleq \min\{0, W\}.$$

The aim is to solve problem (1.1), i.e. to construct an *optimal* control strategy.

Sometimes it is assumed that action a can take values only in subsets $\mathbf{A}(x)$ depending on the previous state $x \in \mathbf{X}$. In such cases, one can modify the loss function $c_t(\cdot)$, putting $c_t(x, a) = \infty$ if $a \notin \mathbf{A}(x)$. Another possibility is to put $p_t(dy|x, a) = p_t(dy|x, \hat{a})$ and $c_t(x, a) = c_t(x, \hat{a})$ for all $a \notin \mathbf{A}(x)$, where $\hat{a} \in \mathbf{A}(x)$ is a fixed point.

As mentioned in future chapters, all similar definitions and constructions hold also for infinite-horizon models with $T = \infty$.

1.3 Dynamic Programming Approach

Bellman formulated his famous *principle of optimality (the Bellman principle)* as follows: "An optimal policy has the property that whatever the initial state and initial decision are, the remaining decisions must constitute an optimal policy with regard to the state resulting from the first decision." [Bellman(1957), Section 3.3].

The Bellman principle leads to the following equation

$$\begin{cases} v_T(x) = C(x); \\ v_{t-1}(x) = \inf_{a \in \mathbf{A}} \left\{ c_t(x, a) + \int_{\mathbf{X}} v_t(y) p_t(dy|x, a) \right\}, \quad t = T, T-1, \ldots, 1. \end{cases} \tag{1.4}$$

Suppose that this *optimality* equation has a measurable solution $v_t(x)$ called the *Bellman function*, and loss functions $c_t(\cdot)$ and $C(\cdot)$ are simultaneously bounded below or above. Then, a control strategy π^* is optimal in problem (1.1) if and only if, for all $t = 1, 2, \ldots, T$, the following equality holds $P_{P_0}^{\pi^*}$ – a.s.

$$v_{t-1}(X_{t-1}) = \int_{\mathbf{A}} \left\{ c_t(X_{t-1}, a) + \int_{\mathbf{X}} v_t(y) p_t(dy|X_{t-1}, a) \right\} \pi_t^*(da|H_{t-1}) \tag{1.5}$$

(here $H_{t-1} = (X_0, A_1, X_1, \ldots, A_{t-1}, X_{t-1})$ is a random history).

$$v^{\pi^*} = \inf_\pi v^\pi = \int_{\mathbf{X}} v_0(x) P_0(dx). \tag{1.6}$$

The following simple example based on [Bäuerle and Rieder(2011), Ex. 2.3.10] confirms that it is not necessary for A_t to provide the infimum in (1.4) for all t and all $x \in \mathbf{X}$.

Let $T = 2$, $\mathbf{X} = \{0,1\}$, $\mathbf{A} = \{0,1\}$, $p_t(a|x,a) \equiv 1$, $c_t(x,a) \equiv -a$, $C(x) = 0$ (see Fig. 1.1).

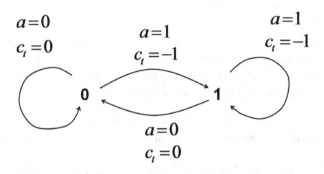

Fig. 1.1 Selector $\varphi_t(x) = I\{t = 1\} + I\{t > 1\}x$ is optimal but not uniformly optimal.

Equation (1.4) has a solution $v_2(x) = 0$; $v_1(x) = -1$; $v_0(x) = -2$ and all actions providing the minima equal 1. Thus, the selector $\varphi_t^1(x) \equiv 1$ is optimal for any initial distribution. But the selector

$$\varphi_t(x) = \begin{cases} 1, & \text{if } t = 1; \\ 1, & \text{if } t = 2 \text{ and } x = 1; \\ 0, & \text{if } t = 2 \text{ and } x = 0 \end{cases}$$

is also optimal because state 0 will not be visited at time 1 and we do not meet it at decision epoch 2. Incidentally, selector φ^1 is *uniformly* optimal whereas the above selector φ is not (see the definitions below).

Suppose a history $h_\tau \in \mathbf{H}_\tau$, $0 \leq \tau \leq T$ is fixed. Then we can consider the controlling process A_t and the controlled process X_t as developing on the time interval $\{\tau+1, \tau+2, \ldots, T\}$ which is empty if $\tau = T$. If a control strategy π (in the initial model) is fixed, then one can build the strategic measure on $\mathbf{H}$, denoted as $P_{h_\tau}^\pi$, in the same way as was described on p.4, satisfying the "initial condition" $P_{h_\tau}^\pi(h_\tau \times (\mathbf{A} \times \mathbf{X})^{T-\tau}) = 1$. The most important case is $\tau = 0$; then we have just $P_{x_0}^\pi$. Note that $P_{x_0}^\pi$ is another notation for $P_{P_0}^\pi$ in the case where $P_0(\cdot)$ is concentrated at point x_0. In reality, $P_{h_\tau}^\pi(\cdot) = P_{P_0}^\pi(\cdot|\mathcal{F}_\tau)$ coincides with the conditional probability for $P_{P_0}^\pi$-almost all h_τ if measure $P_{P_0}^\pi$ on $\mathbf{H}_\tau$ has full support: $\mathbf{Supp}\, P_{P_0}^\pi = \mathbf{H}_\tau$.

We introduce $v_{h_\tau}^\pi \triangleq E_{h_\tau}^\pi \left[\sum_{t=\tau+1}^T c_t(X_{t-1}, A_t) + C(X_T) \right]$ and call a control strategy π^* uniformly optimal if

$$v_{h_\tau}^{\pi^*} = \inf_\pi v_{h_\tau}^\pi \triangleq v_{h_\tau}^* \text{ for all } h_\tau \in \bigcup_{t=0}^T \mathbf{H}_t.$$

In this connection, function

$$v_x^* \triangleq \inf_\pi v_x^\pi$$

represents the minimum possible loss, if started from $X_0 = x$; this is usually also called a Bellman function because it coincides with $v_0(x)$ under weak conditions; see Lemma 1.1 below. Sometimes, if it is necessary to underline T, the time horizon, we denote $V_x^T \triangleq \inf_\pi v_x^\pi$.

Suppose function $v_{h_\tau}^* \neq \pm\infty$ is finite. We call a control strategy π uniformly ε-optimal if $v_{h_\tau}^\pi \leq v_{h_\tau}^* + \varepsilon$ for all $h_\tau \in \bigcup_{t=0}^T \mathbf{H}_t$. Similarly, in the case where $\inf_\pi v^\pi \neq \pm\infty$, we call a strategy π ε-optimal if $v^\pi \leq \inf_\pi v^\pi + \varepsilon$ (see (1.1)). Uniformly (ε)-optimal strategies are sometimes called *persistently ε-optimal* [Kertz and Nachman(1979)].

The dynamic programming approach leads to the following statement: a control strategy π^* is uniformly optimal if and only if equality

$$v_{t-1}(x_{t-1}) = \int_\mathbf{A} \left\{ c_t(x_{t-1}, a) + \int_\mathbf{X} v_t(y) p_t(dy|x_{t-1}, a) \right\} \pi_t^*(da|h_{t-1}) \quad (1.7)$$

holds for all $t = 1, 2, \ldots, T$ and $h_{t-1} \in \mathbf{H}_{t-1}$. In this case,

$$v_{h_\tau}^{\pi^*} = v_\tau(x_\tau). \quad (1.8)$$

Very often, the infimum in (1.4) is provided by a mapping $a = \varphi_t(x)$, so that Markov selectors form a sufficient class for solving problem (1.1). Another general observation is that a uniformly optimal strategy is usually also optimal, but not *vice versa*.

If loss functions $c_t(\cdot)$ and $C(\cdot)$ are not bounded (below or above), the situation becomes more complicated. The following lemma can be helpful.

Lemma 1.1. *Suppose function $c_t(x, a)$ takes only finite values and the optimality equation (1.4) has a measurable solution. Then, for any control strategy π, $\forall h_t = (x_0, a_1, \ldots, x_t) \in \mathbf{H}_t$, $t = 0, 1, \ldots, T$, inequality $v_{h_t}^\pi \geq v_t(x_t)$ is valid. (Note that function $v_t(x)$ can take values $\pm\infty$.)*

In the case where strategy π^* satisfies equality (1.7) and $v_{h_t}^{\pi^*} < +\infty$ for all $h_t \in \mathbf{H}_t$, $t = 0, 1, \ldots, T$, we have equality

$$v_{h_t}^{\pi^*} \equiv v_t(x_t) = \inf_\pi v_{h_t}^\pi,$$

so that π^* is uniformly optimal.

Corollary 1.1. *Under the conditions of Lemma 1.1, $\forall \pi$*

$$v^\pi \geq \int_{\mathbf{X}} v_0(x_0) P_0(dx_0),$$

so that π^* is optimal if $v^{\pi^*} = \int_{\mathbf{X}} v_0(x_0) P_0(dx_0)$.

Corollary 1.2. *Under the conditions of Lemma 1.1, if a strategy π^* satisfies equality (1.7), $v^{\pi^*} < +\infty$, and $v_{h_t}^{\pi^*} < +\infty$ for all $h_t \in \mathbf{H}_t$, $t = 0, 1, \ldots, T$, then control strategy π^* is optimal and uniformly optimal.*

The proof can be found in [Piunovskiy(2009a)].

Even if equality (1.5) (or (1.7)) holds, it can happen that strategy π^* is not (uniformly) optimal. The above lemma and corollaries provide sufficient conditions of optimality.

We mainly study minimization problems. If one considers $v^\pi \to \sup_\pi$ instead of (1.1), then all the statements remain valid if min and inf are replaced with max and sup. Simultaneously, the convention about the infinities should be modified: " $+\infty$" + " $-\infty$" $\stackrel{\triangle}{=}$ " $-\infty$". Lemma 1.1 and Corollaries 1.1 and 1.2 should be also modified in the obvious way; they provide the upper boundary for $v_{h_\tau}^\pi$ and sufficient conditions for the (uniform) optimality of a control strategy.

1.4 Examples

1.4.1 *Non-transitivity of the correlation*

Let $\mathbf{X} = \{-1, 0, 1\}$, $\mathbf{A} = \{0, 1\}$, $T = 1$, $p_1(y|x, a) \equiv p_1(y|a)$ does not depend on x;

$$p_1(y|0) = \begin{cases} 1/3 - \varepsilon_0 - \varepsilon_-, & \text{if } y = 1; \\ 1/3 + \varepsilon_0, & \text{if } y = 0; \\ 1/3 + \varepsilon_-, & \text{if } y = -1, \end{cases}$$

where ε_0 and ε_- are small positive numbers; $p_1(y|1) \equiv 1/3$. Finally, put

$$c_1(x, a) \equiv 0, \quad C(x) = \begin{cases} 1, & \text{if } x = 1; \\ 1 + \delta, & \text{if } x = 0; \\ -1, & \text{if } x = -1, \end{cases} \quad \text{where } \delta > 0 \text{ is a small constant}$$

(see Fig. 1.2).

The random variables X_1 and $W = C(X_1)$ do not depend on the initial distribution. One can check that for an arbitrary distribution of action A_1,

$$\mathrm{Cov}(X_1, W) = 2/3 + O(\varepsilon_0) + O(\varepsilon_-) + O(\delta),$$

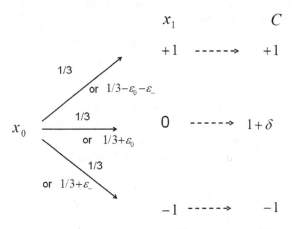

Fig. 1.2 Example 1.4.1.

meaning that X_1 and W are positively correlated for small $\varepsilon_0, \varepsilon_-$ and δ. Under any randomized strategy $\pi_1(a|x_0)$, random variables A_1 and X_1 are also positively correlated for small $\varepsilon_0, \varepsilon_-$ and δ. In other words, if $P\{A_1 = 1\} = p \in (0,1)$ then
$$\mathrm{Cov}(A_1, X_1) = (p - p^2)(\varepsilon_0 + 2\varepsilon_-).$$
One might think that it is reasonable to minimize A_1 in order to obtain $\inf_\pi v^\pi = \inf_\pi E_{P_0}^\pi[W]$, but it turns out that A_1 and W are *negatively* correlated if $\delta > 2\varepsilon_-/\varepsilon_0$: if $P\{A_1 = 1\} = p \in (0,1)$ then
$$\mathrm{Cov}(A_1, W) = (p - p^2)(2\varepsilon_- - \delta\varepsilon_0).$$
In this case,
$$\begin{aligned} v_0(x_0) = \min\{ & (1/3 - \varepsilon_0 - \varepsilon_-) + (1/3 + \varepsilon_0)(1 + \delta) - (1/3 + \varepsilon_-); \\ & 1/3 + 1/3(1 + \delta) - 1/3\} \\ = \ & 1/3 + \delta/3, \end{aligned}$$
and the minimum is provided by $a_1^* = \varphi_1(x_0) \equiv 1$.

The property of being positively correlated is not necessarily transitive. This question was studied in [Langford et al.(2001)].

1.4.2 *The more frequently used control is not better*

Let $\mathbf{X} = \{-2, -1, +1, +2\}$, $\mathbf{A} = \{0, 1\}$, $T = 2$, the transition probability
$$p_1(y|x, a) \equiv p_1(y) = \begin{cases} p, & \text{if } y = +1; \\ q = 1 - p, & \text{if } y = -2; \\ 0 & \text{otherwise} \end{cases}$$

does not depend on x and a;

$$p_2(y|x,a) = \begin{cases} 1, & \text{if } y = x + a; \\ 0 & \text{otherwise.} \end{cases}$$

Finally, put $c_1(x,a) \equiv 0$, $c_2(x,a) \equiv 0$, $C(x) = \begin{cases} b, & \text{if } x = +1; \\ d, & \text{if } x = -1; \\ 0 & \text{otherwise,} \end{cases}$

where b and d are positive numbers (see Fig. 1.3).

Clearly, $\pi_1(a|x_0)$ can be arbitrary, and the control

$$a_2 = \varphi_2^*(x_1) = \begin{cases} 1, & \text{if } x = +1; \\ 0, & \text{if } x = -2 \end{cases}$$

is optimal. ($\pi_2(a|x_1)$ can be arbitrary at $x_1 = +2$ or -1.) Also, $\min_\pi v^\pi = 0$. When $p > q$, control $a_2 = 1$ is applied with higher probability, i.e. more frequently.

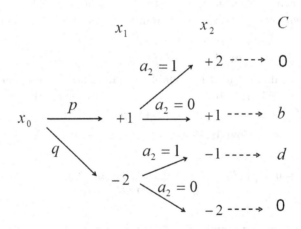

Fig. 1.3 Example 1.4.2.

Now suppose the decision maker cannot observe state x_1, but still has to choose action a_2. It turns out that $a_2 = 0$ is optimal if $b < qd/p$. In reality we deal with another MDP with the Bellman equation

$$v(x_0) = \min\{pb, qd\} = pb,$$

where the first (second) expression in the parentheses corresponds to $a_2 = 0$ (1).

1.4.3 Voting

Suppose three magistrates investigate an accused person who is actually guilty. When making their decisions, the magistrates can make a mistake. To be specific, let p_i, $i = 1, 2, 3$ be the probability that magistrate i decides that the accused is guilty; $q_i \stackrel{\triangle}{=} 1 - p_i$. The final decision is in accordance with the majority among the three opinions. Suppose $p_1 > p_3 > p_2$. Is it not better for the less reliable magistrate 2 to share the opinion of the most reliable magistrate 1, instead of voting independently? Such a problem was discussed in [Szekely(1986), p.171].

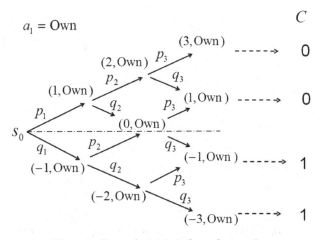

Fig. 1.4   Example 1.4.3: independent voting.

To describe the situation mathematically, we make several assumptions. First of all, the magistrates make their decisions in sequence, one after another. We accept that magistrates 1 and 3 vote according to their personal estimates of the accused's guilt; magistrate 2 either follows the same rule (see Fig. 1.4) or shares the opinion of magistrate 1 (see Fig. 1.5), and he makes his general decision at the very beginning. Put $T = 3$; $\mathbf{X} = \{(y, z), s_0\}$, where component $y \in \{-3, -2, \ldots, +3\}$ represents the current algebraic sum of decisions in favour of finding the accused guilty; $z \in \{\text{Own}, \text{Sh}\}$ denotes the general decision of magistrate 2 made initially; s_0 is a fictitious initial state. $\mathbf{A} = \{\text{Own}, \text{Sh}\}$, and action Own(Sh) means that magistrate 2 will make his own decision (will share the opinion of magistrate 1).

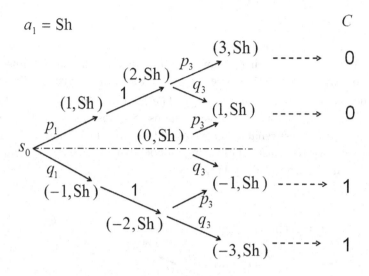

Fig. 1.5 Example 1.4.3: sharing opinion.

Now $X_0 = s_0$, i.e. $P_0(x) = I\{x = s_0\}$;

$$p_1((\hat{y}, \hat{z})|x, a) = I\{\hat{z} = a\} \times \begin{cases} p_1, & \text{if } \hat{y} = 1; \\ q_1, & \text{if } \hat{y} = -1; \\ 0 & \text{otherwise,} \end{cases}$$

$$p_2((\hat{y}, \hat{z})|(y, Own), a) = I\{\hat{z} = \text{Own}\} \times \begin{cases} p_2, & \text{if } \hat{y} = y + 1; \\ q_2, & \text{if } \hat{y} = y - 1; \\ 0 & \text{otherwise,} \end{cases}$$

$$p_2((\hat{y}, \hat{z})|(y, Sh), a) = I\{\hat{z} = \text{Sh}\} \times \begin{cases} 1, & \text{if } \hat{y} = y + 1,\ y > 0; \\ 1, & \text{if } \hat{y} = y - 1,\ y < 0; \\ 0 & \text{otherwise,} \end{cases}$$

$$p_3((\hat{y}, \hat{z})|(y, z), a) = I\{\hat{z} = z\} \times \begin{cases} p_3, & \text{if } \hat{y} = y + 1; \\ q_3, & \text{if } \hat{y} = y - 1; \\ 0 & \text{otherwise.} \end{cases}$$

We have presented only the significant values of the transition probability. Other transition probabilities are zero. If a state x cannot be reached by step t, then there is no reason to pay attention to $p_t(\hat{x}|x, a)$.

$$c_t(x, a) \equiv 0, \quad C(x) = C((y, z)) = \begin{cases} 0, & \text{if } y \geq 0; \\ 1, & \text{if } y < 0. \end{cases}$$

The dynamic programming approach results in the following:

$$v_3(x) = C(x); \quad v_2((y,z)) = \begin{cases} 0, & \text{if } y = +2; \\ q_3, & \text{if } y = 0; \\ 1, & \text{if } y = -2 \end{cases}$$

(other values of y are of no interest).

$$v_1((1, \text{Own})) = q_2 q_3; \quad v_1((-1, \text{Own})) = p_2 q_3 + q_2;$$

$$v_1((1, \text{Sh})) = 0; \quad v_1((-1, \text{Sh})) = 1;$$

$$v_0(s_0) = \min\{p_1 q_2 q_3 + q_1(p_2 q_3 + q_2); \quad q_1\}.$$

The first (second) expression in parentheses corresponds to $a = \text{Own (Sh)}$. Let $p_1 = 0.7$, $p_2 = 0.6$, $p_3 = 0.65$. Then

$$v_0(s_0) = p_1 q_2 q_3 + q_1(p_2 q_3 + q_2) = 0.281 < 0.3 = q_1.$$

Even the less reliable magistrate plays his role. If he shares the opinion of the most reliable magistrate 1 then the total probability of making a mistake increases. Of course the situation changes if p_2 is too small.

1.4.4 The secretary problem

The classical secretary problem was studied in depth in, for example, [Puterman(1994), Section 4.6.4]. See also [Ross(1983), Chapter 1, Section 5] and [Suhov and Kelbert(2008), Section 1.11]. We shall consider only a very simple version here.

An employer seeks to hire an individual to fill a vacancy for a secretarial position. There are two candidates for this job, from two job centres. It is known that the candidate from the first centre is better with probability 0.6. The employer can, of course, interview the first candidate; however, immediately after this interview, he must decide whether or not to make the offer. If the employer does not offer the job, that candidate seeks employment elsewhere and is no longer eligible to receive an offer, so that the employer has to accept the second candidate, from the second job centre. As there is no reason for such an interview, the employer should simply make the offer to the first candidate. The aim is to maximize the probability of accepting the best candidate.

Now, suppose there is a third candidate from a third job centre. For simplicity, assume that the candidates can be ranked only in three ways:

- 1 is better than 2 and 2 is better than 3, the probability of this event being 0.3;

- 3 is better than 1 and 1 is better than 2, the probability is 0.3;

- 2 is better than 1 and 1 is better than 3, making the probability 0.4.

The first candidate is better than the second with probability of 0.6, but to maximize the probability of accepting the best candidate without interviews, the employer has to offer the job to the second candidate. There could be the following conversation between the employers:
– We have candidates from job centres 1 and 2. Who do you prefer?
– Of course, we'll hire the first one.
– Stop. Here is another application from job centre 3.
– Hm. In that case I prefer the candidate from the second job centre.

The employer can interview the candidates sequentially: the first one, the second and the third. At each step he can make an offer; if the first two are rejected then the employer has to hire the third one. Now the situation is similar to the classical case, and the problem can be formulated as an MDP [Puterman(1994), Section 4.6.4]. The dynamic programming approach results in the following optimal strategy: reject the first candidate and accept the second one only if he is better than the first. The probability of hiring the best candidate equals 0.7.

Consider another sequence of interviews: the second job centre, the first and the third. Then the optimal control strategy prescribes the acceptance of candidate 2 (which can be done even without interviews). The probability of success equals 0.4. One can also investigate other sequences of interviews, and the optimal control strategy and probability of success can change again.

1.4.5 *Constrained optimization*

Suppose we have two different loss functions $^1c_t(x,a)$ and $^2c_t(x,a)$, along with $^1C(x)$ and $^2C(x)$. Then every control strategy π results in two performance functionals $^1v^\pi$ and $^2v^\pi$ defined similarly to (1.1). Usually, objectives $^1v^\pi$ and $^2v^\pi$ are inconsistent, so that there does not exist a control strategy providing $\min_\pi {}^1v^\pi$ and $\min_\pi {}^2v^\pi$ simultaneously. One can construct the Pareto set corresponding to non-dominated control strategies. Another approach sets out the passage to a *constrained* problem:

$$^1v^\pi \to \inf_\pi; \quad ^2v^\pi \leq d, \qquad (1.9)$$

where d is a chosen number. Strategies satisfying the inequality in (1.9) are called admissible. In such cases, the method of Lagrange multipliers has proved to be effective, although constructing an optimal strategy becomes much more complicated [Piunovskiy(1997)]. The dynamic programming approach can also be useful here [Piunovskiy and Mao(2000)], but only after some preliminary work.

We present an example similar to [Haviv(1996)] showing that the Bellman principle fails to hold and the optimal control strategy can look strange.

Let $\mathbf{X} = \{1, 2, 3, 4\}$; $\mathbf{A} = \{1, 2\}$; $T = 1$, $P_0(1) = P_0(2) = 1/2$;

$$p_1(y|1, a) = I\{y = 1\}; \quad p_1(y|2, a) = I\{y = 2 + a\}.$$

Other transition probabilities play no role.

Put

$$^1c_1(x, a) = {}^2c_1(x, a) \equiv 0;$$

$$^1C(x) = \begin{cases} 0, & \text{if } x = 1 \text{ or } 2; \\ 20, & \text{if } x = 3; \\ 10, & \text{if } x = 4; \end{cases} \quad {}^2C(x) = \begin{cases} 0.2, & \text{if } x = 1 \text{ or } 2; \\ 0.05, & \text{if } x = 3; \\ 0.1, & \text{if } x = 4; \end{cases}$$

and consider problem (1.9) with $d = 0.125$ (see Fig. 1.6).

Intuition says that as soon as $X_0 = 2$ is realized, it is worth applying action $a = 2$ because that leads to the admissible value $^2C(X_1) = {}^2C(4) = 0.1 \leq 0.125$ and simultaneously to the minimal loss $^1C(X_1) = {}^1C(4) = 10$, when compared with $^1C(3) = 20$.

On the other hand, for such a control strategy we have, after taking into account the initial distribution:

$$^1v^\pi = 1/2 \cdot 10 = 5; \quad {}^2v^\pi = 1/2 \cdot 0.2 + 1/2 \cdot 0.1 = 0.15 > 0.125 = d,$$

meaning that the control strategy mentioned is not admissible. One can see that the only admissible control strategy is $\varphi_1^*(2) = 1$ when $^1v^{\varphi^*} = 1/2 \cdot 20 = 10$, $^2v^{\varphi^*} = 1/2 \cdot 0.2 + 1/2 \cdot 0.05 = 0.125$. Therefore, in state 2 the decision maker should take into account not only the future dynamics, but also other trajectories ($X_0 = X_1 = 1$) that have already no chance of being realized; this means that the Bellman principle does not hold.

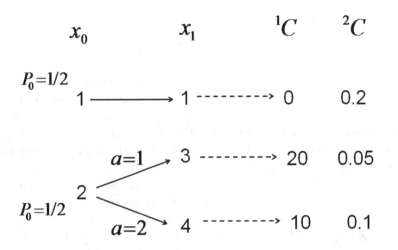

Fig. 1.6 Example 1.4.5.

Suppose now that $^2C(1) = 0.18$. Then, action 2 in state 2 can be used but only with small probability. One should maximize that probability, and the solution to problem (1.9) is then given by

$$\pi_1^*(1|2) = 0.6, \qquad \pi_1^*(2|2) = 1 - \pi_1^*(1|2) = 0.4.$$

Remark 1.1. In the latter case, $^1v^{\pi^*} = 8$ and $^2v^{\pi^*} = 0.125$. Note that selector $\varphi_1(2) = 1$ provides $^1v^{\varphi_1} = 10$, $^2v^{\varphi_1} = 0.115$ and selector $\varphi_2(2) = 2$ provides $^1v^{\varphi_2} = 5$, $^2v^{\varphi_2} = 0.14$. We see that no individual selector results in the same performance vector $(^1v, {}^2v)$ as π^*. In a general MDP with finite horizon and finite spaces $\mathbf{X}$ and $\mathbf{A}$, performance space $\mathcal{V} \triangleq \{(^1v^\pi, {}^2v^\pi), \pi \in \Delta^{\text{All}}\}$ coincides with the (closed) convex hull of performance space $\mathcal{V}^N \triangleq \{(^1v^\varphi, {}^2v^\varphi), \varphi \in \Delta^{\text{MN}}\}$. This statement can be generalized in many directions. Here, as usual, Δ^{All} is the set of all strategies; Δ^{MN} is the set of all Markov selectors.

In the case where $^2C(1) \leq 0.15$, constraint $^2v^\pi \leq d = 0.125$ becomes inessential and $^1v^\pi$ is minimized by the admissible control strategy $\varphi_1^*(2) = 2$.

Note that, very often, the solution to a constrained MDP is given by a randomized Markov control strategy; however, there is still no reason to consider past-dependent strategies. Example 1 in [Frid(1972)], in the framework of a constrained discounted MDP, also shows that randomization is necessary.

One can impose a constraint in a different way:

$$^1v^\pi \to \inf_\pi; \quad ^2W = \sum_{t=1}^{T} {}^2c_t(X_{t-1}, A_t) + {}^2C(X_T) \le d \quad P_{P_0}^\pi \text{ - a.s.}$$

After introducing artificial random variables Z_t:

$$Z_0 = 0, \quad Z_t = Z_{t-1} + {}^2c_t(X_{t-1}, A_t), \quad t = 1, 2, \ldots, T,$$

one should modify the final loss:

$$^1\tilde{C}(x, z) = \begin{cases} {}^1C(x), & \text{if } z + {}^2C(x) \le d; \\ +\infty & \text{otherwise.} \end{cases}$$

In this new model, the Bellman principle holds. On the other hand, an optimal control strategy will depend on the initial distribution. In the example considered (with the original data), there are no solutions if $X_0 = 1$ and $\varphi_1^*(2) = 2$ in the case where $X_0 = 2$. Quite formally, in the new 'tilde'-model $\tilde{v}_0(1) = +\infty$, $\tilde{v}(2) = 10$.

1.4.6 Equivalent Markov selectors in non-atomic MDPs

Consider a one-step MDP with state and action spaces $\mathbf{X}$ and $\mathbf{A}$, initial distribution $P_0(dx)$, and zero final loss, so that the transition probability plays no role. Suppose we have a finite collection of loss functions $\{{}^kc(x,a)\}_{k=1,2,\ldots,K}$. In Remark 1.1 we saw that performance spaces

$$\mathcal{V} \triangleq \{\{{}^kv^\pi\}_{k=1,2,\ldots,K}, \pi \in \Delta^{\text{All}}\} \text{ and } \mathcal{V}^N \triangleq \{\{{}^kv^\varphi\}_{k=1,2,\ldots,K}, \varphi \in \Delta^{\text{MN}}\}$$

can be different: if $\mathbf{X}$ and $\mathbf{A}$ are finite then $\mathcal{V}^N$ is finite because Δ^{MN} is finite, but $\mathcal{V}$ is convex compact [Piunovskiy(1997), Section 3.2.2]. On the other hand, according to [Feinberg and Piunovskiy(2002), Th. 2.1], $\mathcal{V} = \mathcal{V}^N$ if the initial distribution $P_0(dx)$ is non-atomic. See also [Feinberg and Piunovskiy(2010), Th. 3.1]. In other words, for any control strategy π, there exists a selector φ such that their performance vectors coincide

$$\{{}^kv^\pi\}_{k=1,2,\ldots,K} = \{{}^kv^\varphi\}_{k=1,2,\ldots,K}.$$

We shall call such strategies π and φ equivalent w.r.t. $\{{}^kc(\cdot)\}_{k=1,2,\ldots,K}$.

Recall that

$$\mathcal{D} = \{P_{P_0}^\pi(\cdot), \pi \in \Delta^{\text{All}}\} \text{ and } \mathcal{D}^{\text{MN}} = \{P_{P_0}^\varphi(\cdot), \varphi \in \Delta^{\text{MN}}\}$$

are the strategic measures spaces. In the general case, if $\mathbf{A}$ is compact and all transition probabilities $p_t(dy|x, a)$ are continuous, then $\mathcal{D}$ is convex compact [Schäl(1975a), Th. 5.6]. This sounds a little strange; however, in spite of equality $\mathcal{V} = \mathcal{V}^N$ being valid in the non-atomic case, the space $\mathcal{D}^{MN}$ is usually not closed (in the weak topology), as the following example shows [Piunovskiy(1997), p.170].

Let $\mathbf{X} = [0,1]$, $\mathbf{A} = \{1,2\}$, $P_0(dx) = dx$, the Lebesgue measure, and put $\varphi^n(x) = I\{x \in \Gamma_n\} + 2 \cdot I\{x \in \Gamma_n^c\}$, where Γ_n is the set consisting of 2^{n-1} segments of the same length $\delta_n = \left(\frac{1}{2}\right)^n$:

$$\Gamma_n = [\delta_n, 2\delta_n] \cup [3\delta_n, 4\delta_n] \cup \ldots \cup [(2^n-1)\delta_n, 2^n\delta_n = 1]$$

(see Fig. 1.7). Take an arbitrary bounded continuous function $c(x,a)$.

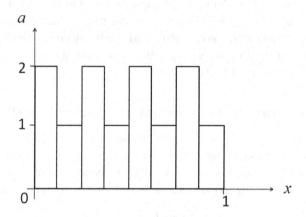

Fig. 1.7   Example 1.4.6: selector φ^n when $n=3$.

Then

$$\sum_{a=1}^{2} \int_0^1 c(x,a) I\{\varphi^n(x) = a\} dx = \int_{\Gamma_n} c(x,1) dx + \int_{\Gamma_n^c} c(x,2) dx$$

$$= \frac{1}{2} \int_0^1 c(x,1) dF_1^n(x) + \frac{1}{2} \int_0^1 c(x,2) dF_2^n(x),$$

where $F_1^n(\cdot)$ and $F_2^n(\cdot)$ are the cumulative distribution functions of uniform random variables on Γ_n and Γ_n^c respectively. Obviously, $\forall x \in \mathbf{X}$, $F_1^n(x)$, $F_2^n(x) \to F(x) = x$ as $n \to \infty$. Hence

$$\int_0^1 \hat{c}(x) dF_1^n(x), \quad \int_0^1 \hat{c}(x) dF_2^n(x) \to \int_0^1 \hat{c}(x) dx$$

for any continuous function $\hat{c}(\cdot)$. In particular,

$$\sum_{a=1}^{2}\int_{0}^{1} c(x,a)I\{\varphi^{n}(x)=a\}dx \to \frac{1}{2}\int_{0}^{1}c(x,1)dx + \frac{1}{2}\int_{0}^{1}c(x,2)dx$$

$$=\sum_{a=1}^{2}\int_{0}^{1}c(x,a)\pi^{*}(a|x)dx,$$

where $\pi^{*}(1|x) = \pi^{*}(2|x) = \frac{1}{2}$ (space $\mathcal{D}$ is weakly closed). But the strategic measure $P_{P_0}^{\pi^*}$ cannot be generated by a selector. As proof of this, suppose such a selector $\varphi(x)$ exists. Then, for function $g_1(x) \stackrel{\triangle}{=} I\{\varphi(x)=1\}$, the following integrals must coincide:

$$\int_{0}^{1} g_1(x)\pi^{*}(1|x)dx = \frac{1}{2}\int_{0}^{1}I\{\varphi(x)=1\}dx,$$

$$\int_{0}^{1} g_1(x)I\{\varphi(x)=1\}dx = \int_{0}^{1}I\{\varphi(x)=1\}dx,$$

meaning that $\int_{0}^{1}I\{\varphi(x)=1\}dx = 0$. Similarly, for $g_2(x) \stackrel{\triangle}{=} I\{\varphi(x)=2\}$, we obtain $\int_{0}^{1}I\{\varphi(x)=2\}dx = 0$. This contradiction implies that such a selector φ does not exist.

The following example shows that in the non-atomic case, the performance spaces $\mathcal{V}$ and $\mathcal{V}^N$ can be different if the collection of loss functions $\{{}^kc(x,a)\}_{k=1,2,\ldots}$ is not finite.

Let $\mathbf{X}$ be an arbitrary Borel space and $\mathbf{A} = \{1,2\}$. It is known [Parthasarathy(2005), Th. 6.6] that there exists a sequence $\{{}^kf(x)\}_{k=1,2,\ldots}$ of bounded uniformly continuous functions on $\mathbf{X}$ such that if

$$\forall k=1,2,\ldots \quad \int_{\mathbf{X}}{}^kf(x)\mu_1(dx) = \int_{\mathbf{X}}{}^kf(x)\mu_2(dx),$$

then the measures μ_1 and μ_2 coincide. Now put ${}^kc(x,a) \stackrel{\triangle}{=} I\{a=1\}\,{}^kf(x)$, take $\pi(1|x) = \pi(2|x) \equiv 1/2$ and suppose there exists a selector φ equivalent to π w.r.t. $\{{}^kc(\cdot)\}_{k=1,2,\ldots}$ meaning that

$$\int_{\mathbf{X}}{}^kf(x)\pi(1|x)P_0(dx) = \int_{\mathbf{X}}{}^kf(x)I\{\varphi(x)=1\}P_0(dx).$$

We see that measures on $\mathbf{X}$ $\mu_1(dx) \stackrel{\triangle}{=} \pi(1|x)P_0(dx)$ and $\mu_2(dx) \stackrel{\triangle}{=} I\{\varphi(x)=1\}P_0(dx)$ must coincide. But for function $g_1(x) \stackrel{\triangle}{=} I\{\varphi(x)=1\}$ we have

$$\int_{\mathbf{X}}g_1(x)\mu_1(dx) = \frac{1}{2}\int_{\mathbf{X}}I\{\varphi(x)=1\}P_0(dx)$$

and
$$\int_{\mathbf{X}} g_1(x)\mu_2(dx) = \int_{\mathbf{X}} I\{\varphi(x) = 1\}P_0(dx),$$

so that $\int_{\mathbf{X}} I\{\varphi(x) = 1\}P_0(dx) = 0$. Similarly, when considering function $g_2(x) \stackrel{\triangle}{=} I\{\varphi(x) = 2\}$ we obtain $\int_{\mathbf{X}} I\{\varphi(x) = 2\}P_0(dx) = 0$. This contradiction shows that such a selector φ does not exist.

1.4.7 Strongly equivalent Markov selectors in non-atomic MDPs

If we have an arbitrary collection $\{\,^\alpha c(x,a)\}_{\alpha \in \mathcal{A}}$ of loss functions, but of special form $^\alpha c(x,a) = \,^\alpha \rho(a) \cdot f(x)$ where all functions $^\alpha \rho(\cdot)$ are bounded (arbitrary in case $f(\cdot) \geq 0$ or $f(\cdot) \leq 0$), then, as before, in the non-atomic case, for any control strategy π there exists a selector φ such that $^\alpha v^\pi = \,^\alpha v^\varphi$ for all $\alpha \in \mathcal{A}$ (see Lemma B.1). The latter statement can be reformulated as follows: if the function $f(x)$ is fixed then, for any control strategy π, there exists a selector φ such that the measures on $\mathbf{A}$

$$\nu^\pi(\Gamma) \stackrel{\triangle}{=} \int_{\mathbf{X}} \pi(\Gamma|x)f(x)P_0(dx) \text{ and } \nu^\varphi(\Gamma) \stackrel{\triangle}{=} \int_{\mathbf{X}} I\{\Gamma \ni \varphi(x)\}f(x)P_0(dx)$$

coincide (here $\Gamma \in \mathcal{B}(\mathbf{A})$). We call such strategies π and φ *strongly equivalent* w.r.t. $f(\cdot)$. This notion is important in the theory of mass transportation and for so-called Monge–Kantorovich problems [Ball(2004)]; [Magaril-Il'yaev and Tikhomirov(2003), Section 12.2]. The generalized definition reads:

Definition 1.1. Suppose a collection of functions $\{\,^k f(x,a)\}_{k=1,2,\ldots,K}$ is given. Two strategies π^1 and π^2 are called *strongly equivalent* w.r.t. $\{\,^k f(\cdot)\}_{k=1,2,\ldots,K}$ if for an arbitrary bounded real measurable function $\rho(a)$ on $\mathbf{A}$, $\forall k = 1, 2, \ldots, K$,

$$^k\nu^{\pi^1} \stackrel{\triangle}{=} \int_{\mathbf{X}} \int_{\mathbf{A}} \rho(a) \,^k f(x,a)\pi^1(da|x)P_0(dx)$$

$$= \,^k\nu^{\pi^2} \stackrel{\triangle}{=} \int_{\mathbf{X}} \int_{\mathbf{A}} \rho(a) \,^k f(x,a)\pi^2(da|x)P_0(dx).$$

If π and φ are equivalent w.r.t. $\{\,^k c(\cdot)\}_{k=1,2,\ldots,K}$ then they may not be strongly equivalent w.r.t. $\{\,^k c(\cdot)\}_{k=1,2,\ldots,K}$. On the other hand, if π

and φ are strongly equivalent w.r.t. $\{{}^k f(\cdot)\}_{k=1,2,\ldots,K}$ then they are equivalent w.r.t. any collection of loss functions of the form $\{{}^\alpha c(x,a)\}_{\alpha \in \mathcal{A}} = \{{}^\gamma \rho(a) \cdot {}^k f(x,a)\}_{\gamma \in \Gamma; \, k=1,2,\ldots,K}$, where $\alpha = (\gamma, k)$, Γ is an arbitrary set and $\forall \gamma \in \Gamma$ function ${}^\gamma \rho(\cdot)$ is bounded (arbitrary in case every function ${}^k f(\cdot)$ is either non-negative or non-positive).

Theorem 1.1. *Let initial distribution $P_0(dx)$ be non-atomic and suppose one of the following conditions is satisfied:*

(a) Action space $\mathbf{A}$ is finite and collection $\{{}^k f(x,a)\}_{k=1,2,\ldots,K}$ is finite.
(b) Action space $\mathbf{A}$ is arbitrary, $K = 1$ and a single function $f(x) = {}^1 f(x)$ is given (independent of a).

Then, for any control strategy π, there exists a selector φ, strongly equivalent to π w.r.t. $\{{}^k f(\cdot)\}_{k=1,2,\ldots,K}$.

Part (a) follows from [Feinberg and Piunovskiy(2002)]; see also [Feinberg and Piunovskiy(2010), Th. 1]. For part (b), see Lemma B.1.

If the collection $\{{}^k f(x,a)\}_{k=1,2,\ldots}$ is not finite then assertion (a) is not valid (see Example 1.4.6). Now we want to show that all conditions in Theorem 1.1(b) are important.

Independence of function $f(\cdot)$ of a. This example is due to [Feinberg and Piunovskiy(2010), Ex. 2].
Let $\mathbf{X} = [0,1]$, $\mathbf{A} = [-1, +1]$, $P_0(dx) = dx$, the Lebesgue measure, and $f(x,a) \stackrel{\triangle}{=} 2x - |a|$. Consider a strategy $\pi(\Gamma|x) \stackrel{\triangle}{=} \frac{1}{2}[I\{\Gamma \ni x\} + I\{\Gamma \ni -x\}]$ to be a mixture of two Dirac measures (see Fig. 1.8), and suppose there exists a selector φ strongly equivalent to π w.r.t. f. Then, for any measurable non-negative or non-positive function $\rho(a)$, we must have

$$\int_0^1 \int_{-1}^1 \rho(a) f(x,a) \pi(da|x) dx = \frac{1}{2} \int_0^1 [\rho(x) \cdot x + \rho(-x) \cdot x] dx$$
$$= \int_0^1 \rho(\varphi(x)) I\{\varphi(x) > 0\}[2x - \varphi(x)] dx \quad (1.10)$$
$$+ \int_0^1 \rho(\varphi(x)) I\{\varphi(x) \le 0\}[2x + \varphi(x)] dx.$$

Consider $\rho(a) \stackrel{\triangle}{=} a \cdot I\{a > 0\}$. Then

$$\frac{1}{2} \int_0^1 x^2 dx = \int_0^1 \varphi(x) I\{\varphi(x) > 0\}[2x - \varphi(x)] dx.$$

Hence
$$\int_0^1 I\{\varphi(x) > 0\}[x - \varphi(x)]^2 dx = \int_0^1 I\{\varphi(x) > 0\} x^2 \, dx - \frac{1}{2} \int_0^1 x^2 \, dx. \quad (1.11)$$

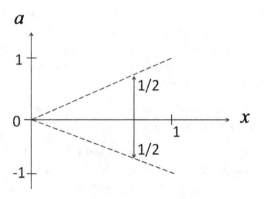

Fig. 1.8 Example 1.4.7: description of the strategy π.

Consider $\rho(a) \triangleq a \cdot I\{a \leq 0\}$. Then

$$-\frac{1}{2}\int_0^1 x^2 dx = \int_0^1 \varphi(x)I\{\varphi(x) \leq 0\}[2x + \varphi(x)]dx.$$

Hence

$$\int_0^1 I\{\varphi(x) \leq 0\}[x + \varphi(x)]^2 dx = \int_0^1 I\{\varphi(x) \leq 0\}x^2\, dx - \frac{1}{2}\int_0^1 x^2\, dx. \tag{1.12}$$

If we add together the right-hand parts of (1.11) and (1.12), we obtain zero. Therefore,

$$\int_0^1 I\{\varphi(x) > 0\}[x - \varphi(x)]^2 dx = \int_0^1 I\{\varphi(x) \leq 0\}[x + \varphi(x)]^2 dx = 0$$

and

$$\left.\begin{array}{l}\varphi(x) = x, \quad \text{if } \varphi(x) > 0;\\ \varphi(x) = -x, \quad \text{if } \varphi(x) \leq 0\end{array}\right\} \text{ almost surely.} \tag{1.13}$$

Consider

$$\rho(a) \triangleq I\{\varphi(a) = a\} \cdot I\{a > 0\}/a.$$

Equality (1.10) implies

$$\frac{1}{2}\int_0^1 I\{\varphi(x) = x\}dx = \int_0^1 I\{\varphi(x) > 0\}\frac{I\{\varphi(\varphi(x)) = \varphi(x)\}}{\varphi(x)}[2x - \varphi(x)]dx$$

$$= \int_0^1 I\{\varphi(x) > 0\}I\{\varphi(\varphi(x)) = \varphi(x)\}\left[\frac{2x}{\varphi(x)} - 1\right]dx.$$

If $\varphi(x) = x$, then $\varphi(\varphi(x)) = \varphi(x)$. Hence,

$$\frac{1}{2}\int_0^1 I\{\varphi(x) = x\}dx \geq \int_0^1 I\{\varphi(x) > 0\}I\{\varphi(x) = x\}\left[\frac{2x}{\varphi(x)} - 1\right]dx$$

$$= \int_0^1 I\{\varphi(x) = x\}dx,$$

meaning that

$$\int_0^1 I\{\varphi(x) = x\}dx = 0. \qquad (1.14)$$

Consider

$$\rho(a) \stackrel{\triangle}{=} I\{\varphi(-a) = a\} \cdot I\{a < 0\}/(-a).$$

Equality (1.10) implies

$$\frac{1}{2}\int_0^1 I\{\varphi(x) = -x\}dx = \int_0^1 I\{\varphi(x) < 0\}\frac{I\{\varphi(-\varphi(x)) = \varphi(x)\}}{-\varphi(x)}[2x+\varphi(x)]dx$$

$$= \int_0^1 I\{\varphi(x) < 0\}I\{\varphi(-\varphi(x)) = \varphi(x)\}\left[\frac{2x}{-\varphi(x)} - 1\right]dx.$$

If $\varphi(x) = -x$, then $\varphi(-\varphi(x)) = \varphi(x)$. Hence

$$\frac{1}{2}\int_0^1 I\{\varphi(x) = -x\}dx \geq \int_0^1 I\{\varphi(x) < 0\}I\{\varphi(x) = -x\}\left[\frac{2x}{-\varphi(x)} - 1\right]dx$$

$$= \int_0^1 I\{\varphi(x) = -x\}dx,$$

meaning that

$$\int_0^1 I\{\varphi(x) = -x\}dx = 0. \qquad (1.15)$$

The contradiction obtained in (1.13), (1.14), (1.15) shows that the selector φ does not exist.

One cannot have more than one function $f(x)$. This example is due to [Loeb and Sun(2006), Ex. 2.7].

Let $\mathbf{X} = [0,1]$, $\mathbf{A} = [-1,+1]$, $P_0(dx) = dx$, the Lebesgue measure, and $^1f(x) \equiv 1$, $^2f(x) = 2x$. Consider the strategy $\pi(\Gamma|x) \stackrel{\triangle}{=} \frac{1}{2}[I\{\Gamma \ni x\}+I\{\Gamma \ni -x\}]$ to be a mixture of two Dirac measures (see Fig. 1.8), and suppose there exists a selector φ strongly equivalent to π w.r.t. $\{\,^1f(\cdot),\,^2f(\cdot)\}$. Then, for any bounded function $\rho(a)$, we must have

$$\int_0^1 \int_{-1}^1 \rho(a)\pi(da|x)\,^1f(x)dx = \frac{1}{2}\int_0^1 [\rho(x)+\rho(-x)]dx = \int_0^1 \rho(\varphi(x))dx; \tag{1.16}$$

$$\int_0^1 \int_{-1}^1 \rho(a)\pi(da|x)\,^2f(x)dx = \frac{1}{2}\int_0^1 [\rho(x)+\rho(-x)]2x\,dx = \int_0^1 \rho(\varphi(x))2x\,dx. \tag{1.17}$$

Consider $\rho(a) \triangleq a^2$. Then (1.16) implies

$$\int_0^1 [\varphi(x)]^2 dx = \int_0^1 x^2\,dx = 1/3.$$

Consider $\rho(a) \triangleq |a|$. Then (1.17) implies

$$\int_0^1 |\varphi(x)|2x\,dx = \int_0^1 x\cdot 2x\,dx = 2/3.$$

Therefore,

$$\int_0^1 [x - |\varphi(x)|]^2 dx = \int_0^1 x^2 dx - 2/3 + 1/3 = 0,$$

meaning that

$$\left.\begin{array}{l}\varphi(x) = x, \quad \text{if } \varphi(x) > 0;\\ \varphi(x) = -x, \quad \text{if } \varphi(x) \le 0\end{array}\right\} \text{ almost surely.} \tag{1.18}$$

Consider $\rho(a) \triangleq I\{\varphi(a) = a\}I\{a > 0\}$. Equality (1.16) implies

$$\frac{1}{2}\int_0^1 I\{\varphi(x) = x\}dx = \int_0^1 I\{\varphi(\varphi(x)) = \varphi(x)\}I\{\varphi(x) > 0\}dx.$$

If $\varphi(x) = x$, then $\varphi(\varphi(x)) = \varphi(x)$. Hence

$$\frac{1}{2}\int_0^1 I\{\varphi(x) = x\}dx \ge \int_0^1 I\{\varphi(x) = x\}dx,$$

meaning that

$$\int_0^1 I\{\varphi(x) = x\}dx = 0. \tag{1.19}$$

Consider $\rho(a) \triangleq I\{\varphi(-a) = a\}I\{a \le 0\}$. Equality (1.16) implies

$$\frac{1}{2}\int_0^1 I\{\varphi(x) = -x\}dx = \int_0^1 I\{\varphi(-\varphi(x)) = \varphi(x)\}I\{\varphi(x) \le 0\}dx.$$

If $\varphi(x) = -x$, then $\varphi(-\varphi(x)) = \varphi(x)$. Hence
$$\frac{1}{2}\int_0^1 I\{\varphi(x) = -x\}dx \geq \int_0^1 I\{\varphi(x) = -x\}dx,$$
meaning that
$$\int_0^1 I\{\varphi(x) = -x\}dx = 0. \tag{1.20}$$

The contradiction obtained in (1.18), (1.19), (1.20) shows that selector φ does not exist.

1.4.8 Stock exchange

Suppose we would like to buy shares and we can choose from two different types. In a one-year period, the ith share ($i = 1,2$) yields Y^i times as much profit as our initial capital was at the beginning of the year. Suppose, for simplicity, $Y^i \in \{+1, -1\}$ can take only two values. We can either double the capital or lose it. Put
$$p_{++} = P\{Y^1 = Y^2 = +1\}, \quad p_{--} = P\{Y^1 = Y^2 = -1\},$$
$$p_{+-} = P\{Y^1 = +1, Y^2 = -1\}, \quad p_{-+} = P\{Y^1 = -1, Y^2 = +1\}.$$

An action is a way to split the initial capital into the two parts to be invested in the first and second shares, namely $\mathbf{A} = \{(a^1, a^2) : a^i \geq 0, \, a^1 + a^2 \leq 1\}$. Since the profit is proportional to the initial capital, we can assume it equals 1. Now $T = 1$,
$$\mathbf{X} = \{s_0, (a^1, a^2, y^1, y^2), \, a^i \geq 0, a^1 + a^2 \leq 1, \, y^i \in \{+1, -1\}\},$$
where s_0 is a fictitious initial state,
$$p((a^1, a^2, y^1, y^2)|s_0, a) = I\{(a^1, a^2) = a\} \times \begin{cases} p_{++}, & \text{if } y^1 = y^2 = +1; \\ p_{--}, & \text{if } y^1 = y^2 = -1; \\ p_{+-}, & \text{if } y^1 = +1, y^2 = -1; \\ p_{-+}, & \text{if } y^1 = -1, y^2 = +1, \end{cases}$$
$c_1(s_0, a) \equiv 0$. If we intend to maximize the expected profit we put
$$^1C(a^1, a^2, y^1, y^2) = (y^1 a^1 + y^2 a^2)$$
(see Fig. 1.9).

The solution is as follows. If $p_{+-} > p_{-+}$ then $\varphi_1^*(s_0) = (1, 0)$ when $p_{++} - p_{--} + p_{+-} - p_{-+} > 0$, and $\varphi_1^*(s_0) = (0, 0)$ otherwise. It is better to invest all the capital in shares 1 if they are at all profitable and if their price

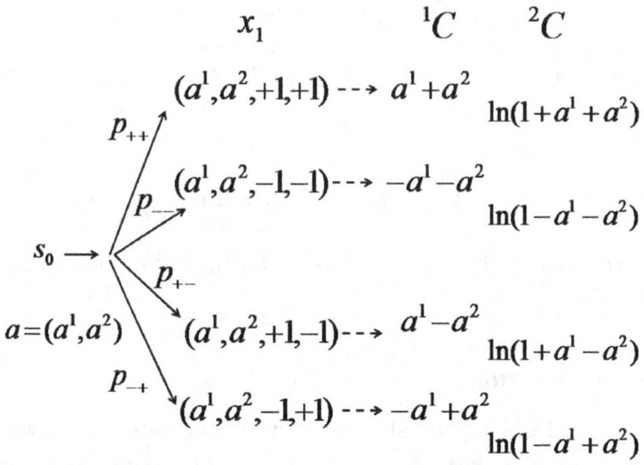

$$
\begin{array}{ccc}
 & x_1 & {}^1C \quad {}^2C
\end{array}
$$

```
                                x_1              ¹C    ²C
                  (a¹,a²,+1,+1) --→  a¹+a²
              p++                              ln(1+a¹+a²)
             /
            /   (a¹,a²,-1,-1) --→ -a¹-a²
           / p-- 
          /                                    ln(1-a¹-a²)
   s₀ →  ⟨  p+-
         \
a=(a¹,a²) \   (a¹,a²,+1,-1) --→   a¹-a²
          p-+                                 ln(1+a¹-a²)
            \
             (a¹,a²,-1,+1) --→ -a¹+a²
                                              ln(1-a¹+a²)
```

Fig. 1.9 Example 1.4.8.

doubles with higher probability than the price of shares 2. This control strategy is probably fine in a one-step process, but if $p_{--} > 0$ then, in the long run, the probability of loosing all the capital approaches 1, and that is certainly not good. (At the same time, the expected total capital approaches infinity!)

Financial specialists use another loss function leading to a different solution. They use ln as a utility function, ${}^2C = \ln({}^1C + 1)$. If the profit per unit of capital approaches -1, then the reward ${}^2C$ goes to $-\infty$, i.e. the investor will make every effort to avoid losing all the capital. In particular, $a^1 + a^2$ will be strictly less than 1 if $p_{00} > 0$. In this case, one should maximize the following expression

$$p_{++}\ln(1+a^1+a^2) + p_{--}\ln(1-a^1-a^2) + p_{+-}\ln(1+a^1-a^2) + p_{-+}\ln(1-a^1+a^2)$$

with respect to a^1 and a^2. Suppose

$$p_{++} > p_{--}, \quad \frac{p_{++}}{p_{--}} > \max\left\{\frac{p_{-+}}{p_{+-}}; \frac{p_{+-}}{p_{-+}}\right\},$$

and all these probabilities are non-zero. Then the optimal control $\varphi_1^*(s_0)$ is given by the equations

$$a^1 + a^2 = \frac{p_{++} - p_{--}}{p_{++} + p_{--}}; \qquad a^1 - a^2 = \frac{p_{+-} - p_{-+}}{p_{+-} + p_{-+}}.$$

Even if the shares exhibit identical properties ($p_{+-} = p_{-+}$), it is better to invest equal parts of the capital in both of them. A similar example was considered in [Szekely(1986), Section 3.6]. Incidentally, if $p_{+-} = p_{-+}$ and $p_{++} > p_{--}$, then using the first terminal reward ^1C, we conclude that all actions (a^1, a^2) satisfying the equality $a^1 + a^2 = 1$ are optimal. In this case, it is worth paying attention to the variance $\sigma^2[\,^1C(X_1)]$, which is minimal when $a^1 = a^2$.

1.4.9 Markov or non-Markov strategy? Randomized or not? When is the Bellman principle violated?

A lemma [Piunovskiy(1997), Lemma 2] says that for every control strategy π, there exists a Markov strategy π^m such that $\forall t = 1, 2, \ldots, T$

$$P_{P_0}^{\pi^m}\{X_{t-1} \in \Gamma^X, A_t \in \Gamma^A\} = P_{P_0}^{\pi}\{X_{t-1} \in \Gamma^X, A_t \in \Gamma^A\}$$

and (obviously)

$$P_{P_0}^{\pi^m}\{X_0 \in \Gamma^X\} = P_{P_0}^{\pi}\{X_0 \in \Gamma^X\}$$

for any $\Gamma^X \in \mathcal{B}(\mathbf{X})$ and $\Gamma^A \in \mathcal{B}(\mathbf{A})$. Therefore,

$$v^\pi = \sum_{t=1}^{T} E_{P_0}^{\pi}[c_t(X_{t-1}, A_t)] + E_{P_0}^{\pi}[C(X_T)]$$

$$= \sum_{t=1}^{T} E_{P_0}^{\pi^m}[c_t(X_{t-1}, A_t)] + E_{P_0}^{\pi^m}[C(X_T)] = v^{\pi^m}$$

in the event that sums of the type "$+\infty$" + "$-\infty$" do not appear. That is why the optimization in the class of all strategies can be replaced by the optimization in the class of Markov strategies, assuming the initial distribution is fixed.

The following example, published in [Piunovskiy(2009a)] shows that the requirement concerning the infinities is essential.

Let $\mathbf{X} = \{0, \pm 1, \pm 2, \ldots\}$, $\mathbf{A} = \{0, -1, -2, \ldots\}$, $T = 3$, $P_0(0) = 1$,

$$p_1(y|x, a) = \begin{cases} \frac{3}{|y|^2 \pi^2}, & \text{if } y \neq 0; \\ 0, & \text{if } y = 0, \end{cases} \quad p_2(0|x, a) = p_3(0|x, a) \equiv 1,$$

$c_1(x, a) \equiv 0$, $c_2(x, a) = x$, $c_3(x, a) = a$, $C(x) = 0$ (see Fig. 1.10).

Since actions A_1 and A_2 play no role, we shall consider only A_3.

The dynamic programming approach results in the following

$$v_3(x) = 0, \quad v_2(x) = -\infty, \quad v_1(x) = -\infty, \quad v_0(x) = -\infty.$$

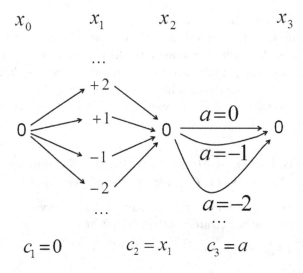

Fig. 1.10 Example 1.4.9: only a non-Markov randomized strategy can satisfy equalities (1.5) and (1.7) and be optimal and uniformly optimal.

Consider the Markov control strategy π^* with $\pi_3^*(0|x_2) = 0$, $\pi_3^*(a|x_2) = \frac{6}{|a|^2\pi^2}$ for $a < 0$. Here equalities (1.7) hold because

$$\sum_{i=1}^{\infty} \frac{(-i) \times 6}{i^2 \pi^2} = -\infty = v_2(x), \quad x + v_2(0) = -\infty = v_1(x),$$

$$0 + \sum_{|y|=1}^{\infty} \frac{3}{|y|^2 \pi^2} \cdot \text{``} - \infty\text{''} = -\infty = v_0(x).$$

On the other hand, for any Markov strategy π^m, $v^{\pi^m} = +\infty$. Indeed, let $\hat{a} = \max\{j : \pi_3^m(j|0) > 0\}$; $0 \geq \hat{a} > -\infty$, and consider random variable $W^+ = (X_1 + A_3)^+$. It takes values $1, 2, 3, \ldots$ with probabilities not smaller than

$$p_1(-\hat{a}+1|0, a)\pi_3^m(\hat{a}|0) = \frac{3\pi_3^m(\hat{a}|0)}{|-\hat{a}+1|^2 \pi^2},$$

$$p_1(-\hat{a}+2|0, a)\pi_3^m(\hat{a}|0) = \frac{3\pi_3^m(\hat{a}|0)}{|-\hat{a}+2|^2 \pi^2},$$

$$p_1(-\hat{a}+3|0, a)\pi_3^m(\hat{a}|0) = \frac{3\pi_3^m(\hat{a}|0)}{|-\hat{a}+3|^2 \pi^2},$$

$$\ldots$$

The expressions come from trajectories ($x_0 = 0, x_1 = -\hat{a} + i, a_1, x_2 = 0, a_2 = \hat{a}, x_3 = 0$). That means

$$E_{P_0}^{\pi^m}[W^+] \geq \pi_3^m(\hat{a}|0) \sum_{i=1}^{\infty} \frac{3i}{|-\hat{a}+i|^2 \pi^2} = +\infty$$

and $v^{\pi^m} = E_{P_0}^{\pi^m}[W] = +\infty$. In particular, $v^{\pi^*} = +\infty$.

At the same time, there exist optimal non-Markov strategies providing $v^\pi = -\infty$. For example, put

$$a_3 = \varphi_3(x_1) = \begin{cases} -x_1, & \text{if } x_1 > 0; \\ 0, & \text{if } x_1 < 0. \end{cases} \quad (1.21)$$

Then, $W = X_1 + A_3 = X_1^- \leq 0$ and $E_{P_0}^\varphi[W] = -\infty$. Note that $x_0 = 0$; so $\inf_\pi v_{x_0}^\pi = \inf_\pi v^\pi = -\infty$, meaning that no one Markov control strategy (including π^*) can be optimal or uniformly optimal.

The optimal control strategy φ presented does not satisfy either equalities (1.5), or (1.7). Indeed, $v_2(0) = -\infty$, and, for example, for history $\hat{h}_2 = (0, a_1, 1, a_2, 0)$ having positive $P_{P_0}^\varphi$ probability, on the right-hand side of (1.5) and (1.7) we have

$$c_3(x_2 = 0, a_3 = \varphi_3(1)) + 0 = \varphi_3(1) = -1.$$

Since for this history $v_{\hat{h}_2}^\varphi = -1$ and $\inf_\pi v_{\hat{h}_2}^\pi = -\infty$, the optimal control strategy φ is not uniformly optimal. This reasoning is correct for an arbitrary selector, meaning that non-randomized strategies cannot satisfy equalities (1.5) and (1.7) and cannot be uniformly optimal.

Therefore, only a non-Markov randomized strategy can satisfy the equalities (1.5) and (1.7) and be optimal and uniformly optimal. As an example, take

$$\pi_3(j|x_1) = \begin{cases} \dfrac{6}{(x_1+j-1)^2\pi^2}, & \text{if } j \leq -x_1 \text{ and } x_1 > 0; \\ \dfrac{6}{j^2\pi^2}, & \text{if } j < 0 \text{ and } x_1 < 0; \\ 0 & \text{otherwise.} \end{cases}$$

In the model investigated, for every optimal control strategy π we have $v_{x_0}^\pi = v_0(x_0)$. It can happen that this statement is false. Consider the following modification of the MDP being studied (see Fig. 1.11):

$$\mathbf{A} = \{0\}, \quad p_3(y|x,a) = \begin{cases} \dfrac{6}{|y|^2\pi^2}, & \text{if } y < 0; \\ 0 & \text{otherwise,} \end{cases} \quad c_3(x,a) = 0, \quad C(x) = x.$$

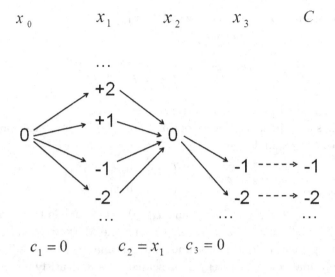

Fig. 1.11   Example 1.4.9: $v_0(x_0) = -\infty < +\infty = v_{x_0}^\varphi = \inf_\pi v_{x_0}^\pi$.

Actually the process is not controlled and can be interpreted as the previous MDP under a fixed Markov control strategy with distribution $\pi_3(\cdot|x) = p_3(\cdot|x,a)$. We know that the total expected loss here equals $+\infty$. Thus, in this modified model for the optimal control strategy (which is unique: $\varphi_t(x) \equiv 0$) we have $v_{x_0}^\varphi = +\infty$. At the same time, the optimality equation (1.4) still gives $v_2(x) = -\infty$, $v_1(x) = -\infty$, and $v_0(x_0) = -\infty$.

Another similar example, illustrating that $v_0(x_0) = -\infty$ and $\inf_\pi v_{x_0}^\pi = +\infty$ at some x_0, is presented in [Bertsekas and Shreve(1978), Section 3.2, Ex. 3].

Finally, let us present a very simple one-step model ($T = 1$) with a negative loss function, where only a randomized strategy is optimal. Let $\mathbf{X} = \{1\}$ be a singleton; $\mathbf{A} = \{1, 2, \ldots\}$; $c_1(x,a) = -2^{-a}$; $C(x) = 0$. For any selector $a = \varphi_1(x)$, $v_x^\varphi = -2^a > -\infty$, but $\inf_\pi\{v_x^\pi\} = -\infty$, and this infimum is attained, e.g. by the randomized strategy $\pi_1^*(a|x) = \left(\frac{1}{2}\right)^a$:

$$v_x^{\pi^*} = \sum_{a=1}^\infty -2^a \left(\frac{1}{2}\right)^a = -\infty.$$

Compare with Example 2.2.18 with the infinite horizon.

1.4.10 Uniformly optimal, but not optimal strategy

We can slightly modify Example 1.4.9 (Fig. 1.10) by ignoring the initial step and putting

$$P_0(x) = \begin{cases} \dfrac{3}{x^2\pi^2}, & \text{if } x \neq 0; \\ 0 & \text{otherwise}. \end{cases}$$

The number of time moments decreases by 1 and $T = 2$. We still have that, for any Markov strategy π^m, $v^{\pi^m} = +\infty$, so that none of them is optimal. Simultaneously, the non-optimal strategy π^* is now uniformly optimal. In the example below, function $v_t(x)$ is finite.

Let $\mathbf{X} = \{\pm 1, \pm 2, \ldots\}$, $\mathbf{A} = \{0, 1\}$, $T = 1$,

$$P_0(x) = \begin{cases} \dfrac{6}{|x|^2\pi^2}, & \text{if } x > 0; \\ 0 & \text{otherwise}, \end{cases}$$

$$p_1(y|x,1) = I\{y = -x\}, \qquad p_1(y|x,0) = \begin{cases} 1/4, & \text{if } y = 2x; \\ 3/4, & \text{if } y = -2x; \\ 0 & \text{otherwise}. \end{cases}$$

$c_1(x,a) = x$, $C(x) = x$ (see Fig. 1.12).

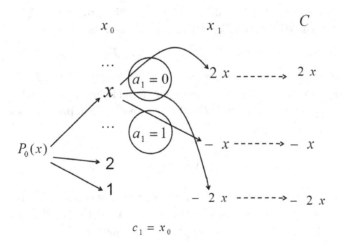

Fig. 1.12 Example 1.4.10.

The dynamic programming approach results in the following: $v_1(x) = x$, $v_0(x) = 0$, and both actions provide the equality in equation (1.4).

Consider action 1: $\varphi_1^1(x) = 1$. This control strategy φ^1 is uniformly optimal, since
$$v_{x_0}^{\varphi^1} = 0 = v_0(x_0) = \inf_\pi v_{x_0}^\pi = v_{x_0}^*.$$
It is also optimal, because only trajectories $(x_0, a_1 = 1, x_1 = -x_0)$ are realized, for which $W = X_0 - X_0 = 0$, so that $v^{\varphi^1} = 0$.

Consider now action 0: $\varphi_1^0(x) = 0$. This control strategy φ^0 is also uniformly optimal, since
$$v_{x_0}^{\varphi^0} = x_0 + \frac{1}{4}(2x_0) + \frac{3}{4}(-2x_0) = 0 = v_0(x_0) = \inf_\pi v_{x_0}^\pi = v_{x_0}^*.$$
It also satisfies equality (1.7). However, it is not optimal, because
$$E_{P_0}^{\varphi^0}[W^+] = \sum_{i=1}^\infty 3i \cdot 1/4 \cdot \frac{6}{i^2 \pi^2} = +\infty, \quad E_{P_0}^{\varphi^0}[W^-] = \sum_{i=1}^\infty (-i) \cdot 3/4 \cdot \frac{6}{i^2 \pi^2} = -\infty,$$
so that $v^{\varphi^0} = +\infty > v^{\varphi^1} = 0$.

This example, first published in [Piunovskiy(2009a)], shows that the condition $v^{\pi^*} < +\infty$ in Corollary 1.2 is important.

1.4.11 Martingales and the Bellman principle

Recall that an adapted real-valued stochastic process $Y_0, Y_1, \ldots$ on a stochastic basis $(\Omega, \mathcal{B}(\Omega), \{\mathcal{F}_t\}, P)$ is called a *martingale* if
$$\forall t \geq 0 \quad Y_t = E[Y_{t+1} | \mathcal{F}_t].$$

In the case where a control strategy π is fixed, we have the $\mathcal{F}_t$-adapted stochastic process X_t (the basis was constructed in Section 1.2) which allows the building of the following $\mathcal{F}_t$-adapted real-valued process, sometimes called an *estimating* process,
$$Y_t^\pi = \sum_{i=1}^t c_i(X_{i-1}, A_i) + v_t(X_t).$$

Here v is a (measurable) solution to the optimality equation (1.4). If functions c and C are bounded below, then Y_t^π is a martingale if and only if π is optimal [Boel(1977)]; [Piunovskiy(1997), Remark 8].

Example 1.4.9 shows that the latter statement can be false. Take the optimal strategy φ defined by (1.21). Since
$$v_0(x) = v_1(x) = v_2(x) = -\infty,$$

we have
$$Y_2^\varphi = X_1 + v_2(X_2) = -\infty.$$
At the same time, $v_3(x) \equiv 0$ and
$$Y_3^\varphi = X_1 + A_3 = X_1^-,$$
so that $E[Y_3^\varphi | \mathcal{F}_2] = X_1^- \neq Y_2^\varphi$.

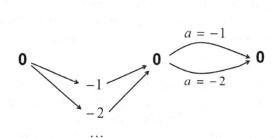

Fig. 1.13 Example 1.4.11: the estimating process is not a martingale.

When we consider the strictly negative modification of Example 1.4.9 presented in Fig. 1.13 with $\mathbf{A} = \{-1, -2\}$, $p_1(y|x, a) = \frac{6}{|y|^2 \pi^2}$, we still see that the optimal selector $\varphi_3(x_1) \equiv -1$ providing $v^\varphi = -\infty$ leads to a process Y_t^φ which is not a martingale:
$$v_3(x) = 0, \quad v_2(x) = -2, \quad v_1(x) = x - 2, \quad v_0(x) = -\infty;$$
$$E[Y_3^\varphi | \mathcal{F}_2] = X_1 - 1 \neq Y_2^\varphi = X_1 - 2.$$

The point is that the Bellman principle is violated: action $a_3 = -1$ is definitely not optimal at state $x_2 = 0$; nevertheless, it is optimal for the whole process on time horizon $t = 0, 1, 2, 3$. The substantial negative loss $c_2(x_1, a_2) = x_1$ on the second step improves the performance up to $-\infty$.

1.4.12 Conventions on expectation and infinities

Some authors [Feinberg(1982), Section 4.1], [Feinberg(1987), Section 4], [Feinberg and Piunovskiy(2002)], [Puterman(1994), Section 5.2] suggest the following formula to calculate the performance criterion:

$$v^\pi = E_{P_0}^\pi \left[\sum_{t=1}^T c_t^+(X_{t-1}, A_t) + C^+(X_T) \right] \\ + E_{P_0}^\pi \left[\sum_{t=1}^T c_t^-(X_{t-1}, A_t) + C^-(X_T) \right] \quad (1.22)$$

still accepting the rule "$+\infty$" + "$-\infty$" = "$+\infty$". (We adjusted the model of maximizing rewards studied in [Feinberg(1982); Feinberg(1987)] to our basic case of minimizing the losses.) In this situation, the value of v^π can only increase, meaning that most of the statements in Examples 1.4.9 and 1.4.11 still hold. On the other hand, in the basic model presented in Fig. 1.10, any control strategy gives $v^\pi = +\infty$ simply because

$$E_{P_0}^\pi \left[c_2^+(X_1, A_2) \right] = E_{P_0}^\pi \left[X_1^+ \right] = +\infty.$$

(The same happens to Example 1.4.10.) Thus, any control strategy can be called optimal! But it seems intuitively clear that selector φ given in (1.21) is better than many other strategies because it compensates positive values of X_1. Similarly, it is natural to call the selector φ^1 optimal in Example 1.4.10.

If we accept (1.22), then it is easy to elaborate an example where the optimality equation (1.4) has a finite solution but where, nevertheless, only a control strategy for which criterion (1.7) is violated, is optimal. The examples below first appeared in [Piunovskiy(2009a)].

Put $\mathbf{X} = \{0, 1, 2, \ldots\}$, $\mathbf{A} = \{0, 1\}$, $T = 2$, $P_0(0) = 1$,

$$p_1(y|x, 0) = I\{y = 0\}, \quad p_1(y|x, 1) = \begin{cases} \dfrac{6}{y^2\pi^2}, & \text{if } y > 0; \\ 0, & \text{if } y = 0, \end{cases}$$

$$p_2(y|x, a) = I\{y = x\},$$

$$c_1(x, a) = 1 - a, \quad c_2(x, a) = x, \quad C(x) = -x.$$

Since action A_2 plays no role, we shall consider only A_1 (see Fig. 1.14).

The dynamic programming approach results in the following:

$$v_2(x) = C(x) = -x, \quad v_1(x) = x - x = 0, \quad v_0(x) = \min\{1 + 0,\ 0 + 0\} = 0,$$

and action $a_1 = 1$ provides this minimum. At the same time, for control strategy $\varphi_1(x_0) = 1$ we have

$$E_{P_0}^\varphi \left[c_2^+(X_1, A_2) \right] = E_{P_0}^\varphi [X_1] = +\infty,$$

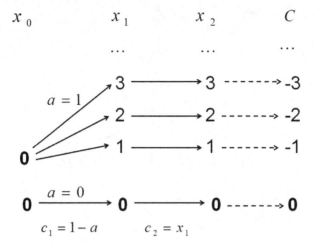

Fig. 1.14 Example 1.4.12: $v_t(x)$ is finite, but $\inf_\pi v_{x_0}^\pi = v_{x_0}^{\varphi^*} = 1 > v_0(x_0) = 0$.

so that (1.22) gives $v^\varphi = +\infty$. Hence, control strategy $\varphi_1^*(x_0) = 0$, resulting in $v^{\varphi^*} = 1$, must be called optimal. In the opposite case, $\varphi_1(x_0) = 1$ is optimal if we accept the formula

$$v^\pi = E_{P_0}^\pi \left[W^+\right] + E_{P_0}^\pi \left[W^-\right], \qquad (1.23)$$

where $W = \sum_{t=1}^T c_t(X_{t-1}, A_t) + C(X_T)$ is the total realized loss. The big loss X_1 on the second step is totally compensated by the final (negative) loss C.

The following example from [Feinberg(1982), Ex. 4.1] shows that a control strategy, naturally uniformly optimal under convention (1.23), is not optimal if we accept formula (1.22).

Let $\mathbf{X} = \{0, 1, 2, \ldots\}$, $\mathbf{A} = \{1, 2, \ldots\}$, $T = 2$, $P_0(0) = 1$, $p_1(y|x, a) = I\{y = a\}$, $p_2(y|x, a) = I\{y = 0\}$, $c_1(x, a) = -a$, $c_2(x, a) = x/2$, $C(x) = 0$ (see Fig. 1.15).

The dynamic programming approach results in the following:
$$v_2(x) = C(x) = 0, \quad v_1(x) = x/2, \quad v_0(x) = \inf_{a \in \mathbf{A}} \{-a + a/2\} = -\infty.$$
No one selector is optimal, and the randomized (Markov) strategy
$$\pi_1^*(a|x_0) = \frac{6}{a^2 \pi^2}, \quad \pi_2^*(a|x_2) \text{ is arbitrary}$$
is uniformly optimal if we accept formula (1.23):
$$W = -\frac{a_1}{2}; \quad v_{x_0}^{\pi^*} = E_{x_0}^{\pi^*}[-a/2] = -\infty.$$

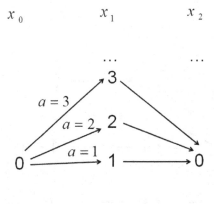

Fig. 1.15 Example 1.4.12: no optimal strategies under convention (1.22).

On the other hand, if we accept convention (1.22), then
$$v_{x_0}^{\pi^*} = E_{x_0}^{\pi^*}[c_1(x_0,a_1)] + E_{x_0}^{\pi^*}[c_2(x_1,a_2)] = -\infty + \infty = +\infty,$$
so that π^* (as well as any other control strategy) is not optimal. If $E_{P_0}^{\pi^*}[c_1(x_0,a_1)] = -\infty$, then $E_{P_0}^{\pi^*}[c_2(x_1,a_2)] = +\infty$, and hence, $v^\pi = +\infty$.

In the current chapter, we accept formula (1.23).

We now discuss the possible conventions for infinity. Basically, if in (1.23) the expression " $+\infty$ " + " $-\infty$ " appears, then the random variable W is said to be not integrable. We have seen in Examples 1.4.9, 1.4.10, and 1.4.11 that the convention
$$\text{`` } +\infty \text{'' } + \text{`` } -\infty \text{'' } = +\infty \tag{1.24}$$
leads to violation of the Bellman principle and to other problems. One can show that all those principal difficulties also appear if we put " $+\infty$ " + " $-\infty$ " $= -\infty$. But convention (1.24) is still better.

Assume for a moment that " $+\infty$ " + " $-\infty$ " $= -\infty$. Then in Example 1.4.9 (Fig. 1.10), any Markov strategy π^m provides $v^\pi = -\infty$, so that all of them are equally optimal, in common with all the other control strategies. But again, selector φ given by (1.21) seems better, and we want this to be mathematically confirmed. In a nutshell, if we meet " $+\infty$ " + " $-\infty$ " in (1.23), it is better to say that all such strategies are equally bad than to accept that they are equally good.

Lemma 1.1 and Corollary 1.1 provided the lower boundary for the performance functional. That will not be the case if " $+\infty$ " + " $-\infty$ " $= -\infty$, as the following example shows (compare with Example 1.4.10).

Let $\mathbf{X} = \{0, \pm 1, \pm 2, \ldots\}$, $\mathbf{A} = \{0, 1\}$, $T = 1$,

$$P_0(x) = \begin{cases} \frac{6}{x^2 \pi^2}, & \text{if } x > 0; \\ 0 & \text{otherwise,} \end{cases}$$

$$p_1(y|x, 0) = \begin{cases} 1/4, & \text{if } y = 2x; \\ 3/4, & \text{if } y = -2x; \\ 0 & \text{otherwise,} \end{cases} \qquad p_1(y|x, 1) = I\{y = -x\},$$

$$c_1(x, a) = x + a, \qquad C(x) = x$$

(see Fig. 1.16).

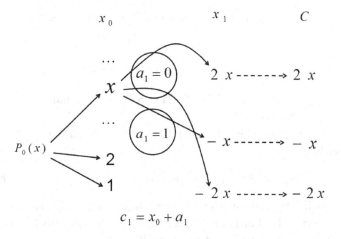

Fig. 1.16 Example 1.4.12: no boundaries for v^π.

The dynamic programming approach results in the following:

$$v_1(x) = x, \qquad v_0(x) = 0,$$

and $a_1 = 0$ provides the minimum in the formula

$$v_0(x_0) = \min_a \{x_0 + a - x_0\} = 0.$$

At the same time, for control strategy $\varphi_1^0(x_1) = 0$, we have for $W = X_0 + X_1$

$$E_{P_0}^{\varphi^0}[W^+] = \sum_{i=1}^{\infty} 3i \cdot 1/4 \cdot \frac{6}{i^2 \pi^2} = +\infty, \quad E_{P_0}^{\varphi^0}[W^-] = \sum_{i=1}^{\infty} (-i) \cdot 3/4 \cdot \frac{6}{i^2 \pi^2} = -\infty,$$

so that $v^{\varphi^0} = -\infty < v_0(x_0) = 0$. Incidentally, for the selector $\varphi_1^1(x_1) = 1$ we have

$$W = X_0 + 1 + X_1 = X_0 + 1 - X_0 = 1,$$

and $v^{\varphi^1} = 1 > v_0(x_0) = 0$. Thus, the solution to the optimality equation (1.4) provides no boundaries to the performance functional.

Everywhere else in this book we accept formula (1.24).

Remark 1.2. All the pathological situations in Sections 1.4.9–1.4.12 appear only because we encounter expressions "$+\infty$" + "$-\infty$" when calculating expectations. That is why people impose the following conditions: for every strategy π, $\forall x_0$, $\forall t$

$$E_{x_0}^\pi[r_t^+(x_{t-1}, a_t)] < +\infty \text{ and } E_{x_0}^\pi[C^+(x_T)] < +\infty$$

or

$$E_{x_0}^\pi[r_t^-(x_{t-1}, a_t)] > -\infty \text{ and } E_{x_0}^\pi[C^-(x_T)] > -\infty.$$

(See [Bertsekas and Shreve(1978), Section 8.1] and [Hernandez-Lerma and Lasserre(1999), Section 9.3.2].)

To guarantee this, one can restrict oneself to "negative" or "positive" models with

$$c_t(x, a) \leq K, \quad C(x) \leq K, \text{ or with } c_t(x, a) \geq -K, \quad C(x) \geq -K.$$

Another possibility is to consider "contracting" models [Altman(1999), Section 7.7]; [Hernandez-Lerma and Lasserre(1999), Section 8.3.2], where, for some positive function $\nu(x)$ and constant K,

$$\int_\mathbf{X} \nu(y) p_t(dy|x, a) \leq K\nu(x) \text{ and } \frac{|c_t(x, a)|}{\nu(x)} \leq K, \quad \frac{C(x)}{\nu(x)} \leq K$$

for all t, x, a.

1.4.13 Nowhere-differentiable function $v_t(x)$; discontinuous function $v_t(x)$

It is known that a functional series can converge (absolutely) to a continuous, but nowhere-differentiable function. As an example, take

$$f(x) = \sum_{i=0}^\infty \left(\frac{1}{2}\right)^i [\cos(7^i \cdot \pi x) + 1], \quad x \in \mathbb{R}$$

[Gelbaum and Olmsted(1964), Section 3.8]. Note that $f_n(x) \geq f_{n-1}(x)$, where

$$f_n(x) \triangleq \sum_{i=0}^{n} \left(\frac{1}{2}\right)^i [\cos(7^i \cdot \pi x) + 1] \in \mathbf{C}^\infty.$$

On the other hand, we also know that the function

$$g(a) = \begin{cases} \exp\left[-\dfrac{1}{a^2} e^{-\frac{1}{(1-a)^2}}\right], & \text{if } 0 < a < 1; \\ 0, & \text{if } a = 0; \\ 1, & \text{if } a = 1 \end{cases} \tag{1.25}$$

is strictly increasing on $[0,1]$ and belongs to $\mathbf{C}^\infty$, and $\left.\frac{dg}{da}\right|_{a=0} = \left.\frac{dg}{da}\right|_{a=1} = 0$.

Now put

$$c_1(x,a) = -f_{n-1}(x) - \left(\frac{1}{2}\right)^n [\cos(7^n \cdot \pi x) + 1]g(a - n + 1) \text{ if } a \in [n-1, n],$$

where $n \in \mathbb{N}$, $x \in \mathbf{X} = [0,2]$, $a \in \mathbf{A} = \mathbb{R}^+ = [0, \infty)$.

In the MDP $\{\mathbf{X}, \mathbf{A}, T = 1, p, c, C \equiv 0\}$ with an arbitrary transition probability $p_1(y|x,a)$, we have

$$v_1(x) = C(x) = 0 \text{ and } v_0(x) = \inf_{a \in \mathbf{A}} c_1(x,a) = f(x),$$

so that $c_1(\cdot), C(\cdot) \in \mathbf{C}^\infty$ but function $v_0(x)$ is continuous and nowhere differentiable (see Figs 1.17–1.20).

One can easily construct a similar example where $v_0(x)$ is discontinuous, although $c_1(\cdot) \in \mathbf{C}^\infty$:

$$c_1(x,a) = h_n(x) + [h_{n+1}(x) - h_n(x)]g(a - n + 1), \text{ if } a \in [n-1, n], \ n \in \mathbb{N},$$

where

$$h_n(x) = \begin{cases} 0, & \text{if } x \leq 1 - \frac{1}{n}; \\ g(1 - n + nx), & \text{if } 1 - \frac{1}{n} \leq x \leq 1; \\ 1, & \text{if } x \geq 1, \end{cases}$$

and function $g(\cdot)$ is defined as in (1.25). Now

$$v_1(x) = C(x) = 0 \text{ and } v_0(x) = \begin{cases} 1, & \text{if } x \geq 1; \\ 0, & \text{if } x < 1 \end{cases} \tag{1.26}$$

(see Figs 1.21 and 1.22).

Remark 1.3. In general, Theorem A.14 proved in [Bertsekas and Shreve(1978), Statements 7.33 and 7.34] can be useful. Incidentally, if A is compact then function $\inf_{a \in \mathbf{A}} C(x,a)$ is continuous provided C is continuous.

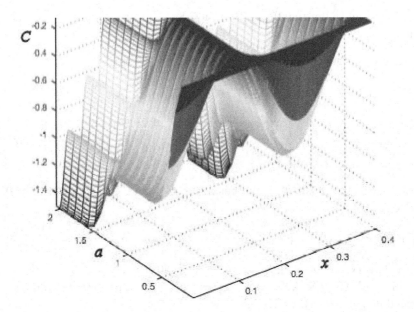

Fig. 1.17   Example 1.4.13: function $c_1(x,a)$ on a small area $x \in [0,\ 0.4]$, $a \in [0,2]$.

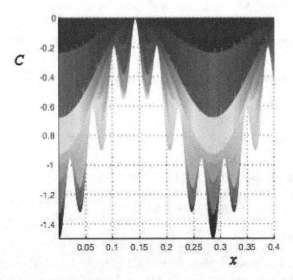

Fig. 1.18   Example 1.4.13: projection on the plane $x \times c$ of the function from Fig. 1.17.

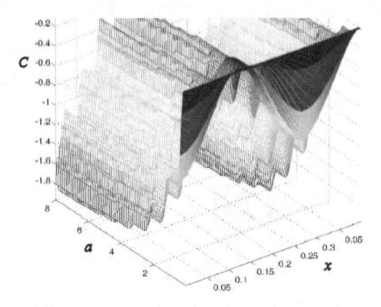

Fig. 1.19 Example 1.4.13: function $c_1(x, a)$ on a greater area $x \in [0,\ 0.4]$, $a \in [0, 8]$.

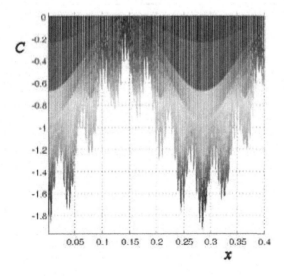

Fig. 1.20 Example 1.4.13: projection on the plane $x \times c$ of the function from Fig. 1.19.

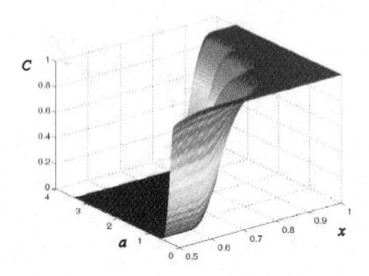

Fig. 1.21 Example 1.4.13: graph of function $c_1(x,a)$ on subset $0.5 \leq x \leq 1$, $a \in [0,3]$ (discontinuous function $v_t(x)$).

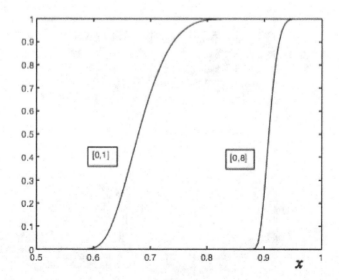

Fig. 1.22 Example 1.4.13: graphs of functions $\inf_{a \in [0,1]} c_1(x,a)$ and $\inf_{a \in [0,8]} c_1(x,a)$ for $c_1(x,a)$ shown in Fig. 1.21 (discontinuous function $v_t(x)$).

1.4.14 The non-measurable Bellman function

The dynamic programming approach is based on the assumption that optimality equation (1.4) has a measurable solution. The following example [Bertsekas and Shreve(1978), Section 8.2, Example 1] shows that the Bellman function $v_t(x)$ may be not Borel-measurable even in the simplest case having $T = 1$, $C(x) \equiv 0$ with a measurable loss function $c_1(x, a)$.

Let $\mathbf{X} = [0, 1]$ and $\mathbf{A} = \mathcal{N}$, the Bair null space; $c_1(x, a) = -I\{(x, a) \in B\}$, where B is a closed (hence Borel-measurable) subset of $\mathbf{X} \times \mathbf{A}$ with projection $B^1 = \{x \in \mathbf{X} : \exists a \in \mathbf{A} : (x, a) \in B\}$ that is not Borel (see Fig. 1.23).

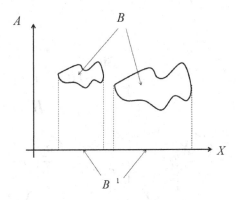

Fig. 1.23 Example 1.4.14: non-measurable Bellman function.

Now the function

$$v_0(x) = \inf_{a \in \mathbf{A}} \{c_1(x, a)\} = -I\{x \in B^1\}$$

is not Borel-measurable. A similar example can be found in [Mine and Osaki(1970), Section 6.3].

Incidentally, the Bellman function $v_t(x)$ is lower semi-analytical if the loss functions $c_t(x, a)$ and $C(x)$ are all Borel-measurable and bounded below (or above) [Bertsekas and Shreve(1978), Corollary 8.2.1].

1.4.15 *No one strategy is uniformly ε-optimal*

This example was first published in [Blackwell(1965)] and in [Strauch(1966), Ex. 4.1].

Let $T = 2$; $\mathbf{X} = B \cup [0,1] \cup \{x_\infty\}$, where, similarly to Example 1.4.14, B is a Borel subset of the rectangle $Y_1 \times Y_2 = [0,1] \times [0,1]$ with projection $B^1 = \{y_1 \in Y_1 : \exists y_2 \in Y_2 : (y_1, y_2) \in B\}$ that is not Borel. One can find the construction of such a set in [Dynkin and Yushkevich(1979), Appendix 2, Section 5]. We put $\mathbf{A} = [0,1]$, $p_1(B \cup \{x_\infty\}|(y_1, y_2), a) \equiv 0$; $p_1(\Gamma|(y_1, y_2), a) = I\{\Gamma \ni y_1\}$ for all $a \in \mathbf{A}$, $x = (y_1, y_2) \in B$, $\Gamma \subseteq [0,1]$; $p_1(\Gamma|x, a) = I\{\Gamma \ni x_\infty\}$ for all $a \in \mathbf{A}$, $x \in [0,1] \cup \{x_\infty\}$, $\Gamma \in \mathcal{B}(\mathbf{X})$. The transition probability p_2 does not play any role since we put $C(x) \equiv 0$. The loss function is $c_1(x, a) \equiv 0$ for all $x \in \mathbf{X}, a \in \mathbf{A}$; $c_2(x, a) \equiv 0$ if $x \in B \cup \{x_\infty\}$; $c_2(x, a) = -I\{(x, a) \in B\}$ for $x \in [0,1]$. See Fig. 1.24.

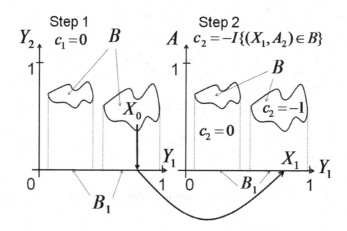

Fig. 1.24 Example 1.4.15: no uniformly ε-optimal Markov strategies.

Action A_1 plays no role.

For any $x \in [0,1] \cup \{x_\infty\}$, for any π $v_x^\pi = 0$ and $v_x^* = 0$. But for any $x = (y_1, y_2) \in B$, on step 2, one can choose $a_2 = y_2$ (or any other action $a \in \mathbf{A}$ such that $(y_1, a) \in B$). The total loss will equal -1. Hence, $v_x^* = -1$ for $x \in B$. Similarly, for a realized history (x_0, a_1, x_1),

$$v^*_{(x_0, a_1, x_1)} = \begin{cases} 0, & \text{if } x_1 = x_\infty \text{ (case } x_0 \in [0,1] \cup \{x_\infty\}); \\ -1, & \text{if } x_1 \in [0,1] \cap B^1 \text{ (case } x_0 \in B), \end{cases}$$

and $v^*_{(x_0,a_1,x_1,a_2,x_2)} \equiv 0$. (Note that x_1 cannot belong to B or to $[0,1] \setminus B^1$ with probability 1.) Incidentally, the Bellman equation (1.4) has an obvious solution

$$v_2(x) \equiv 0; \quad v_1(x) = \begin{cases} 0, & \text{if } x \in B \cup \{x_\infty\} \\ & \text{or } x \in [0,1] \setminus B^1; \\ -1, & \text{if } x \in [0,1] \cap B^1; \end{cases}$$

$$v_0(x) = \begin{cases} -1, & \text{if } x \in B; \\ 0 & \text{otherwise.} \end{cases}$$

Note that function $v_1(x)$ is not measurable because B^1 is not Borel-measurable.

Any strategy π^*, such that $\pi_2^*(\Gamma|x_0, a_1, x_1) = I\{\Gamma \ni y_2\}$ for $\Gamma \in \mathcal{B}(\mathbf{A})$, in the case $x_0 = (y_1, y_2) \in B$, is optimal for all $x_0 \in B$. On the other hand, suppose $\pi_t^m(\Gamma|x)$ is an arbitrary Markov strategy. Then

$$\{x \in B : v_x^{\pi^m} < 0\} = \{(y_1, y_2) \in B : E^{\pi^m}_{(y_1, y_2)}[c_2(X_1, A_2)] < 0\}$$

$$= \{(y_1, y_2) \in B : \int_{\mathbf{A}} I\{(y_1, a) \in B\} \pi_2^m(da|y_1) < 0\}.$$

But the set

$$\{y_1 \in [0,1] : \int_{\mathbf{A}} I\{(y_1, a) \in B\} \pi_2^m(da|y_1) < 0\}$$

is a measurable subset of B^1, and hence different from B^1; thus there is $\hat{y}_1 \in B^1$ such that

$$\int_{\mathbf{A}} I\{(\hat{y}_1, a) \in B\} \pi_2^m(da|\hat{y}_1) = 0.$$

Therefore, for each $x \in B$ of the form $(\hat{y}_1, y_2)$ we have $v_x^{\pi^m} = 0$. A Markov strategy, ε-optimal simultaneously for all $x_0 \in B$, does not exist if $\varepsilon < 1$.

In this model, there are no uniformly ε-optimal strategies for $\varepsilon < 1$ because, similarly to the above reasoning, for any measurable stochastic kernel $\pi_1(da|x)$ on $\mathbf{A}$ given $[0,1]$, there is $\hat{x} \in [0,1] \cap B^1$ such that

$$\int_{\mathbf{A}} I\{(\hat{x}, a) \in B\} \pi_1(da|\hat{x}) = 0,$$

i.e. $v_{\hat{x}}^\pi = 0 > v_{\hat{x}}^* = -1$.

Remark 1.4. Both in Examples 1.4.14 and 1.4.15, we have a "two-dimensional" Borel set B whose projection B^1 is not Borel. Note that in the first case, B is closed, but $\mathbf{A}$ is not compact. On the other hand, in Example 1.4.15 Y_2 is compact, but B is only measurable. We emphasize that one cannot have simultaneously closed B and compact $\mathbf{A}$ (or Y_2), because in this case projection B^1 would have been closed: it is sufficient to apply Theorem A.14 to function $-I\{(x,a) \in B\}$ (we use the notation of Example 1.4.14).

Another example, in which there are no uniformly ε-optimal selectors, but the Bellman function is well defined, can easily be built using the construction of Example 3.2.8.

1.4.16 Semi-continuous model

MDP is called *semi-continuous* if the following condition is satisfied.

Condition 1.1.

(a) The action space $\mathbf{A}$ is compact,
(b) The transition probability $p_t(dy|x,a)$ is a continuous stochastic kernel on $\mathbf{X}$ given $\mathbf{X} \times \mathbf{A}$, and
(c) The loss functions $c_t(x,a)$ and $C(x)$ are lower semi-continuous and bounded below.

In such models, the function $v_t(x)$ is also lower semi-continuous and bounded below. Moreover, there exists a (measurable) selector $\varphi_t^*(x)$ providing the required minimum:

$$v_{t-1}(x) = c_t(x, \varphi_t^*(x)) + \int_{\mathbf{X}} v_t(y) p_t(dy|x, \varphi_t^*(x))$$

[Hernandez-Lerma and Lasserre(1996a), Section 3.3.5],[Bertsekas and Shreve(1978), Statement 8.6], [Piunovskiy(1997), Section 1.1.4.1].

If the action space is not compact, or the transition probability is not continuous, or the loss functions are not lower semi-continuous, then trivial examples show that the desired selector φ^* may not exist. The following example based on [Feinberg(2002), Ex. 6.2] confirms that the boundedness below of the loss functions is also important.

Let $T = 1$, $\mathbf{X} = \{\Delta, 0, 1, 2, \ldots\}$ with the discrete topology. Suppose $\mathbf{A} = \{\infty, 1, 2, \ldots\}$, action ∞ being the one-point compactification of the sequence $1, 2, \ldots$, that is, $\lim_{a \to \infty} a = \infty$. We put $P_0(0) = 1$;

$$\text{for } a \neq \infty, \quad p_1(y|x,a) = \begin{cases} 1/a, & \text{if } y = a; \\ (a-1)/a, & \text{if } y = \Delta; \\ 0 & \text{otherwise;} \end{cases} \quad p_1(\Delta|x,\infty) = 1,$$

$$c_1(x,a) \equiv 0, \qquad C(x) = \begin{cases} 1-x, & \text{if } x \neq \Delta; \\ 0, & \text{if } x = \Delta \end{cases}$$

(see Fig. 1.25).

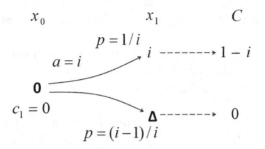

Fig. 1.25 Example 1.4.16: not a semi-continuous model.

This MDP satisfies all the conditions 1.1 apart from the requirement that the loss functions be bounded below. One can easily calculate

$$v_1(x) = C(x) = \begin{cases} 1-x, & \text{if } x \neq \Delta; \\ 0, & \text{if } x = \Delta \end{cases}$$

$$v_0(x) = \min\left\{\inf_{a\in\{1,2,\ldots\}}\left\{\frac{1}{a}[1-a]\right\};\ 0\right\} = -1$$

(in the last formula, zero corresponds to action ∞). But for any $a \in \mathbf{A}$,

$$c_1(0,a) + \sum_{y\in\mathbf{X}} v_1(y)p_1(y|0,a) > v_0(0) = -1,$$

and no one strategy is optimal.

One can reformulate this model in the following way. $T = 1$, $\mathbf{X} = \{0, \Delta\}$, $\mathbf{A} = \{\infty, 1, 2, \ldots\}$, $P_0(0) = 1$, $p_1(\Delta|x,a) \equiv 1$, $c_1(\Delta,a) = 0$, $c_1(0,a) = \frac{1}{a} - 1$ if $a \neq \infty$, $c_1(0,\infty) = 0$, $C(x) \equiv 0$. Now the loss function is bounded below, but it ceases to be lower semi-continuous: $\lim_{a\to\infty} c_1(0,a) = -1 < c_1(0,\infty) = 0$.

Using Remark 2.6, one can make the loss functions uniformly bounded; however, the time horizon will be infinite. According to Theorem 2.6, one can guarantee the existence of an optimal stationary selector only under additional conditions (e.g. if the state space is finite). See also [Bertsekas and Shreve(1978), Corollary 9.17.2]: an optimal stationary selector exists in semicontinuous *positive* homogeneous models with the total expected loss.

The next example shows that the Bellman function $v_t(\cdot)$ may not necessarily be lower semi-continuous if the action space $\mathbf{A}$ is not compact, even when the infimum in the optimality equation (1.4) is attained at every $x \in \mathbf{X}$.

If the space $\mathbf{A}$ is not compact, it is convenient to impose the following additional condition: the loss function $c_t(x, a)$ is *inf-compact* for any $x \in \mathbf{X}$, i.e. the set $\{a \in \mathbf{A}: c_t(x,a) \leq r\}$ is compact for each $r \in \mathbb{R}^1$.

Now the infimum in equation

$$v_{t-1}(x) = \inf_{a \in \mathbf{A}} \left\{ c_t(x,a) + \int_{\mathbf{X}} v_t(y) p_t(dy|x,a) \right\}$$

is provided by a measurable selector $\varphi_t^*(x)$, if function $v_t(\cdot)$ is bounded below and lower semi-continuous. The function in the parentheses is itself bounded below, lower semi-continuous and inf-compact for any $x \in \mathbf{X}$ [Hernandez-Lerma and Lasserre(1996a), Section 3.3.5]. To apply the mathematical induction method, what remains is to find conditions which guarantee that the function

$$v(x) = \inf_{a \in \mathbf{A}} \{f(x,a)\}$$

is lower semi-continuous for a lower semi-continuous inf-compact (for any $x \in \mathbf{X}$) function $f(\cdot)$. This problem was studied in [Luque-Vasquez and Hernandez-Lerma(1995)]; it is, therefore, sufficient to require in addition that the multifunction

$$x \to \mathbf{A}^*(x) = \{a \in \mathbf{A}: \ v(x) = f(x,a)\} \tag{1.27}$$

is *lower semi-continuous*; that is, the set $\{x \in \mathbf{X}: \ \mathbf{A}^*(x) \cap \Gamma \neq \emptyset\}$ is open for every open set $\Gamma \subseteq \mathbf{A}$.

The following example from [Luque-Vasquez and Hernandez-Lerma(1995)] shows that the last requirement is essential.

Let $\mathbf{X} = \mathbb{R}^1$, $\mathbf{A} = [0, \infty)$, and let

$$f(x,a) = \begin{cases} 1+a, & \text{if } x \leq 0 \text{ or } x > 0, \ 0 \leq a \leq \frac{1}{2x}; \\ (2+\frac{1}{x}) - (2x+1)a, & \text{if } x > 0, \ \frac{1}{2x} \leq a \leq \frac{1}{x}; \\ a - \frac{1}{x}, & \text{if } x > 0, \ a \geq \frac{1}{x} \end{cases}$$

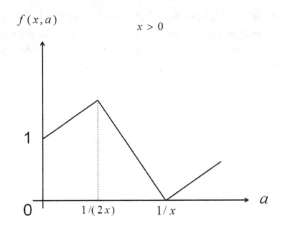

Fig. 1.26   Example 1.4.16: function $f(x,a)$.

(see Fig. 1.26).

Function $f(\cdot)$ is non-negative, continuous and inf-compact for any $x \in$ **X**. But multifunction (1.27) (it is actually a function) has the form

$$\mathbf{A}^*(x) = \begin{cases} 0, & \text{if } x \le 0; \\ \frac{1}{x}, & \text{if } x > 0, \end{cases}$$

and is not lower semi-continuous because the set $\Gamma \stackrel{\triangle}{=} [0, r)$ is open in **A** for $r > 0$ and the set

$$\{x \in \mathbf{X}: \mathbf{A}^*(x) \in \Gamma\} = (-\infty, 0] \cup (1/r, \infty)$$

is not open in **X**. The function $v(x) = \inf_{a \in \mathbf{A}} \{f(x,a)\} = \begin{cases} 1, & \text{if } x \le 0; \\ 0, & \text{if } x > 0 \end{cases}$

is not lower semi-continuous; it is upper semi-continuous (see Theorem A.14).

As for the classical semi-continuous MDP, we mention Example 2.2.15. Any finite-horizon model can be considered as an infinite-horizon MDP with expected total loss. It is sufficient to introduce an artificial cemetery state and to make the process absorbed there at time T without any future loss. (The model becomes homogeneous if we incorporate the number of the decision epoch as the second component of the state: (x,t).) The definition of a semi-continuous model given in Section 2.2.15 differs from the one presented at the beginning of the current section, however, in a similar way to Example 2.2.16, one can usually introduce different topologies, so that one or two of the requirements (a,b,c) will be satisfied. Discussions of several slightly different semi-continuity conditions which guarantee the existence of (uniformly) optimal selectors can be found in [Hernandez-Lerma and Lasserre(1996a), Section 3.3] and [Schäl(1975a)].

Chapter 2

Homogeneous Infinite-Horizon Models: Expected Total Loss

2.1 Homogeneous Non-discounted Model

We now assume that the time horizon $T = \infty$ is not finite. The definitions of (Markov, stationary) strategies and selectors are the same as in Chapter 1. For a given initial distribution $P_0(\cdot)$ and control strategy π, the *strategic measure* $P_{P_0}^\pi(\cdot)$ is built in a similar way to that in Chapter 1; the rigorous construction is based on the Ionescu Tulcea Theorem [Bertsekas and Shreve(1978), Prop. 7.28], [Hernandez-Lerma and Lasserre(1996a), Prop. C10], [Piunovskiy(1997), Th. A1.11]. The goal is to find an optimal control strategy π^* solving the problem

$$v^\pi = E_{P_0}^\pi \left[\sum_{t=1}^\infty c(X_{t-1}, A_t) \right] \to \inf_\pi . \qquad (2.1)$$

As usual, v^π is called the *performance functional*.

In the current chapter, the following condition is always assumed to be satisfied.

Condition 2.1. For any control strategy π, either $E_{P_0}^\pi \left[\sum_{t=1}^\infty c^+(X_{t-1}, A_t) \right]$ or $E_{P_0}^\pi \left[\sum_{t=1}^\infty c^-(X_{t-1}, A_t) \right]$ is finite.

Note that the loss function $c(x,a)$ and the transition probability $p(dy|x,a)$ do not depend on time. Such models are called *homogeneous*. As before, we write $P_{x_0}^\pi$ and $v_{x_0}^\pi$ if the initial distribution is concentrated at a single point $x_0 \in \mathbf{X}$. In this connection,

$$v_x^* \stackrel{\triangle}{=} \inf_\pi v_x^\pi$$

is the Bellman function. We call a strategy π (uniformly) ε-optimal if (for all x) $v_x^\pi \leq \begin{cases} v_x^* + \varepsilon, & \text{if } v_x^* > -\infty; \\ -\frac{1}{\varepsilon}, & \text{if } v_x^* = -\infty; \end{cases}$ a (uniformly) 0-optimal strategy is called (uniformly) optimal; here $-\frac{1}{0}$ should be replaced by $-\infty$. Note that this definition of uniform optimality is slightly different from the one given in Chapter 1.

Remark 2.1. As mentioned at the beginning of Section 1.4.9, Markov strategies are sufficient for solving optimization problems if the initial distribution is fixed (Lemma 2 in [Piunovskiy(1997)] holds for $T = \infty$, too). Since the uniform optimality concerns all possible initial states, it can happen that only a semi-Markov strategy is uniformly (ε-) optimal: see Sections 2.2.12, 2.2.13 and Theorems 2.1 and 2.7.

The Bellman function v_x^* satisfies the optimality equation

$$v(x) = \inf_{a \in \mathbf{A}} \left\{ c(x,a) + \int_{\mathbf{X}} v(y) p(dy|x,a) \right\}, \qquad (2.2)$$

except for pathological situations similar to those described in Chapter 1, where expressions " $+\infty$ " + " $-\infty$ " appear. In the finite-horizon case, the minimal expected loss coincides with the solution of (1.4), except for those pathological cases. If $T = \infty$, that is not the case, because it is obvious that if $v(\cdot)$ is a solution of (2.2) then $v(\cdot) + r$ is also a solution for any $r \in \mathbb{R}$. Moreover, as Example 2.2.2 below shows, there can exist many other non-trivial solutions of (2.2). Thus, generally speaking, a solution of the optimality equation (2.2) provides no boundaries for the Bellman function.

A stationary selector φ is called *conserving* (or *thrifty*) if

$$v_x^* = c(x, \varphi(x)) + \int_{\mathbf{X}} v_y^* p(dy|x, \varphi(x));$$

it is called *equalizing* if

$$\forall x \in \mathbf{X} \quad \lim_{T \to \infty} E_x^\varphi \left[v_{X_T}^* \right] \geq 0$$

[Puterman(1994), Section 7.1.3].

It is known that a conserving and equalizing stationary selector φ is (uniformly) optimal, i.e. $v_x^\varphi = v_x^*$, under very general conditions [Puterman(1994), Th. 7.1.7], in fact as soon as the following representation holds:

$$E_x^\varphi \left[\sum_{t=1}^T c(X_{t-1}, A_t) \right] = v_x^* + E_x^\varphi \left[\sum_{t=1}^T \{ c(X_{t-1}, A_t) \right.$$
$$\left. + \int_{\mathbf{X}} v_y^* p(dy|x, \varphi(X_{t-1})) - v_{X_{t-1}}^* \} \right] - E_x^\varphi [v_{X_T}^*].$$

Below, we provide several general statements that are proved, for example, in [Puterman(1994)]. That book is mainly devoted to *discrete* models, where the spaces **X** and **A** are countable or finite.

Suppose the following conditions are satisfied:

Condition 2.2. [Puterman(1994), Section 7.2].

(a) $\forall x \in \mathbf{X}, \forall \pi \in \Delta^{\text{All}}$

$$E_x^\pi \left[\sum_{t=1}^\infty c^-(X_{t-1}, A_t) \right] > -\infty;$$

(b) $\forall x \in \mathbf{X}, \exists a \in \mathbf{A} \quad c(x, a) \leq 0$.

Then, $v_x^* \leq 0$ is the maximal non-positive solution to (2.2).

In *positive* models, where $c(x, a) \geq 0$, any strategy is equalizing and v_x^* is the minimal non-negative solution to (2.2) provided $\exists \pi \in \Delta^{\text{All}}: \forall x \in \mathbf{X} \; v_x^\pi < +\infty$ [Puterman(1994), Th. 7.3.3]. We call a model *negative* if $c(x, a) \leq 0$.

Theorem 2.1. *[Bertsekas and Shreve(1978), Props 9.19 and 9.20]. In a positive (negative) model, for each $\varepsilon > 0$, there exists a uniformly ε-optimal Markov (semi-Markov) selector.*

Recall that an MDP is called *absorbing* if there is a state (say 0 or Δ), for which the controlled process is absorbed at time $T_0 \stackrel{\Delta}{=} \min\{t \geq 0 : X_t = 0\}$ and $\forall \pi \; E_{P_0}^\pi[T_0] < \infty$. All the future loss is zero: $c(0, a) \equiv 0$. Absorbing models are considered in Sections 2.2.2, 2.2.7, 2.2.10, 2.2.13, 2.2.16, 2.2.17, 2.2.19, 2.2.20, 2.2.21, 2.2.24, 2.2.28.

The examples in Sections 2.2.3, 2.2.4, 2.2.9, 2.2.13, 2.2.18 are from the area of *optimal stopping* in which, on each step, there exists the possibility of putting the controlled process in a special absorbing state (say 0, or Δ), sometimes called *cemetery*, with no future loss. Note that optimal stopping problems are not always about absorbing MDP: the absorption may be indefinitely delayed, as in the examples in Sections 2.2.4, 2.2.9, 2.2.18.

Many examples from Chapter 1, for example the conventions on the infinities, can be adjusted for the infinite-horizon case.

2.2 Examples

2.2.1 *Mixed Strategies*

The space of all probability measures on the space of trajectories $\mathbf{H} = \mathbf{X} \times (\mathbf{A} \times \mathbf{X})^\infty$ is a Borel space (Theorems A.5, A.9). In this connection, under a fixed initial distribution P_0, the spaces of strategic measures $\mathcal{D}^\Delta = \{P_{P_0}^\pi, \pi \in \Delta\}$ are known to be measurable and hence Borel for $\Delta = \Delta^{\text{All}} = \{\text{all strategies}\}$, $\Delta = \Delta^{\text{M}} = \{\text{Markov strategies}\}$, $\Delta = \Delta^{\text{S}} = \{\text{stationary strategies}\}$, and for Δ^{AllN}, Δ^{MN}, Δ^{SN}, where letter N corresponds to non-randomized strategies (see [Feinberg(1996), Sections 2 and 3]). Below, we use the notation $\mathcal{D}$ for $\Delta = \Delta^{\text{All}}$, $\mathcal{D}^{\text{N}}$ for $\Delta = \Delta^{\text{AllN}}$ and so on.

Now, we say that a strategy π is *mixed* if, for some probability measure $\nu(dP)$ on $\mathcal{D}^{\text{N}}$,

$$P_{P_0}^\pi = \int_{\mathcal{D}^{\text{N}}} P \nu(dP). \tag{2.3}$$

Similarly, a Markov (stationary) strategy π is called *mixed* if

$$P_{P_0}^\pi = \int_{\mathcal{D}^{\text{MN}}} P \nu(dP) \qquad \left(P_{P_0}^\pi = \int_{\mathcal{D}^{\text{SN}}} P \nu(dP) \right).$$

Incidentally, the space $\mathcal{D}$ is convex and, for any probability measure ν on $\mathcal{D}$,

$$P^\nu \stackrel{\triangle}{=} \int_{\mathcal{D}} P\nu(dP) \in \mathcal{D}. \tag{2.4}$$

According to [Feinberg(1996), Th. 5.2], any general strategy $\pi \in \Delta^{\text{All}}$ is mixed, and any Markov strategy $\pi \in \Delta^{\text{M}}$ is mixed as well. (Examples 5.3 and 5.4 in [Feinberg(1996)] confirm that measure ν here is not necessarily unique.) The following example from [Feinberg(1987), Remark 3.1] shows that the equivalent statement does not hold for stationary strategies.

Let $\mathbf{X} = \{0\}$, $\mathbf{A} = \{0, 1\}$, $p(0|0, a) \equiv 1$, and consider the stationary randomized strategy $\pi^s(a|0) = 0.5$, $a \in \mathbf{A}$. In this model, we have only two non-randomized stationary strategies $\varphi^0(0) = 0$ and $\varphi^1(0) = 1$. If

$$P_{P_0}^{\pi^s} = \int_{\mathcal{D}^{\text{SN}}} P\nu(dp) = \alpha P_{P_0}^{\varphi^0} + (1-\alpha) P_{P_0}^{\varphi^1} \quad \text{for } \alpha \in [0,1],$$

then measure $P_{P_0}^{\pi^s}$ would have been concentrated on two trajectories ($x_0 = 0, a_1 = 0, x_1 = 0, a_2 = 0, \ldots$) and ($x_0 = 0, a_1 = 1, x_1 = 0, a_2 = 1, \ldots$) only, which is not the case. At the same time, for each $t = 1, 2, \ldots$

$$P_{P_0}^{\pi^s}(A_t = 0) = \frac{1}{2} P_{P_0}^{\varphi^0}(A_t = 0) + \frac{1}{2} P_{P_0}^{\varphi^1}(A_t = 0) = \frac{1}{2}.$$

The following example from [Piunovskiy(1997), Remark 34] shows that if φ^1 and φ^2 are two stationary selectors, then it can happen that the equality

$$P^\pi_{P_0}(X_{t-1} \in \Gamma^X, A_t \in \Gamma^A) = \frac{1}{2} P^{\varphi^1}_{P_0}(X_{t-1} \in \Gamma^X, A_t \in \Gamma^A)$$
$$+ \frac{1}{2} P^{\varphi^2}_{P_0}(X_{t-1} \in \Gamma^X, A_t \in \Gamma^A), \quad \Gamma^X \in \mathcal{B}(\mathbf{X}), \Gamma^A \in \mathcal{B}(\mathbf{A}), \ t = 1, 2, \ldots$$
(2.5)

holds for no one stationary strategy π.

Let $\mathbf{X} = \{1, 2\}$, $\mathbf{A} = \{1, 2\}$, $p(1|1,1) = 1$, $p(2|1,2) = 1$, $p(2|2,a) \equiv 1$, with the other transition probabilities zero (see Fig. 2.1).

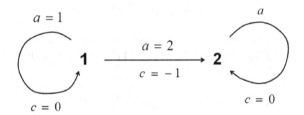

Fig. 2.1 Examples 2.2.1 and 2.2.3.

Suppose $P_0(1) = 1$, $\varphi^1(x) \equiv 1$, $\varphi^2(x) \equiv 2$. If we take $t = 1$, $\Gamma^X = \{1\}$ and $\Gamma^A = \{1\}$, then (2.5) implies $\pi_1(1|1) = \frac{1}{2}$ and

$$P^\pi_{P_0}(X_1 = 1) = \frac{1}{2} = \frac{1}{2} P^{\varphi^1}_{P_0}(X_1 = 1). \tag{2.6}$$

Now

$$\frac{1}{2} P^{\varphi^1}_{P_0}(X_1 = 1, \ A_2 = 1) + \frac{1}{2} P^{\varphi^2}_{P_0}(X_1 = 1, \ A_2 = 1)$$

$$= \frac{1}{2} P^{\varphi^1}_{P_0}(X_1 = 1, \ A_2 = 1) = \frac{1}{2},$$

and it follows from (2.6) that we must put $\pi_2(1|1) = 1$ in order to have (2.5) for $t = 2$, $\Gamma^X = \{1\}$, $\Gamma^A = \{1\}$. Since $\pi_2(1|1) \neq \pi_1(1|1)$, equality (2.5) cannot hold for a stationary strategy π. At the same time, the equality does hold for some Markov strategy $\pi = \pi^m$ [Dynkin and Yushkevich(1979), Chapter 3, Section 8], [Piunovskiy(1997), Lemma 2]; also see Lemma 3.1.

One can consider the performance functional v^π as a real measurable functional on the space of strategic measures $\mathcal{D}$: $v^\pi = V(P_{P_0}^\pi)$. It is concave in the following sense: for each probability measure ν on $\mathcal{D}$,

$$V(P^\nu) \geq \int_\mathcal{D} V(P)\nu(dP),$$

where P^ν is defined in (2.4) [Feinberg(1982)]. In fact, if V is the total loss, then we have equality in the last formula.

Theorem 2.2. *[Feinberg(1982), Th. 3.2] If the performance functional is concave then, for any $P \in \mathcal{D}$, $\forall \varepsilon > 0$, there exists $P^N \in \mathcal{D}^N$ such that*

$$V(P^N) \leq \begin{cases} V(P), & \text{if } V(P) > -\infty; \\ -\frac{1}{\varepsilon}, & \text{if } V(P) = -\infty. \end{cases}$$

Note that the given definition of a concave functional differs from the standard definition: a mapping $V : \mathcal{D} \to \mathbb{R}^1$ is usually called concave if, for any $P^1, P^2 \in \mathcal{D}$, $\forall \alpha \in [0,1]$,

$$V(\alpha P^1 + (1-\alpha)P^2) \geq \alpha V(P^1) + (1-\alpha)P^2.$$

The following example [Feinberg(1982), Ex. 3.1] shows that, if the mapping V is concave (in the usual sense), then Theorem 2.2 can fail.

Let $\mathbf{X} = \{0\}$ be a singleton (there is no controlled process). Put $\mathbf{A} = [0,1]$ and let

$$V(P) = \begin{cases} -1, & \text{if the marginal distribution of } A_1, \text{ i.e. } \pi_1(da|0), \text{ is} \\ & \text{absolutely continuous w.r.t. the Lebesgue measure;} \\ 0 & \text{otherwise.} \end{cases}$$

The mapping V is concave, but for each $P^N \in \mathcal{D}^N$ we have $V(P^N) = 0$, whereas $V(P_{P_0}^\pi) = -1$, if $\pi_1(\cdot|0)$ is absolutely continuous.

2.2.2 Multiple solutions to the optimality equation

Consider a discrete-time queueing model. During each time period t, there may be an arrival of a customer with probability λ or a departure from the queue with probability μ; $\lambda + \mu \leq 1$. State X means there are X customers in the queue. There is no control here, and we wish to compute the expected time for the queue to empty. A similar example was presented in [Altman(1999), Ex. 9.1].

Let $\mathbf{X} = \{0, 1, 2, \ldots\}$, $\mathbf{A} = \{0\}$ (a dummy action), $p(0|0, a) = 1$,

$$\forall x > 0 \quad p(y|x, a) = \begin{cases} \lambda, & \text{if } y = x + 1; \\ \mu, & \text{if } y = x - 1; \\ 1 - \lambda - \mu, & \text{if } y = x; \\ 0 & \text{otherwise;} \end{cases} \quad c(x, a) = I\{x > 0\}.$$

The process is absorbing at zero, and the one-step loss equals 1 at all positive states (see Fig. 2.2).

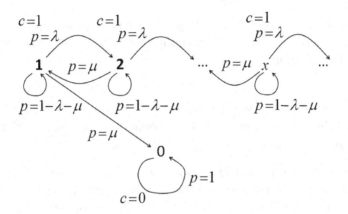

Fig. 2.2 Example 2.2.2: multiple solutions to the optimality equation.

Equation (2.2) has the form

$$v(x) = 1 + \mu v(x - 1) + \lambda v(x + 1) + (1 - \lambda - \mu)v(x), \quad x > 0. \qquad (2.7)$$

In the case where $\lambda \neq \mu$, the general solution to (2.7) is as follows:

$$v(x) = k_1 \left(\frac{\mu}{\lambda}\right)^x + k_2 + \frac{x}{\mu - \lambda}.$$

The Bellman function must satisfy the condition $v_x^* = 0$ at $x = 0$, so that one should put $k_2 = -k_1$. And even now, constant k_1 can be arbitrary. In fact, v_x^* is the *minimal* non-negative solution to (2.7), i.e. $k_1 = k_2 = 0$ in the case where $\mu > \lambda$.

If $\mu < \lambda$ then equation (2.7) has no finite non-negative solutions. Here $v_x^* \equiv \infty$ for $x > 0$.

2.2.3 Finite model: multiple solutions to the optimality equation; conserving but not equalizing strategy

Let $\mathbf{X} = \{1,2\}$, $\mathbf{A} = \{1,2\}$, $p(1|1,1) = 1$, $p(2|1,2) = 1$, $p(2|2,a) \equiv 1$, with other transition probabilities zero; $c(1,1) = 0$, $c(1,2) = -1$, $c(2,a) \equiv 0$ (see Fig. 2.1; similar examples were presented in [Dynkin and Yushkevich(1979), Chapter 4, Section 7], [Puterman(1994), Ex. 7.2.3 and 7.3.1] and [Kallenberg(2010), Ex. 4.1].)

The optimality equation (2.2) is given by

$$\left.\begin{array}{l} v(1) = \min\{v(1); \quad -1 + v(2)\}; \\ v(2) = v(2). \end{array}\right\} \quad (2.8)$$

Any pair of numbers satisfying $v(1) \leq v(2) - 1$ provides a solution. Conditions 2.2 are satisfied, so the Bellman function coincides with the maximal non-positive solution:

$$v_1^* = -1, \quad v_2^* = 0.$$

Any control strategy is conserving, but the stationary selector $\varphi^1(x) \equiv 1$ is not equalizing; $v_1^{\varphi^1} = 0$, $v_2^{\varphi^1} = 0$. In the opposite case, selector $\varphi^2(x) \equiv 2$ is equalizing and hence optimal; $v_1^{\varphi^2} = -1$, $v_2^{\varphi^2} = 0$.

Remark 2.2. For a discounted model with discount factor $\beta \in (0,1)$ (Chapter 3), the optimality equation is given by (3.2). In that case, if the loss function c is bounded, it is known [Puterman(1994), Th. 6.2.2] that $v(x) \geq v_x^*$ (or $v(x) \leq v_x^*$) if function $v(x)$ satisfies the inequality

$$v(x) \geq (\text{ or } \leq) \inf_{a \in \mathbf{A}} \left\{ c(x,a) + \beta \int_{\mathbf{X}} v(y) p(dy|x,a) \right\}.$$

In the current example (with the discount factor $\beta = 1$) this statement does not hold: one can take $v(1) = 0$; $v(2) = 2$ or $v(1) = v(2) = -3$.

2.2.4 The single conserving strategy is not equalizing and not optimal

Let $\mathbf{X} = \{0,1,2,\ldots\}$, $\mathbf{A} = \{1,2\}$, $p(0|0,a) \equiv 1$, $c(0,a) \equiv 0$, $\forall x > 0$ $p(0|x,1) \equiv 1$, $p(x+1|x,2) \equiv 1$, with all other transition probabilities zero; $c(x,1) = 1/x - 1$, $c(x,2) \equiv 0$ (see Fig. 2.3; similar examples were presented in [Bäuerle and Rieder(2011), Ex. 7.4.4], [Bertsekas(2001), Ex. 3.4.4], [Dynkin and Yushkevich(1979), Chapter 6, Section 3], [Puterman(1994), Ex. 7.2.4], [Strauch(1966), Ex. 4.2] and in [van der Wal and Wessels(1984), Ex. 3.4].)

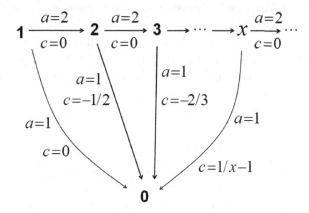

Fig. 2.3   Example 2.2.4: no optimal strategies.

The optimality equation (2.2) is given by
$$v(0) = v(0);$$
for $x > 0$,
$$v(x) = \min\{1/x - 1 + v(0),\ v(x+1)\}. \tag{2.9}$$

We are interested only in the solutions with $v(0) = 0$. Conditions 2.2 are satisfied and v_x^* is the maximal non-positive solution to (2.9), i.e. $v_x^* \equiv -1$ for $x > 0$.

The stationary selector $\varphi^*(x) \equiv 2$ is the single conserving strategy at $x > 0$, but
$$\lim_{t \to \infty} E_x^{\varphi^*}\left[v_{X_t}^*\right] \equiv -1,$$
so that it is not equalizing.

There exist no optimal strategies in this model. The stationary selector φ^* indefinitely delays absorption in state 0, so that the decision maker receives no negative loss: $v_x^{\varphi^*} \equiv 0$.

Note that $\forall \varepsilon > 0$ the stationary selector
$$\varphi^\varepsilon(x) = \begin{cases} 2, & \text{if } x \leq \varepsilon^{-1}; \\ 1, & \text{if } x > \varepsilon^{-1} \end{cases}$$
is (uniformly) ε-optimal: $\forall x > 0\ v_x^{\varphi^\varepsilon} < \varepsilon - 1$.

Suppose now that there is an additional action 3 leading directly to state 0 with cost -1: $p(0|x, 3) \equiv 1$, $c(x, 3) \equiv -1$. Now the stationary selector $\varphi(x) \equiv 3$ is conserving and equalizing, and hence is uniformly optimal.

This example can also be adjusted for a discounted model; see Example 3.2.4.

Consider the following example motivated by [Feinberg(2002), Ex. 6.3]. Let $\mathbf{X} = \{0, 1, 2, \ldots\}$; $\mathbf{A} = \{1, 2\}$; $p(0|0, a) \equiv 1$, $p(0|x, 1) \equiv 1$, $p(x+1|x, 2) = 1$ for $x \geq 1$, with other transition probabilities zero; $c(0, a) \equiv 0$, for $x > 0$ $c(x, 1) = 2^{-x} - 1$, $c(x, 2) = -2^{-x}$ (see Fig. 2.4).

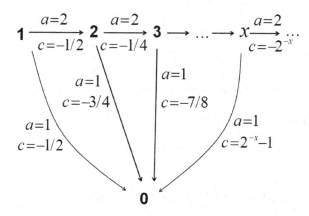

Fig. 2.4 Example 2.2.4: no optimal strategies; stationary selectors are not dominating.

The maximal non-positive solution to the optimality equation (2.2) can be found by value iteration – see (2.10):

$$v_x^* = \begin{cases} 0, & \text{if } x = 0; \\ -2^{-x+1} - 1, & \text{if } x > 0. \end{cases}$$

The stationary selector $\varphi^*(x) \equiv 2$ is the only conserving strategy at $x > 0$, but it is not equalizing and not optimal because $v_x^{\varphi^*} = -2^{-x+1}$ for $x > 0$. There exist no optimal strategies in this model.

Consider the following stationary randomized strategy: $\pi^s(1|x) = \pi^s(2|x) = 1/2$. One can compute $v_x^{\pi^s} \equiv -1$ (for all $x > 0$), e.g. again using the value iteration. Now, if φ is an arbitrary stationary selector, then either $\varphi = \varphi^*$ and $v_2^{\varphi} = -1/2 > v_2^{\pi^s} = -1$, or $\varphi(\hat{x}) = 1$ for some $\hat{x} > 0$, so that $v_{\hat{x}}^{\varphi} = 2^{-\hat{x}} - 1 > v_{\hat{x}}^{\pi^s} = -1$. We conclude that the strategy π^s cannot be dominated by any stationary selector.

2.2.5 When strategy iteration is not successful

If the model is positive, i.e. $c(x,a) \geq 0$, and function c is bounded, then the following theorem holds.

Theorem 2.3.

(a) [Strauch(1966), Th. 4.2] Let π and σ be two strategies and suppose $\exists n_0 \colon \forall n > n_0$ $v_x^{\pi^n \sigma} \leq v_x^{\sigma}$ for all $x \in \mathbf{X}$. Here $\pi^n \sigma = \{\pi_1, \pi_2, \ldots, \pi_n, \sigma_{n+1} \ldots\}$ is the natural combination of the strategies π and σ. Then $v_x^{\pi} \leq v_x^{\sigma}$.

(b) [Strauch(1966), Cor. 9.2] Let φ^1 and φ^2 be two stationary selectors and put

$$\hat{\varphi}(x) \triangleq \begin{cases} \varphi^1(x), & \text{if } v_x^{\varphi^1} \leq v_x^{\varphi^2}; \\ \varphi^2(x), & \text{otherwise.} \end{cases}$$

Then, for all $x \in \mathbf{X}$, $v_x^{\hat{\varphi}} \leq \min\{v_x^{\varphi^1}, v_x^{\varphi^2}\}$.

Example 2.2.4 (Fig. 2.3) shows that this theorem can fail for *negative* models, where $c(x,a) \leq 0$ (see [Strauch(1966), Examples 4.2 and 9.1]).

Statement (a). Let $\pi_t(2|h_{t-1}) \equiv 2$ and $\sigma_t(1|h_{t-1}) \equiv 1$ be stationary selectors. Then $v_0^{\sigma} = 0$; for $x > 0$ $v_x^{\sigma} = 1/x - 1$; $v_0^{\pi^n \sigma} = 0$ and $v_x^{\pi^n \sigma} = 1/(x+n) - 1$ for $x > 0$. Therefore, $v_x^{\pi^n \sigma} \leq v_x^{\sigma}$ for all n, but $v_x^{\pi} = 0 > v_x^{\sigma}$ for $x > 1$.

Statement (b). For $x > 0$ let

$$\varphi^1(x) = \begin{cases} 1, & \text{if } x \text{ is odd;} \\ 2, & \text{if } x \text{ is even;} \end{cases} \qquad \varphi^2(x) = \begin{cases} 1, & \text{if } x \text{ is even;} \\ 2, & \text{if } x \text{ is odd.} \end{cases}$$

Then, for positive odd $x \in \mathbf{X}$,

$$v_x^{\varphi^1} = \frac{1}{x} - 1; \qquad v_x^{\varphi^2} = \frac{1}{x+1} - 1,$$

so that $\hat{\varphi}(x) = \varphi^2(x) = 2$.

For positive even $x \in \mathbf{X}$,

$$v_x^{\varphi^1} = \frac{1}{x+1} - 1; \qquad v_x^{\varphi^2} = \frac{1}{x} - 1,$$

so that $\hat{\varphi}(x) = \varphi^1(x) = 2$ (for $x = 0$, $v_0^{\pi} = 0$ for any strategy π).

Now, for all $x > 0$, we have $\hat{\varphi}(x) = 2$ and $v_x^{\hat{\varphi}} \equiv 0 > \min\{v_x^{\varphi^1}, v_x^{\varphi^2}\} = \frac{1}{x+1} - 1$.

The basic strategy iteration algorithm constitutes a paraphrase of [Puterman(1994), Section 7.2.5].

1. Set $n = 0$ and select a stationary selector φ^0 arbitrarily enough.

2. Obtain $w^n(x) \triangleq v_x^{\varphi^n}$.

3. Choose $\varphi^{n+1} : \mathbf{X} \to \mathbf{A}$ such that
$$c(x, \varphi^{n+1}(x)) + \int_{\mathbf{X}} w^n(y) p(dy|x, \varphi^{n+1}(x))$$
$$= \inf_{a \in \mathbf{A}} \left\{ c(x, a) + \int_{\mathbf{X}} w^n(y) p(dy|x, a) \right\},$$
setting $\varphi^{n+1}(x) = \varphi^n(x)$ whenever possible.

4. If $\varphi^{n+1} = \varphi^n$, stop and set $\varphi^* = \varphi^n$. Otherwise, increment n by 1 and return to step 2.

This is proven to stop in a finite number of iterations and return an optimal strategy φ^* in negative finite models, i.e. if $c(x, a) \leq 0$ and all the spaces $\mathbf{X}$ and $\mathbf{A}$ are finite [Puterman(1994), Th. 7.2.16].

Theorem 2.4. *[Puterman(1994), Prop. 7.2.14] For discrete negative models with finite values of v_x^π at any π and $x \in \mathbf{X}$, if the strategy iteration algorithm terminates, then it returns an optimal strategy.*

Example 2.2.4 (Fig. 2.3) shows that this algorithm does not always converge even if the action space is finite. Indeed, choose $\varphi^0(x) \equiv 2$; then $w^0(x) = v_x^{\varphi^0} \equiv 0$. Now $\varphi^1(x) = 1$ if $x > 1$ and $\varphi^1(0) = \varphi^1(1) = 2$. Therefore,
$$w^1(x) = v_x^{\varphi^1} = \begin{cases} \frac{1}{x} - 1, & \text{if } x > 1; \\ 0, & \text{if } x \leq 1. \end{cases}$$
Now, for $x \geq 1$, we have
$$c(x, 2) + w^1(x+1) = \frac{1}{x+1} - 1 < c(x, 1) + w^1(0) = \frac{1}{x} - 1,$$
so that $\varphi^2(x) \equiv 2$ for all $x \in \mathbf{X}$, and the strategy iteration algorithm will cycle between these two stationary selectors φ^0 and φ^1. This is not surprising, because Example 2.2.4 illustrates that there are no optimal strategies at all.

Now we modify the model from Example 2.2.3: we put $c(1, 2) = +1$. See Fig. 2.5; this is a simplified version of Example 7.3.4 from [Puterman(1994)].

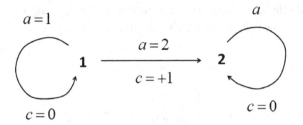

Fig. 2.5 Example 2.2.5: the strategy iteration returns a sub-optimal strategy.

The optimality equation

$$v(1) = \min\{v(1),\ 1 + v(2)\};$$
$$v(2) = v(2)$$

– compare with (2.8) – again has many solutions. We deal now with the positive model, in which the minimal non-negative solution $v(1) = 0$, $v(2) = 0$ coincides with the Bellman function v_x^*, and the stationary selector $\varphi^*(x) \equiv 1$ is conserving, equalizing and optimal. If we apply the strategy iteration algorithm to selector $\varphi^0(x) \equiv 2$, we see that

$$w^0(x) = \begin{cases} +1, & \text{if } x = 1; \\ 0, & \text{if } x = 2. \end{cases}$$

Hence, φ^0 provides the minimum on step 3 in the following expressions corresponding to $x = 1$ and $x = 2$:

$$\min\{w^0(1),\ +1 + w^0(2)\};$$
$$\min\{w^0(2),\ w^0(2)\}.$$

Thus, the strategy iteration algorithm terminates and returns a stationary selector $\varphi^0(x) \equiv 2$ which is not optimal. Condition $c(x, a) \leq 0$ is important in Theorem 2.4. Note that, in the discounted case, the strategy iteration algorithm is powerful much more often [Puterman(1994), Section 6.4].

2.2.6 When value iteration is not successful

This algorithm works as follows:

$$v^0(x) \equiv 0;$$
$$v^{n+1}(x) = \inf_{a \in \mathbf{A}} \left\{ c(x, a) + \int_{\mathbf{X}} v^n(y) p(dy|x, a) \right\}, \quad n = 0, 1, 2, \ldots$$

(we leave aside the question of the measurability of v^{n+1}). It is known that, e.g., in negative models, there exists the limit

$$\lim_{n\to\infty} v^n(x) \stackrel{\triangle}{=} v^\infty(x), \qquad (2.10)$$

which coincides with the Bellman function v_x^*, [Bertsekas and Shreve(1978), Prop. 9.14] and [Puterman(1994), Th. 7.2.12]. The same statement holds for discounted models, if e.g., $\sup_{x\in\mathbf{X}} \sup_{a\in\mathbf{A}} |c(x,a)| < \infty$ (see [Bertsekas and Shreve(1978), Prop. 9.14], [Puterman(1994), Section 6.3]). Some authors call a Markov Decision Process *stable* if the limit

$$\lim_{n\to\infty} v^n(x) = v_x^*$$

exists and coincides with the Bellman function [Kallenberg(2010), p. 112]. Below, we present three MDPs which are not stable. See also Remark 2.5.

Let $\mathbf{X} = \{0, 1, 2\}$, $\mathbf{A} = \{1, 2\}$, $p(0|0, a) = p(0|1, a) \equiv 1$, $p(2|2, 1) = 1$, $p(1|2, 2) = 1$, $c(0, a) \equiv 0$, $c(1, a) \equiv 1$, $c(2, 1) = 0$, $c(2, 2) = -2$ (see Fig. 2.6; a similar example was presented in [Kallenberg(2010), Ex. 4.1]).

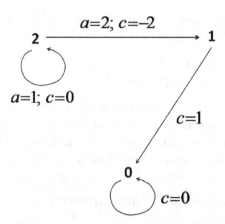

Fig. 2.6 Example 2.2.6: unstable MDP.

Obviously, $v_0^* = 0$, $v_1^* = 1$, $v_2^* = -1$, but value iterations lead to the following:

$$\begin{array}{lll} v^0(0) = 0 & v^0(1) = 0 & v^0(2) = 0 \\ v^1(0) = 0 & v^1(1) = 1 & v^1(2) = -2 \\ v^2(0) = 0 & v^2(1) = 1 & v^2(2) = -2 \\ \cdots & \cdots & \cdots \end{array}$$

In the following examples, the limit (2.10) does not exist at all.

Let $\mathbf{X} = \{\Delta, 0, 1, 2, \ldots\}$, $\mathbf{A} = \{0, 1, 2, \ldots\}$, $p(\Delta|\Delta, a) \equiv 1$, $p(2a+1|0, a) = 1$, $p(\Delta|1, 0) = 1$, for $a > 0$ $p(\Delta|1, a) = p(0|1, a) \equiv 1/2$, for $x > 1$ $p(x-1|x, a) \equiv 1$. All the other transition probabilities are zero. Let $c(\Delta, a) \equiv 0$, $c(0, a) \equiv 12$, $c(1, 0) = 1$, for $a > 1$ $c(1, a) \equiv -4$, for $x > 1$ $c(x, a)$

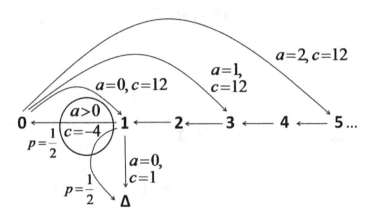

Fig. 2.7 Example 2.2.6: unstable MDP, no limits.

This MDP is absorbing, the performance functional v^π and the Bellman function v_x^* are well defined and finite. Since any one cycle $1 \to 0 \to 2a+1 \to 2a \to \cdots \to 1$ leads to an increment of the performance, one should put $\varphi^*(1) = 0$, so that

$$v_0^* = 13, \qquad \text{for } x > 0 \;\; v_x^* = 1, \qquad \text{and } v_\Delta^* = 0.$$

On the other hand, the value iteration gives the following values:

x	0	1	2	3	4	5	...
$v^0(x)$	0	0	0	0	0	0	
$v^1(x)$	12	−4	0	0	0	0	
$v^2(x)$	8	1	−4	0	0	0	
$v^3(x)$	12	0	1	−4	0	0	
$v^4(x)$	8	1	0	1	−4	0	
$v^5(x)$	12	0	1	0	1	−4	...
...						...	

The third example is based on [Whittle(1983), Chapter 25, Section 5]. Let $\mathbf{X} = \{0,1,2\}$, $\mathbf{A} = \{1,2\}$, $p(0|0,a) \equiv 1$, $p(0|1,1) = 1$, $p(2|1,2) = 1$, $p(1|2,a) \equiv 0$, $c(1,1) =$

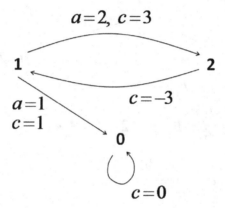

Fig. 2.8   Example 2.2.6: unstable MDP, no limits.

Here Condition 2.1 is violated, and one can call a strategy π^* optimal if it minimizes $\limsup_{\beta \to 1^-} v^{\pi,\beta}$ (see Section 3.1; β is the discount factor). In [Whittle(1983)], such strategies, which additionally satisfy equation

$$\lim_{T\to\infty} \left| E_{P_0}^{\pi^*}\left[\sum_{t=1}^{T} c(X_{t-1}, A_t)\right] - E_{P_0}^{\pi^*}\left[\sum_{t=1}^{\infty} c(X_{t-1}, A_t)\right] \right| = 0 \qquad (2.11)$$

are called *transient-optimal*. It is assumed that all the mathematical expectations in (2.11) are well defined.

Since, for any $\beta \in [0,1)$, the stationary selector $\varphi^*(x) \equiv 1$ is (uniformly) optimal in problem (3.1) and obviously satisfies (2.11), it is transient-

optimal. The discounted Bellman function equals
$$v_s^{*,\beta} = \begin{cases} 0, & \text{if } x = 0; \\ 1, & \text{if } x = 1; \\ \beta - 3, & \text{if } x = 2; \end{cases}$$
so that in this example one should put
$$v_x^* = \lim_{\beta \to 1-} v_x^{*,\beta} = \begin{cases} 0, & \text{if } x = 0; \\ 1, & \text{if } x = 1; \\ -2, & \text{if } x = 2. \end{cases}$$

But the value iterations lead to the following:

$v^0(x) \equiv 0;$

$$v^1(x) = \begin{cases} 0, & \text{if } x = 0; \\ 1, & \text{if } x = 1; \\ -3, & \text{if } x = 2, \end{cases} \quad v^2(x) = \begin{cases} 0, & \text{if } x = 0; \\ 0, & \text{if } x = 1; \\ -2, & \text{if } x = 2, \end{cases}$$

$$v^3(x) = \begin{cases} 0, & \text{if } x = 0; \\ 1, & \text{if } x = 1; \\ -3, & \text{if } x = 2, \end{cases}$$

and so on. At every step, the optimal action in state 1 switches from 1 to 2 and back.

2.2.7 When value iteration is not successful: positive model I

The limit (2.10) also exists in positive models, but here
$$v^\infty \leq v^*, \tag{2.12}$$
and $v^\infty \equiv v^*$ if and only if
$$v^\infty(x) = \inf_{a \in A} \left\{ c(x,a) + \int_X v^\infty(y) p(dy|x,a) \right\}$$
for all $x \in \mathbf{X}$ [Bertsekas and Shreve(1978), Prop. 9.16]. The following example shows that inequality (2.12) can be strict. Recall that $v^\infty \equiv v^*$ when the action space $\mathbf{A}$ is finite [Bertsekas and Shreve(1978), Corollary 9.17.1].

Let $\mathbf{X} = \{0, 1, 2, \ldots\}$, $\mathbf{A} = \{1, 2, \ldots\}$, $p(0|0, a) \equiv 1$, $p(0|2, a) \equiv 1$, $p(x-1|x, a) \equiv 1$ for $x > 2$, and
$$p(y|1, a) = \begin{cases} 1, & \text{if } y = 1 + a; \\ 0 & \text{otherwise.} \end{cases}$$

All the other transition probabilities are zero. Let $c(2,a) \equiv 1$, with all other losses zero (see Fig. 2.9). Versions of this example were presented in [Bäuerle and Rieder(2011), Ex. 7.2.4], [Bertsekas and Shreve(1978), Chapter 9, Ex. 1], [Dynkin and Yushkevich(1979), Chapter 4, Section 6], [Puterman(1994), Ex. 7.3.3] and [Strauch(1966), Ex. 6.1].

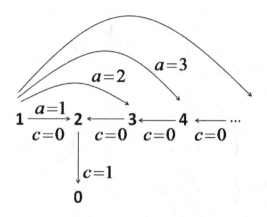

Fig. 2.9 Example 2.2.7: value iteration does not converge to the Bellman function.

The optimality equation (2.2) takes the form

$$v(0) = v(0);$$
$$v(1) = \inf_{a \in \mathbf{A}} \{v(1+a)\};$$
$$v(2) = 1 + v(0);$$
$$v(x) = v(x-1), \text{ if } x > 2.$$

The minimal non-negative solution $v(x)$ coincides with the Bellman function v_x^* and can be built using the following reasoning: $v(0) = 0$, hence $v(2) = 1$ and $v(x) = 1$ for all $x > 2$; therefore, $v(1) = 1$.

The value iteration results in the following sequence

$v^0(x) \equiv 0;$
$v^1(0) = 0;\ v^1(2) = 1;\qquad\qquad v^1(x) \equiv 0$ for $x > 2$ and $v^1(1) = 0$;
$v^2(0) = 0;\ v^2(2) = 1;\ v^2(3) = 1;\ v^2(x) \equiv 0$ for $x > 3$ and $v^2(1) = 0$;
and so on.

Eventually, $\lim_{n \to \infty} v^n(0) = 0$, $\lim_{n \to \infty} v^n(x) = 1$ for $x \geq 2$, but $v^n(1) = 0$ for each n, so that $\lim_{n \to \infty} v^n(1) = 0 < 1 = v_1^*$.

If $x = 1$, we have the strict inequality
$$v^\infty(1) = 0 < 1 = \inf_{a \in \mathbf{A}} \{v^\infty(1+a)\}.$$

2.2.8 When value iteration is not successful: positive model II

Example 2.2.7 showed that the following statement can fail if the model is not negative:

Statement 1. $\exists \lim_{n \to \infty} v^n(x) = v_x^*$.

It is also useful to look at the strategies that provide the infimum in the value iterations. Namely, let

$$\Gamma_n(x) \stackrel{\triangle}{=} \left\{ a \in \mathbf{A}: \ c(x,a) + \int_{\mathbf{X}} v^n(y) p(dy|x,a) = v^{n+1}(x) \right\};$$

$\Gamma_\infty(x) \stackrel{\triangle}{=} \{a \in \mathbf{A}: \ a$ is an accumulation point of some sequence a_n with $a_n \in \Gamma_n(x)\}$ (here we assume that $\mathbf{A}$ is a topological space). Then

$$\Gamma^*(x) = \left\{ a \in \mathbf{A}: \ c(x,a) + \int_{\mathbf{X}} v_y^* p(dy|x,a) = v_x^* \right\}$$

is the set of conserving actions.

Statement 2. $\Gamma_\infty(x) \subseteq \Gamma^*(x)$ for all $x \in \mathbf{X}$.

Sufficient conditions for statements 1 and 2 were discussed in [Schäl(1975b)]. Below, we present a slight modification of Example 7.1 from [Schäl(1975b)] which shows that statements 1 and 2 can fail separately or simultaneously.

Let $\mathbf{X} = [0,2] \times \{1,2,\ldots\}$, $\mathbf{A} = [0,2]$, and we consider the natural topology in $\mathbf{A}$. For $x = (y,k)$ with $k \geq 2$, we put $p((y,k+1)|(y,k),a) \equiv 1$; $p((a,2)|(y,1),a) \equiv 1$ (see Fig. 2.10).

To describe the one-step loss $c(x,a)$, we need to introduce functions $\delta_n(y)$ on $[0,2]$, $n = 1, 2, \ldots$ Suppose positive numbers $c_2 \leq c_3 \leq \cdots \leq d \leq b$ are fixed; $c_\infty \stackrel{\triangle}{=} \lim_{i \to \infty} c_i$. Let $\delta_1(y) \equiv 0$. For $n \geq 2$, we put

$$\delta_n(y) = \begin{cases} b, & \text{if } y = 0; \\ c_n, & \text{if } 0 < y \leq 1/n; \\ b, & \text{if } 1/n < y < 1; \\ d, & \text{if } 1 \leq y \leq 2 \end{cases}$$

(see Fig. 2.11). Now $c((y,1)) \equiv 0$ and, for $x = (y,k)$ with $k \geq 2$, $c(x,a) \equiv \delta_k(y) - \delta_{k-1}(y)$.

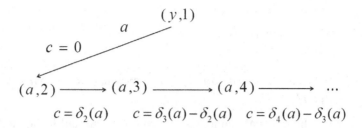

Fig. 2.10 Example 2.2.8: value iteration is unsuccessful.

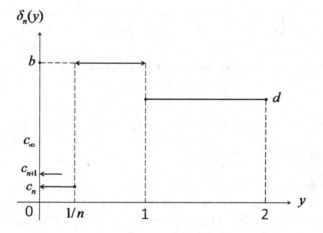

Fig. 2.11 Example 2.2.8: construction of the loss function.

Value iterations give the following table:

x	$(y,1)$	$(y,2)$	$(y,3)$	$(y,4)$	...
$v^0(x)$	0	0	0	0	
$v^1(x)$	0	$\delta_2(y)$	$\delta_3(y) - \delta_2(y)$	$\delta_4(y) - \delta_3(y)$	
$v^2(x)$	$\inf_a \delta_2(a) = c_2$	$\delta_3(y)$	$\delta_4(y) - \delta_2(y)$	$\delta_5(y) - \delta_3(y)$	
$v^3(x)$	$\inf_a \delta_3(a) = c_3$	$\delta_4(y)$	$\delta_5(y) - \delta_2(y)$	$\delta_6(y) - \delta_3(y)$	...
...	...	...	...	...	

For $x = (y, k)$ with $k \geq 2$,
$$v_x^* = \lim_{n \to \infty} v^n(x) = \lim_{i \to \infty} \delta_i(y) - \delta_{k-1}(y)$$
and $\Gamma_\infty(x) = \Gamma^*(x) = \mathbf{A}$.
Since for $x = (y, 2)$,
$$\lim_{n \to \infty} v^n(x) = v_x^* = \lim_{i \to \infty} \delta_i(y) = \begin{cases} b, & \text{if } 0 \leq y < 1; \\ d, & \text{if } 1 \leq y \leq 2, \end{cases}$$
and $d \leq b$, we conclude that $v_{(y,1)}^* = d$ and $\Gamma^*((y,1)) = [1, 2]$ when $d < b$. If $d = b$ then $\Gamma^*((y,1)) = [0, 2]$.

At the same time, $\lim_{n \to \infty} v^n((y,1)) = \lim_{n \to \infty} c_n = c_\infty$ and $\Gamma_\infty((y,1)) = 0$ because $\Gamma_n((y,1)) = (0, 1/n]$.

Therefore,

- f $c_\infty < d < b$ then Statements 1 and 2 both fail.

- f $c_\infty = d = b$ then Statements 1 and 2 both hold.

- If $c_\infty = d < b$ then Statement 1 holds, but Statement 2 fails.

- If $c_\infty < d = b$ then Statement 1 fails, but Statement 2 holds.

2.2.9 Value iteration and stability in optimal stopping problems

The pure optimal stopping problem has the action space $\mathbf{A} = \{s, n\}$, where s (n) means the decision to stop (not to stop) the process. Let Δ be a specific absorbing state (cemetery) meaning that the process is stopped: for $x \neq \Delta$, $p(\Gamma|x, s) = I\{\Gamma \ni \Delta\}$ $p(\mathbf{X} \setminus \{\Delta\}|x, n) = 1$; $p(\Gamma|\Delta, a) \equiv I\{\Gamma \ni \Delta\}$ $c(\Delta, a) \equiv 0$.

Now, equation (2.2) takes the form
$$v(x) = \min\{c(x, n) + \int_{\mathbf{X}} v(y) p(dy|x, n); \ c(x, s)\}, \quad x \in \mathbf{X} \setminus \{\Delta\};$$
$$v(\Delta) = 0.$$

In this framework, the traditional value iteration described in Section 2.2.7 is often replaced with calculation of V_x^n, the minimal expected total cost incurred if we start in state $x \in \mathbf{X} \setminus \{\Delta\}$ and are allowed a maximum of n steps before stopping.

Function V_x^n satisfies the equations

$$V_x^0 = c(x, s);$$
$$V_x^{n+1} = \min\{c(x,n) + \int_X V_y^n p(dy|x,n); \ c(x,s)\}, \quad n = 0, 1, 2, \ldots$$

Definition 2.1. [Ross(1983), p. 53]. The optimal stopping problem is called *stable* if

$$\lim_{n \to \infty} V_x^n = v_x^*.$$

In the following example, published in [Ross(1983), Chapter III, Ex. 2.1a], we present an unstable problem for which the traditional value iteration algorithm provides $v^n(x) \to v^*(x)$ as $n \to \infty$.

Let $\mathbf{X} = \{\Delta, 0, \pm 1, \pm 2, \ldots\}$; $\mathbf{A} = \{s, n\}$; for $x \neq \Delta$, $p(x+1|x,n) = p(x-1|x,n) = 1/2$, $c(x,n) \equiv 0$, $c(x,s) = x$ (see Fig. 2.12).

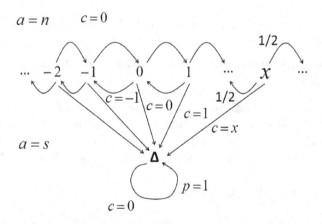

Fig. 2.12 Example 2.2.9: a non-stable optimal stopping problem.

One can check that $V_x^n = x$.

On the other hand, there is obviously no reason to stop the process at positive states: if the chain is never stopped then the total loss equals zero. Hence we can replace $c(x,s)$ with zeroes at $x > 0$ and obtain a negative model, for which $\lim_{n\to\infty} v^n(x) = v^*_x$, where

$$v^0(x) = 0,$$

$$v^1(x) = \min\{x; 0\} = \begin{cases} x, & \text{if } x \le 0; \\ 0, & \text{if } x > 0, \end{cases}$$

$$v^2(x) = \begin{cases} x, & \text{if } x < 0; \\ -1/2, & \text{if } x = 0; \\ 0, & \text{if } x > 0, \end{cases}$$

$$v^3(x) = \begin{cases} x, & \text{if } x < -1; \\ -5/4, & \text{if } x = -1; \\ -1/2, & \text{if } x = 0; \\ -1/4, & \text{if } x = 1; \\ 0, & \text{if } x > 1, \end{cases}$$

and so on, meaning that $v^*_x = -\infty$ for all $x \in \mathbf{X} \setminus \{\Delta\}$.

It is no surprise that $v^*_x = -\infty$. Indeed, for the control strategy

$$\varphi^N(x) = \begin{cases} n, & \text{if } x > -N; \\ s, & \text{if } x \le -N, \end{cases}$$

where $N > 0$, we have $v^{\varphi^N}_x \le -N$ for each $x \in \mathbf{X} \setminus \{\Delta\}$, because the random walk under consideration is (null-)recurrent, so that state $-N$ will be reached from any initial state $x > -N$. Therefore, $\inf_{N>0} v^{\varphi^N}_x = -\infty$.

At the same time, $\lim_{n\to\infty} V^n(x) = x > -\infty$.

2.2.10 A non-equalizing strategy is uniformly optimal

In fact, the model under consideration is uncontrolled (there exists only one control strategy), and we intend to show that the Bellman function v^*_x is finite and, for some values of x,

$$\lim_{t\to\infty} E_x\left[v^*_{X_t}\right] < 0. \tag{2.13}$$

Let $\mathbf{X} = \{0, 1, 1', 1'', 2, 2', 2'', \ldots\}$, $\mathbf{A} = \{0\}$ (a dummy action),

$$\forall x > 0 \ p(y|x, a) = \begin{cases} p, & \text{if } y = x'; \\ 1-p, & \text{if } y = x''; \\ 0 & \text{otherwise,} \end{cases} \quad c(x, a) \equiv 0, \ p(x+1|x', a) = 1,$$

$c(x', a) = p^{-x}$, $p(0|x'', a) = 1$, $c(x'', a) = -\frac{p^{-x+1}}{1-p}$, $p(0|0, a) = 1$, $c(0, a) = 0$, where $p \in (0, 1)$ is a fixed constant (see Fig. 2.13).

The optimality equation is given by

$$v(0) = v(0);$$

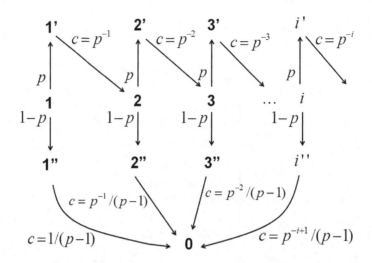

Fig. 2.13 Example 2.2.10: a non-equalizing strategy is optimal.

for $x > 0$

$$v(x) = pv(x') + (1-p)v(x''),$$

$$v(x') = p^{-x} + v(x+1),$$

$$v(x'') = -\frac{p^{-x+1}}{1-p} + v(0).$$

We are interested only in solutions with $v(0) = 0$. If we substitute the second and the third equations into the first one, we obtain

$$v(x) = pv(x+1).$$

The general solution is given by $v(x) = kp^{-x}$, and we intend to show that the Bellman function equals

$$v_x^* = -\frac{p}{1-p} \cdot p^{-x}.$$

Indeed, only the following trajectories are realized, starting from initial state $X_0 = x$:

$$x, x', (x+1), (x+1)', \ldots, (x+n), (x+n)'', 0, 0, \ldots \quad (n = 0, 1, 2, \ldots).$$

The probability equals $p^n(1-p)$, and the associated loss equals

$$W = \sum_{j=0}^{n-1} p^{-(x+j)} - \frac{p^{-(x+n)+1}}{1-p} = -\frac{p^{-x+1}}{1-p}.$$

Therefore,
$$v_x^* = -\sum_{n=0}^{\infty} p^n(1-p)\frac{p^{-x+1}}{1-p} = -\frac{p^{-x+1}}{1-p}$$
and also
$$v_{x'}^* = v_{x''}^* = -\frac{p^{-x+1}}{1-p}.$$
Now, starting from $X_0 = x$,
$$X_{2t} = \begin{cases} x+t & \text{with probability } p^t; \\ 0 & \text{with probability } 1-p^t, \end{cases}$$

$$X_{2t+1} = \begin{cases} (x+t)'' & \text{with probability } p^t(1-p); \\ (x+t)' & \text{with probability } p^{t+1}; \\ 0 & \text{with probability } 1-p^t. \end{cases}$$

Therefore,
$$E_x\left[v_{X_{2t}}^*\right] = E_x\left[v_{X_{2t+1}}^*\right] = -\frac{p^{-x+1}}{1-p}.$$

Similar calculations are valid for $X_0 = x'$. Thus, inequality (2.13) holds.

Note that, in this example, Conditions 2.2 are violated: $c(x, a)$ takes negative and positive values, and
$$E_x\left[\sum_{t=1}^{\infty} r^-(X_{t-1}, A_t)\right] = -\sum_{n=0}^{\infty} p^n(1-p)\frac{p^{-(x+n)+1}}{1-p} = -\infty$$
(one should ignore positive losses $c(x', a) = p^{-x}$).

2.2.11 A stationary uniformly ε-optimal selector does not exist (positive model)

This example was published in [Dynkin and Yushkevich(1979), Chapter 6, Section 6]. Let $\mathbf{X} = \{0, 1, 2\}$, $\mathbf{A} = \{1, 2, \ldots\}$, $p(0|0, a) = p(0|1, a) \equiv 1$, $p(1|2, a) = \left(\frac{1}{2}\right)^a$, $p(2|2, a) = 1 - \left(\frac{1}{2}\right)^a$, with all other transition probabilities zero. We put $c(0, a) \equiv 0$, $c(1, a) \equiv 1$, $c(2, a) \equiv 0$. See Fig. 2.14; a similar example was presented in [Puterman(1994), Ex. 7.3.2].

One should put $v(0) = 0$, and from the optimality equation, we obtain $v(1) = 1$,
$$v(2) = \inf_{a \in \mathbf{A}} \left\{\left(\frac{1}{2}\right)^a + \left(1 - \left(\frac{1}{2}\right)^a\right) v(2)\right\}.$$

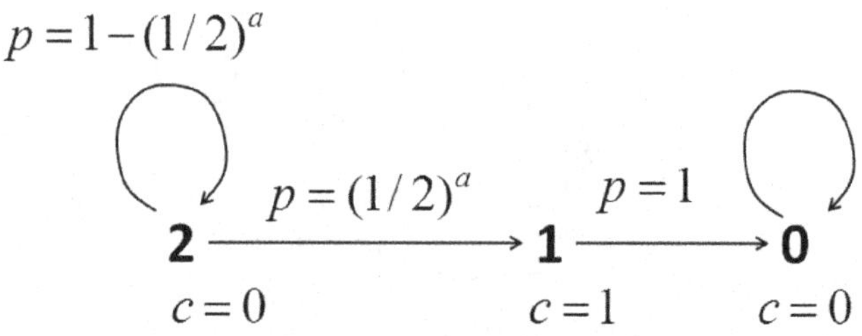

$$p = 1-(1/2)^a$$

$$\overset{\curvearrowleft}{2} \xrightarrow{p=(1/2)^a} 1 \xrightarrow{p=1} \overset{\curvearrowright}{0}$$

$$c=0 \qquad\qquad c=1 \qquad c=0$$

Fig. 2.14 Example 2.2.11: a stationary ε-optimal strategy does not exist.

Any value $v(2) \in [0,1]$ satisfies the last equation, and the minimal solution is $v(2) = 0$. The function $v(x) = v_x^*$ coincides with the Bellman function. For a fixed integer $m \geq 0$, the non-stationary selector $\hat{\varphi}_t^m(2) = m+t$ provides a total loss equal to zero with probability $\prod_{t=1}^{\infty}\left(1-\left(\frac{1}{2}\right)^{m+t}\right)$, and equal to one with the complementary probability, so that

$$v_2^{\hat{\varphi}^m} = 1 - \prod_{t=1}^{\infty}\left(1-\left(\frac{1}{2}\right)^{m+t}\right).$$

The last expression approaches 0 as $m \to \infty$, because $\prod_{t=1}^{\infty}\left(1-\left(\frac{1}{2}\right)^{m+t}\right)$ is the tail of the converging product $\prod_{t=1}^{\infty}\left(1-\left(\frac{1}{2}\right)^t\right) > 0$, and hence approaches 1 as $m \to \infty$.

At the same time, for any stationary selector φ (and also for any stationary randomized strategy) we have $v_2^\varphi = 1$.

We present another simple example for which a stationary ε-optimal strategy does not exist. Let $\mathbf{X} = \{1\}$, $\mathbf{A} = \{1,2,\ldots\}$, $p(1|1,a) \equiv 1$, $c(1,a) = \left(\frac{1}{2}\right)^a$. The model is positive, and the minimal non-negative solution to equation (2.2), which has the form

$$v(1) = \inf_{a \in \mathbf{A}}\left\{\left(\frac{1}{2}\right)^a + v(1)\right\},$$

equals $v(1) = v_1^* = 0$. No one strategy is conserving. Moreover, for any stationary strategy π, there is the same positive loss on each step, equal to $\sum_{a \in \mathbf{A}} \pi(a|1) \left(\frac{1}{2}\right)^a$, meaning that $v_1^\pi = \infty$. At the same time, for each $\varepsilon > 0$, there is a non-stationary ε-optimal selector. For instance, one can put $\varphi_1(1)$ equal any $a_1 \in \mathbf{A}$ such that $c(1, a_1) < \frac{\varepsilon}{2}$, $\varphi_2(1)$ equal any $a_2 \in \mathbf{A}$ such that $c(1, a_2) < \frac{\varepsilon}{4}$, and so on.

The same reasoning remains valid for any positive loss, such that $\inf_{a \in \mathbf{A}} c(1, a) = 0$.

2.2.12 A stationary uniformly ε-optimal selector does not exist (negative model)

Let $\mathbf{X} = \{\ldots, -2, -1, 0, 1, 2, \ldots\}$, $\mathbf{A} = \{1, 2, \ldots\}$. For all $i > 0$, $p(-i+1|-i, a) = p(1|1-i, a) \equiv \frac{1}{2}$, $p(i|i, a) \equiv 1$, $p(a|0, a) = 1$. All other transition probabilities are zero. We put $c(0, a) = -a$, with all other values of the loss function zero. See Fig. 2.15; a similar example was presented in [Ornstein(1969), p. 564], see also [Bertsekas and Shreve(1978), Chapter 8, Ex. 2].

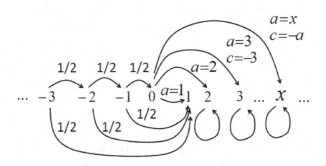

Fig. 2.15 Example 2.2.12: a stationary uniformly ε-optimal selector does not exist.

Obviously, $v_x^* = \begin{cases} 0, & \text{if } x > 0; \\ -\infty & \text{if } x \leq 0. \end{cases}$ However, for any stationary selector φ, if $\hat{a} = \varphi(0)$ then $v_x^\varphi = -\left(\frac{1}{2}\right)^x \hat{a}$ for $x \leq 0$, so that, for any $\varepsilon > 0$, selector φ is not uniformly ε-optimal. If we put $\varphi(0, x_0) \geq 2^{|x_0|}/\varepsilon$, then this semi-Markov selector will be uniformly ε-optimal (see Theorem 2.1).

In the next example (Section 2.2.13) the Bellman function is finite, but again unbounded.

We now present a very special example from [Ornstein(1969)], where $v_x^* \equiv -1$, but, for any stationary selector φ, there is a state $\hat{x}$ such that $v_{\hat{x}}^\varphi > -(1/2)$. In fact, according to the proof, $1/2$ can be replaced here by any positive number.

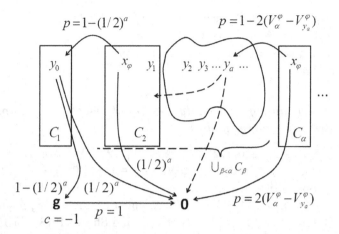

Fig. 2.16 Example 2.2.12: a stationary uniformly ε-optimal selector does not exist.

Let $\mathbf{A} = \{1, 2, \ldots\}$; the state space $\mathbf{X}$ and the transition probability will be defined inductively. Note that the space $\mathbf{X}$ will be not Borel, but it will be clear that the optimization problem (2.1) is well defined.

Let $C_1 \triangleq \{y_0\} \subset \mathbf{X}$ and let 0 and g be two isolated states in $\mathbf{X}$; $p(0|0, a) \equiv p(0|g, a) \equiv 1$, $c(g, a) \equiv -1$, with all the other values of the loss function zero. $p(g|y_0, a) = 1 - (1/2)^a$, $p(0|y_0, a) = (1/2)^a$. Obviously, $v^*_{y_0} = -1$. State g is the "goal", and state 0 is the "cemetery".

Suppose we have built all the subsets $C_\beta \subset \mathbf{X}$ for $\beta < \alpha$, where $\alpha, \beta \in \Omega$ which is the collection of all ordinals up to (and excluding) the first uncountable one. (Or, more simply, Ω is the first uncountable ordinal.) Suppose also that $v^*_x = -1$ for all $x \in \bigcup_{\beta<\alpha} C_\beta$. We shall build C_α such that $C_\alpha \cap C_\beta = \emptyset$ for all $\beta < \alpha$, and eventually we shall put

$$\mathbf{X} \triangleq \bigcup_{\alpha < \Omega} C_\alpha \cup \{0, g\}.$$

Note that, for each $x \in C_\alpha$, there will be a sequence $y_a \in \bigcup_{\beta<\alpha} C_\beta$, $a = 1, 2, \ldots$ such that only transition probabilities $p(y_a|x, a)$ and $p(0|x, a)$ are positive.

For each stationary selector φ on $\bigcup_{\beta<\alpha} C_\beta$, i.e. for each mapping $\varphi : \bigcup_{\beta<\alpha} C_\beta \to \mathbf{A}$, such that

$$V^\varphi_\alpha \triangleq \sup_{y \in \bigcup_{\beta<\alpha} C_\beta} v^\varphi_y \leq -\frac{1}{2},$$

we will introduce one point x_φ in C_α. The transition probabilities from x_φ are defined as follows:

- if $v_{\hat{y}}^\varphi = V_\alpha^\varphi$ for some $\hat{y} \in \bigcup_{\beta<\alpha} C_\beta$, then $p(\hat{y}|x_\varphi, a) = 1 - \left(\frac{1}{2}\right)^a$ and $p(0|x_\varphi, a) = \left(\frac{1}{2}\right)^a$;
- otherwise, pick a sequence $y_a \in \bigcup_{\beta<\alpha} C_\beta$, $a = 1, 2, \ldots$, such that $V_\alpha^\varphi - v_{y_a}^\varphi < 1/2$ and $\lim_{a\to\infty} v_{y_a}^\varphi = V_\alpha^\varphi$, and put $p(y_a|x_\varphi, a) = 1 - 2(V_\alpha^\varphi - v_{y_a}^\varphi)$ and $p(0|x_\varphi, a) = 2(V_\alpha^\varphi - v_{y_a}^\varphi)$.

We consider only initial distributions concentrated on singletons. (One could also consider initial distributions concentrated on countable sets of points.) Now, starting from $x_0 \in \mathbf{X}$, under any control strategy, the probability of each trajectory $x_0, a_1, x_1, \ldots$ equals the product $P_0(x_0)\pi_1(a_1|x_0)p(x_1|x_0, a_1)\cdots$, simply because both π_t and p are atomic. Hence, the optimization problem (2.1) is well defined.

For any $x \in C_\alpha$, for each $\varepsilon > 0$, we can apply a control $a \in \mathbf{A}$ such that some point y from $\bigcup_{\beta<\alpha} C_\beta$ will be reached with a probability bigger than $1 - \varepsilon$. Since, according to the induction supposition, $v_y^* = -1$, we conclude that $v_x^* = -1$. Therefore, $v_x^* \equiv -1$ for all $x \in \mathbf{X}$.

Now, taking any stationary selector φ, we shall prove that $V_\alpha^\varphi > -(1/2)$ for some $\alpha < \Omega$.

Suppose $\forall \alpha < \Omega \; V_\alpha^\varphi \leq -(1/2)$. Then the function

$$h(\alpha) \stackrel{\triangle}{=} \sup_{y \in \bigcup_{\beta \leq \alpha} C_\beta} v_y^\varphi$$

exhibits the following properties: h is (obviously) non-positive and non-decreasing, for each $\alpha \in \Omega$,

$$\sup_{\gamma < \alpha} h(\gamma) = \sup_{y \in \bigcup_{\beta < \alpha} C_\beta} v_y^\varphi = V_\alpha^\varphi < 0,$$

and $h(\alpha) > \sup_{\gamma<\alpha} h(\gamma)$. Indeed, for a fixed α, consider the restriction of φ to $\bigcup_{\beta<\alpha} C_\beta$ and take the point $x_\varphi \in C_\alpha$ as constructed above. We have

- either $v_{x_\varphi}^\varphi = \left[1 - (1/2)^{\varphi(x_\varphi)}\right] v_{\hat{y}}^\varphi > V_\alpha^\varphi$ when $v_{\hat{y}}^\varphi = V_\alpha^\varphi$,
- or $v_{x_\varphi}^\varphi = [1 - 2(V_\alpha^\varphi - v_{y_a}^\varphi)]v_{y_a}^\varphi$ otherwise, where $a = \varphi(x_\varphi)$, $y_a \in \bigcup_{\beta<\alpha} C_\beta$.

In the latter case, $v_{y_a}^\varphi < V_\alpha^\varphi \leq -(1/2)$, so that $2v_{y_a}^\varphi < -1$ and $v_{x_\varphi}^\varphi > v_{y_a}^\varphi + V_\alpha^\varphi - v_{y_a}^\varphi = V_\alpha^\varphi$. Hence, in any case,

$$h(\alpha) \geq v_{x_\varphi}^\varphi > V_\alpha^\varphi = \sup_{\gamma<\alpha} h(\gamma).$$

But, according to Theorem B.1, $h(\alpha) = 0$ for some $\alpha < \Omega$, which contradicts the assumption $V_{\alpha+1}^\varphi \leq -(1/2)$.

Therefore, $\exists \alpha$: $V_\alpha^\varphi > -(1/2)$ and there is $\hat{x} \in \bigcup_{\beta < \alpha} C_\beta$ such that $v_{\hat{x}}^\varphi > -(1/2)$.

The model considered can be called *gambling*, as for each state (e.g. amount of money), when using gamble $a \in \mathbf{A}$, the gambler either reaches his or her goal (state g), loses everything (state 0), or moves to one of a countable number of new states. The objective is to maximize the probability of reaching the goal. We emphasize that the state space $\mathbf{X}$ is uncountable in the current example. For gambling models with countable $\mathbf{X}$, for each $\varepsilon > 0$ there is a stationary selector φ such that $v_x^\varphi \leq (1-\varepsilon) v_x^*$ for all $x \in \mathbf{X}$ [Ornstein(1969), Th. B], see also [Puterman(1994), Th. 7.2.7]. Note that we reformulate all problems as minimization problems.

Other gambling examples are given in Sections 2.2.25 and 2.2.26.

2.2.13 Finite-action negative model where a stationary uniformly ε-optimal selector does not exist

Let $\mathbf{X} = \{0, 1, 2, \ldots\}$, $\mathbf{A} = \{1, 2\}$, $p(0|x, 1) = 1$, $p(0|0, a) \equiv 1$, i.e. state 0 is absorbing; $p(0|x, 2) = p(x+1|x, 2) = 1/2$ for $x > 0$. All the other transition probabilities are zero. We put $c(0, a) = 0$, $c(x, 1) = 1 - 2^x$; $c(x, 2) = 0$ for $x > 0$. See Fig. 2.17; a similar example was presented in [Puterman(1994), Ex. 7.2.5] and in [Altman(1999), Section 9.8]. See also Section 2.2.14.

Remark 2.3. This example can be reformulated as a gamble [Dynkin and Yushkevich(1979), Chapter 6, Section 6]. State x means 2^x units in hand, and if the decision is to continue ($a = 2$) then the capital is either doubled or lost with equal probability. If the decision is to stop ($a = 1$) then the player has to pay one unit. The goal is to maximize the final wealth.

The optimality equation (2.2) is given by

$$v(0) = v(0); \quad \text{for } x > 0 \quad v(x) = \min\{1 - 2^x + v(0); \ 0.5v(x+1) + 0.5v(0)\}.$$

We are interested only in the solutions with $v(0) = 0$; thus if $x > 0$ then

$$v(x) = \min\{1 - 2^x; \ 0.5v(x+1)\}. \tag{2.14}$$

We check Condition 2.2(a). If $\pi \in \Delta^{\text{AllN}}$ (π is non-randomized) then, in the case where control $a = 1$ is never used (i.e. $\pi_t(1|h_{t-1}) \equiv 0$ for all histories h_{t-1} having non-zero probability), we have $E_x^\pi [\sum_{t=1}^{\infty} r^-(X_{t-1}, A_t)] = 0$. Otherwise, control $a = 1$ is actually only applied in a single state $\hat{x} \geq x$, where $x = X_0$. That is, the minimal number $\hat{x}$ for which $\pi(1|x, 2, x+$

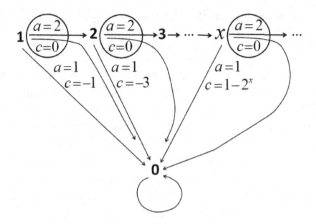

Fig. 2.17 Example 2.2.13: a stationary uniformly ε-optimal selector does not exist.

$1, 2, \ldots, \hat{x}) = 1$. In this case

$$E_x^\pi \left[\sum_{t=1}^\infty r^-(X_{t-1}, A_t) \right] = \left(\frac{1}{2}\right)^{\hat{x}-x} [1 - 2^{\hat{x}}] \geq -2^x.$$

We see that $\inf_{\pi \in \Delta^{\text{AllN}}} v_x^\pi \geq -2^x$. Next, we integrate the random total loss w.r.t. the left-hand side measure in formula (2.3). After applying the Fubini Theorem to the right-hand side, we conclude that

$$\forall \pi \in \Delta^{\text{All}} \quad E_x^\pi \left[\sum_{t=1}^\infty r^-(X_{t-1}, A_t) \right] \geq -2^x,$$

and Condition 2.2 is satisfied.

We want to construct the maximal non-positive solution to equation (2.14) and build a conserving stationary selector. That selector will be not equalizing and not optimal.

Clearly,

$$v(1) \leq -1 = -2 + 1;$$

$$\text{besides } v(1) \leq \frac{1}{2} v(2);$$

$$v(2) \leq -3 = -4 + 1, \text{ so that } v(1) \leq -\frac{3}{2} = -2 + \frac{1}{2}; \qquad (2.15)$$

$$\text{besides } v(2) \leq \frac{1}{2} v(3);$$

$$v(3) \leq -7 = -8 + 1, \text{ so that } v(2) \leq -4 + \frac{1}{2} \text{ and } v(1) \leq -2 + \frac{1}{4};$$

$$\text{besides } v(3) \leq \frac{1}{2}v(4),$$

and so on. Therefore, $v(x) \leq -2^x$, but $v_x^* = -2^x$ is a solution to (2.14), and that is the maximal non-positive solution. For any constant $K \geq 1$, $Kv_x^* = -K2^x$ is also a solution to equation (2.14).

The stationary selector $\varphi^2(x) \equiv 2$ is conserving but not equalizing: $\forall x > 0,\ t\ E_x^{\varphi^2}[v_{X_t}^*] = -2^x$. It is far from being optimal because $v_x^{\varphi^2} \equiv 0 > v_x^* = -2^x$. In a similar way to Example 2.2.4, this selector indefinitely delays the ultimate absorption in state 0. In this model, there are no optimal strategies, because, for such a strategy, we must have the equation $v_x^\pi = 0.5 v_{h_1=(x,2,x+1)}^\pi$ which only holds if $\pi_1(2|x) = 1$ (control $A_1 = 1$ is excluded because it leads to a loss $1 - 2^x > v_x^* = -2^x$). Thus, we must have $v_{(x,2,x+1)}^\pi = 2^{x+1}$. The same reasoning applies to state $x+1$ after history $h_1 = (x, 2, x+1)$ is realized, and so on. Therefore, the only candidate for the optimal strategy is φ^2, but we already know that it is not optimal.

If φ is an arbitrary stationary selector, different from φ^2, then $\varphi(\hat{x}) = 1$ for some $\hat{x} > 0$, and $v_{\hat{x}}^\varphi = 1 - 2^{\hat{x}} > v_{\hat{x}}^* = -2^{\hat{x}}$. Hence, for $\varepsilon < 1$, a stationary uniformly ε-optimal selector does not exist. On the other hand, for any given initial state x, for each $\varepsilon > 0$, there exists a special selector for which $v_x^\varphi \leq -2^x + \varepsilon$. Indeed, we put $\varphi(y) = 2$ for all $y < x + n$ and $\varphi(x+n) = 1$, where $n \in \mathbb{N}$ is such that $n \geq -\frac{\ln \varepsilon}{\ln 2}$. Then

$$v_x^\varphi = \left(\frac{1}{2}\right)^n (1 - 2^{x+n}) \leq -2^x + \varepsilon.$$

The constructed selector is ε-optimal for the given initial state x, but it is not uniformly ε-optimal. To put it another way, we have built a uniformly ε-optimal semi-Markov selector (see Theorem 2.1).

At the same time, for an arbitrary $\varepsilon \in (0,1)$, the stationary randomized strategy $\hat{\pi}(1|x) = \delta = \frac{\varepsilon}{2-\varepsilon};\ \hat{\pi}(2|x) = 1 - \delta = \frac{2(1-\varepsilon)}{2-\varepsilon}$ is uniformly ε-optimal. Indeed, a trajectory of the form $(x, 2, x+1, 2, \ldots, x+n, 1, 0, a_{n+1}, 0, \ldots)$ is realized with probability $\left[\frac{1-\delta}{2}\right]^n \delta$ and leads to a loss $(1 - 2^{x+n})$. All other trajectories result in zero loss. Therefore,

$$v_x^{\hat{\pi}} = \sum_{n=0}^\infty \left(\frac{1-\delta}{2}\right)^n \delta(1 - 2^{x+n}) = -2^x + \frac{2\delta}{1+\delta} = -2^x + \varepsilon = v_x^* + \varepsilon.$$

The MDP thus considered is semi-continuous; more about such models is provided in Section 2.2.15. We also remark that this model is *absorbing*; the corresponding theory is developed, e.g., in [Altman(1999)].

Remark 2.4. We show that the *general uniform Lyapunov function* μ does not exist [Altman(1999), Section 7.2]; that is the inequality

$$\nu(x,a) + 1 + \sum_{y \neq 0} p(y|x,a)\mu(y) \leq \mu(x) \qquad (2.16)$$

cannot hold for positive function μ. (In fact, the function μ must exhibit some additional properties: see [Altman(1999), Def. 7.5].) Here $\nu(x,a)$ is the positive *weight* function, and the theory developed in [Altman(1999)] requires that $\sup_{(x,a) \in \mathbf{X} \times \mathbf{A}} \frac{|c(x,a)|}{\nu(x,a)} < \infty$. Since $c(x,1) = 1 - 2^x$, we have to put at least $\nu(x,1) = 2^x$. Now, for $a = 2$ we have

$$2^x + 1 + \frac{1}{2}\mu(x+1) \leq \mu(x),$$

so that, if $\mu(1) = k$, then

$$\mu(2) \leq 2k - 6; \qquad \mu(3) \leq 4k - 22,$$

and in general

$$\mu(x) \leq k 2^{x-1} + 2 - x 2^x,$$

meaning that $\mu(x)$ becomes negative for any value of k. If $c(x,a)$ were of the order γ^x with $0 < \gamma < 2$, then one could take $\nu(x) = \gamma^x$ and $\mu(x) = 2 + \frac{2\gamma^x}{2-\gamma}$, and a uniformly optimal stationary selector would have existed according to [Altman(1999), Th. 9.2].

2.2.14 Nearly uniformly optimal selectors in negative models

If the state space $\mathbf{X}$ is countable, then the following statement holds.

Theorem 2.5. *[Ornstein(1969), Th. C] Suppose the model is negative and $v_x^* > -\infty$ for all $x \in \mathbf{X}$. Then, for each $\varepsilon > 0$, there is a stationary selector φ such that*

$$v_x^\varphi \leq v_x^* + \varepsilon |v_x^*|. \qquad (2.17)$$

The following example, based on [van der Wal and Wessels(1984), Ex. 7.1], shows that this theorem cannot be essentially improved. In other words, if $\mathbf{X} = \{0, 1, 2, \ldots\}$ and if we replace (2.17) with

$$v_x^\varphi \leq v_x^* + \varepsilon |v_x^*| \delta(x), \qquad (2.18)$$

where $0 < \delta(x) \leq 1$ is a fixed model-independent function, $\lim_{x \to \infty} \delta(x) = 0$, then Theorem 2.5 fails to hold.

Let $\mathbf{X} = \{0, 1, 2, \ldots\}$, $\mathbf{A} = \{1, 2\}$, $p(0|0, a) \equiv 1$, $p(0|x, 1) \equiv 1$, $p(x + 1|x, 2) = \frac{1+\gamma_x}{2(1+\gamma_{x+1})}$, $p(0|x, 2) = \frac{1+2\gamma_{x+1}+\gamma_x}{2(1+\gamma_{x+1})}$ for $x > 0$, where $\gamma_x \leq 1$ is some non-negative sequence; $\liminf_{x \to \infty} \gamma_x = 0$. All the other transition probabilities are zero. We put $c(0, a) = 0$, $c(x, 2) \equiv 0$ and $c(x, 1) = -2^x$ for $x > 0$. See Fig. 2.18, compare with Example 2.2.13.

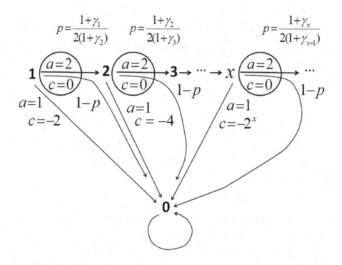

Fig. 2.18 Example 2.2.14.

The optimality equation (2.2) takes the form:

$$v(0) = v(0); \quad \text{clearly, we must put } v(0) = 0;$$

$$v(x) = \min\left\{-2^x, \frac{1+\gamma_x}{2(1+\gamma_{x+1})}v(x+1)\right\} \quad \text{for } x > 0.$$

Now, function $w(x) \stackrel{\triangle}{=} \frac{v(x)}{1+\gamma_x}$ at $x > 0$ satisfies the equation

$$w(x) = \min\left\{-\frac{2^x}{1+\gamma_x}, \frac{1}{2} \cdot w(x+1)\right\}.$$

Following similar reasoning to (2.15), we conclude that the maximal non-positive solution is given by $w(x) = -\frac{2^x}{1+\inf_{i \geq x} \gamma_i} = -2^x$; hence

$$v_x^* = v(x) = -2^x(1+\gamma_x).$$

Suppose the desired sequence $\delta(x)$ exists, and put $\gamma_x \stackrel{\triangle}{=} \sqrt{\delta(x)}$. If a stationary selector φ satisfies (2.18) for some $\varepsilon > 0$, then $\varphi(x) = 1$ for infinitely many $x \in \mathbf{X}$ (otherwise, $v_x^\varphi = 0$ for all sufficiently large x). For those values of x, namely, $x_1, x_2, \ldots$: $\lim_{i\to\infty} x_i = \infty$, we have $v_{x_i}^\varphi = -2^{x_i}$, and

$$\frac{v_{x_i}^\varphi - v_{x_i}^*}{|v_{x_i}^*|\delta(x_i)} = \frac{\gamma_{x_i}}{\delta(x_i)(1 + \gamma_{x_i})} \geq \frac{1}{2\sqrt{\delta_{x_i}}}.$$

The right-hand side cannot remain smaller than $\varepsilon > 0$ for all x_i, meaning that inequality (2.18) is violated and sequence $\delta(x)$ does not exist.

2.2.15 Semi-continuous models and the blackmailer's dilemma

Very powerful results are known for *semi-continuous* models, in which the following conditions are satisfied. See also the discussion at the end of Section 1.4.16.

Condition 2.3.

(a) The action space $\mathbf{A}$ is compact;
(b) for each $x \in \mathbf{X}$ the loss function $c(x, a)$ is lower semi-continuous in a;
(c) for each $x \in \mathbf{X}$ function $\int_{\mathbf{X}} u(y)p(dy|x,a)$ is continuous in a for every (measurable) bounded function u.

Note that this definition, accepted everywhere in the current chapter, is slightly different from that introduced at the beginning of Section 1.4.16.

Suppose for a moment that Condition 1.1 is satisfied. In this case, in *positive* models, there exist uniformly optimal stationary selectors [Bertsekas and Shreve(1978), Corollary 9.17.2]. Example 1.4.16 shows that requirement $c(x, a) \geq 0$ is important.

Theorem 2.6. *[Cavazos-Cadena et al.(2000), Th. 3.1] If the model is semi-continuous (that is, Condition 2.3 is satisfied), the state space $\mathbf{X}$ is finite, Condition 2.2(a) is satisfied, and, for each stationary selector, the controlled process X_t has a single positive recurrent class (unichain model), then there exists a uniformly optimal stationary selector.*

Example 2.2.13 shows that, for countable $\mathbf{X}$, this assertion can fail. Note that model is semi-continuous, and $\{0\}$ is the single positive recurrent class. A more complicated example illustrating the same ideas can be found in [Cavazos-Cadena et al.(2000), Section 4]; see also Section 1.4.16.

We now show that the unichain condition in Theorem 2.6 is important; similar examples were published in [Cavazos-Cadena et al.(2000), Ex. 3.1] and in [van der Wal and Wessels(1984), Ex. 3.5].

Let $\mathbf{X} = \{0,1\}$; $\mathbf{A} = [0,1]$; $p(0|0,a) \equiv 1$, $p(1|0,a) \equiv 0$, $p(y|1,a) = \begin{cases} a, & \text{if } y = 0; \\ 1-a, & \text{if } y = 1, \end{cases}$ $c(0,a) \equiv 0$, $c(1,a) = -a(1-a)$. Here, for $\varphi(x) \equiv 1$, both states are positive recurrent. See Fig. 2.19.

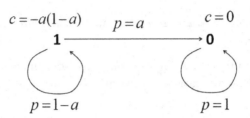

Fig. 2.19 Example 2.2.15: no optimal selectors in a semi-continuous model.

Condition 2.2 is satisfied. Indeed, value iteration converges to the function $v^\infty(0) = 0$, $v^\infty(1) = -1$ and $v^\infty(x) = v_x^* = \inf_\pi v_x^\pi$ [Bertsekas and Shreve(1978), Prop. 9.14]; [Puterman(1994), Th. 7.2.12].

The optimality equation
$$v(0) = v(0);$$
$$v(1) = \inf_{a \in \mathbf{A}} \{-a(1-a) + av(0) + (1-a)v(1)\}$$
has the following maximal non-negative solution:
$$v(0) = v_0^* = 0; \qquad v(1) = v_1^* = -1.$$
The stationary selector $\varphi^*(x) \equiv 0$ is the single conserving strategy at $x = 1$, but
$$\lim_{t \to \infty} E_x^{\varphi^*}[v_{X_t}^*] = v_1^* = -1,$$
so that it is not equalizing and not optimal if $X_0 = 1$. Indeed, $v_1^{\varphi^*} = 0$, and for each stationary selector $\varphi(x) \equiv \hat{a} > 0$,
$$v_1^\varphi = -\hat{a}(1-\hat{a})E_1^\varphi[\tau],$$
where τ, the time to absorption in state 0, has a geometric distribution with parameter $\hat{a}$. Thus, $v_1^\varphi = -(1-\hat{a})$ and $\inf_{\hat{a} \in \mathbf{A}} v_1^\varphi = -1 = v_1^*$. There are no optimal strategies in this model, unless it is semi-continuous.

The next example, based on [Bertsekas(1987), Section 6.4, Ex. 2], also illustrates that the unichain assumption and Condition 2.2(a) in Theorem 2.6 are important. Moreover, it shows that Theorem 2.8 does not hold in negative models.

Let $\mathbf{X} = \{1, 2\}$; $\mathbf{A} = [0, 1]$; $p(2|2, a) \equiv 1$, $p(2|1, a) = 1 - p(1|1, a) = a^2$; $c(2, a) \equiv 0$, $c(1, a) = -a$ (see Fig. 2.20).

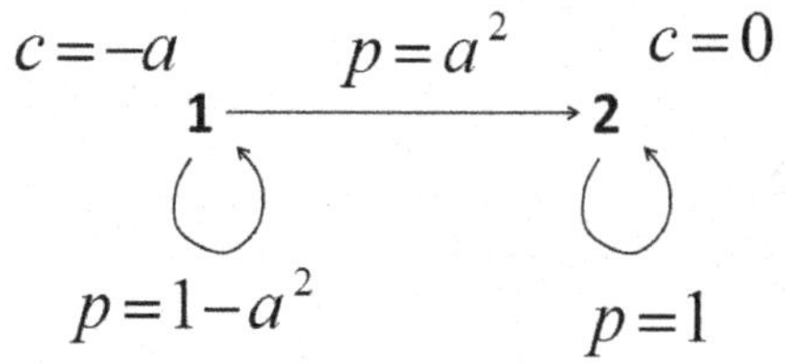

Fig. 2.20 Example 2.2.15: the blackmailer's dilemma.

"We may regard a as a demand made by a blackmailer, and state 1 as the situation where the victim complies. State 2 is the situation where the victim refuses to yield to the blackmailer's demand. The problem then can be seen as one whereby the blackmailer tries to maximize his total gain by balancing his desire for increased demands with keeping his victim compliant." [Bertsekas(1987), p. 254].

Obviously, $v_2^* = 0$, and the optimality equation (2.2) for state $x = 1$ has the form

$$v(1) = \inf_{a \in [0,1]} \{-a + (1 - a^2)v(1)\}.$$

One can check formally that $v(1)$ cannot be positive (or zero). Assuming that $v(1) < 0$ leads to equation $0 = \frac{1}{4v(1)}$ having no finite solutions. (Here the minimum is provided by $a^* = \frac{-1}{2v(1)}$.) Assuming that $\frac{-1}{2v(1)} \notin [0, 1]$ leads to a contradiction.

If $\varphi^0(x) \equiv 0$ then $v_1^{\varphi^0} = 0$, and if $\varphi(x) \equiv a \in (0, 1]$ then $v_1^\varphi = -1/a$, because v_1^φ is a solution to the equation

$$v_1^\varphi = -a + (1 - a^2)v_1^\varphi.$$

Therefore, $v_1^* = -\infty$, but no one stationary selector (or stationary randomized strategy) is optimal. One can also check that the value iteration converges to the function $v^\infty(2) = 0$, $v^\infty(1) = -\infty$ and $v^\infty(x) = v_x^* = \inf_\pi v_x^\pi$ [Bertsekas and Shreve(1978), Prop. 9.14],[Puterman(1994), Th. 7.2.12].

Suppose $X_0 = 1$, and consider the non-stationary selector $\varphi_t^*(x) = \sqrt{1 - e^{-1/t^2}}$. Clearly,

$$v_1^{\varphi^*} = -\varphi_1^*(1) + [1 - (\varphi_1^*(1))^2]\{-\varphi_2^*(1) + [1 - (\varphi_2^*(1))^2]\{\cdots\}\}.$$

First of all, notice that $Q \stackrel{\triangle}{=} \prod_{t=1}^{\infty}[1 - \varphi_t^*(1))^2] > 0$, because

$$\sum_{t=1}^{\infty} \ln[1 - (\varphi_t^*(1))^2] = -\sum_{t=1}^{\infty} \frac{1}{t^2} < \infty.$$

Now, $v_1^{\varphi^*} \leq -Q \cdot \sum_{t=1}^{\infty} \varphi_t^*(1)$, but

$$\sum_{t=1}^{\infty} \varphi_t^*(1) = \sum_{t=1}^{\infty} \sqrt{1 - e^{-1/t^2}} = +\infty,$$

because $\sum_{t=1}^{\infty} \frac{1}{t} = +\infty$ and

$$\lim_{t \to \infty} \frac{\sqrt{1 - e^{-1/t^2}}}{\frac{1}{t}} = \lim_{\delta \to 0} \frac{\sqrt{1 - e^{-\delta^2}}}{\delta} = 1.$$

Therefore, $v_1^{\varphi^*} = -\infty$ and selector φ^* is (uniformly) optimal. Any actions taken in state 2 play no role. Another remark about the blackmailer's dilemma appears at the end of Section 4.2.2.

In the examples presented, the *polytope* condition is satisfied: for each $x \in \mathbf{X}$ the set $\Pi(x) \stackrel{\triangle}{=} \{p(0|x,a), p(1|x,a), \ldots, p(m|x,a)|a \in \mathbf{A}\}$ has a finite number of extreme points. (Here we assume that the state space $\mathbf{X} = \{0, 1, \ldots, m\}$ is finite.) It is known that in such MDPs with average loss, an optimal stationary selector exists, if the model is semi-continuous [Cavazos-Cadena et al.(2000)]. The situation is different for MDPs with expected total loss.

2.2.16 Not a semi-continuous model

If the model is not semi-continuous then one cannot guarantee the existence of optimal strategies. Moreover, the following example [van der Wal and Wessels(1984), Ex. 4.1] shows that no one stationary selector is ε-optimal for all $\varepsilon < 1$.

Let $\mathbf{X} = \{0, 1, 2\}$, $\mathbf{A} = [0, 1)$. Note that $\mathbf{A}$ is not compact. Put $p(0|0,a) = p(0|2,a) \equiv 1$, $p(1|1,a) = a$, $p(2|1,a) = 1 - a$; all the other transition probabilities are zero; $c(2,a) = 1$, $c(0,a) = c(1,a) \equiv 0$. See Fig. 2.21. In fact, this is a slight modification of the example in Section 2.2.11.

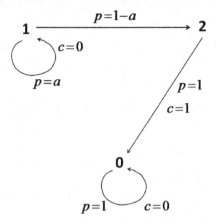

Fig. 2.21 Example 2.2.16: no ε-optimal selectors for $\varepsilon < 1$.

The optimality equation (2.2) takes the form (with, clearly, $v(0) = 0$):

$$v(2) = 1 + v(0);$$
$$v(1) = \inf_{a \in \mathbf{A}} \{av(1) + (1-a)v(2)\}.$$

Hence, $v(0) = 0$, $v(1) = 0$, $v(2) = 1$. But for any stationary selector, $\varphi(1) < 1$, so that $v_1^\varphi = 1$, and φ is not ε-optimal for $\varepsilon < 1$. Here, no one strategy is conserving.

Now we present a unichain model with finite space $\mathbf{X}$ which is not semi-continuous. This example is trivial but can help the understanding of several topological issues.

Let $\mathbf{X} = \{0, 1, 2\}$; $\mathbf{A} = \{a_\infty^1, a_\infty^2, 1, 2, \ldots\}$; $p(0|0, a) \equiv 1$,

$$p(2|1, a) = \begin{cases} 1/2, & \text{if } a = a_\infty^1; \\ 1/7, & \text{if } a = a_\infty^2; \\ 1/2 - (1/3)^a, & \text{if } a = 1, 2, \ldots, \end{cases}$$

$p(0|1, a) = 1 - p(2|1, a)$, $p(1|2, a) \equiv 1$; all the other transition probabilities are zero; $c(0, a) = c(2, a) \equiv 0$,

$$c(1, a) = \begin{cases} D, & \text{if } a = a_\infty^1; \\ -3/2, & \text{if } a = a_\infty^2; \\ -1 + 1/a, & \text{if } a = 1, 2, \ldots \end{cases}$$

(see Fig. 2.22). The optimality equation (2.2) takes the form (with, clearly, $v(0) = 0$):

$$v(2) = v(1);$$
$$v(1) = \min \left\{ -3/2 + (1/7)v(2); \quad D + (1/2)v(2); \right.$$
$$\left. \inf_{a=1,2,3,\ldots} \{-1 + 1/a + (1/2 - (1/3)^a)v(2)\} \right\}.$$

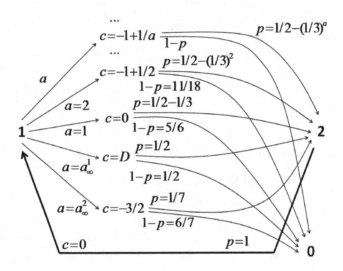

Fig. 2.22 Example 2.2.15: not a semi-continuous model.

Suppose $D > -1$. The maximal non-positive solution can then be obtained by, e.g., using the value iteration:

$$v(0) = v_0^* = 0, \quad v(1) = v_1^* = -2, \quad v(2) = v_2^* = -2.$$

No one stationary selector is conserving. There are no (uniformly) optimal strategies in this model: in state 1, action $a + 1$ is better than a, a_∞^1 and a_∞^2.

One can introduce the topology in space **A** in different ways:

(a) Suppose the topology is *discrete*: all singletons are open sets, hence all subsets are open (and simultaneously closed). Then any function is certainly continuous, Conditions 2.3(b,c) are satisfied, but **A** is not compact. Thus, the model is not semi-continuous.

(b) To make **A** compact, we can say that all singletons $\{a\}$ in **A**, except for a_∞^1, are open, along with their complements $\mathbf{A}\setminus\{a\}$, and consider the coarsest topology containing all those open sets. In other words, we simply accept that sequence $1, 2, \ldots$ in **A** converges to a_∞^1, or one can interpret a_∞^1 as 0, $a = i$ as $1/i$, $a = a_\infty^2$ as 2, and consider the trace on **A** of the standard topology in $\mathbb{R}^1$. Now Condition 2.3(c) is satisfied, but the loss function $c(1, a)$ is not lower semi-continuous in a because $D > -1$. Note that if $D \leq -1$, then this construction leads to a semi-continuous model, and $\varphi(x) \equiv a_\infty^1$ is the (uniformly) optimal stationary selector.

(c) We can make **A** compact in a different way by announcing that all singletons $\{a\}$ in **A**, except for a_∞^2, are open, along with their complements $\mathbf{A}\setminus\{a\}$. Now Condition 2.3(b) is satisfied, but part (c) is violated: in the case where $u(2) = 1$ and $u(0) = 0$ we have $\sum_{y\in\mathbf{X}} u(y)p(y|1,a) = (1/2) - (1/3)^a$ does not converge to $\sum_{y\in\mathbf{X}} u(y)p(y|1,a_\infty^2) = 1/7$ as $a = 1, 2, \ldots$ increases (equivalently, as $a \to a_\infty^2$). Hence, this topology in **A** again does not result in a semi-continuous model.

2.2.17 The Bellman function is non-measurable and no one strategy is uniformly ε-optimal

It was mentioned in Section 2.1 that many examples from Chapter 1 can be modified as infinite-horizon models. In particular, Example 1.4.15 can be adjusted in the following way. The state and action spaces **X** and **A** remain the same, and we put $p(\Gamma|x,a) \equiv p_1(\Gamma|x,a)$ and $c(x,a) = -I\{(x,a) \in B\}I\{x \in [0,1]\}$. Fig. 1.24 is still relevant; if $X_1 \in [0,1]$ then on the next step $X_2 = x_\infty$, and this state is absorbing. The Bellman function has the form
$$v_x^* = \begin{cases} -1, & \text{if } x \in B \text{ or } x \in [0,1] \cap B^1; \\ 0 & \text{otherwise,} \end{cases}$$
and is again not measurable. For any fixed $X_0 = x_0 \in \mathbf{X}$, there exists an optimal stationary selector. For $x_0 \in [0,1] \cap B^1$, it is sufficient to put
$$\varphi^*(x) = \text{ any fixed } a \in \mathbf{A} \text{ such that } (x_0,a) \in B.$$
For $x_0 = (y_1, y_2) \in B$, it is sufficient to put $\varphi^*(x) \equiv y_2$. If $x_0 = x_\infty$ or $x_0 \in [0,1] \setminus B^1$, then $\varphi^*(x)$ does not play any role.

It should be emphasized that this selector cannot be represented as a measurable mapping (semi-Markov selector) $\varphi(x_0, x) : \mathbf{X} \times \mathbf{X} \to \mathbf{A}$.

Moreover, no one strategy is uniformly ε-optimal for $\varepsilon < 1$: all the reasoning given in Example 1.4.15 applies (see also Section 3.2.7). On the other hand, the constructed selector φ^* is optimal simultaneously for all $x_0 \in B$. In this connection, one can show explicitly the dependence of φ^* on $x_0 = (y_1, y_2) \in B$: $\varphi^*(x_0, x) \equiv y_2$. We have built a semi-Markov selector. Remember, no one Markov strategy is as good as φ^* for $x_0 \in B$. In fact, semi-Markov strategies very often form a sufficient class in the following sense.

Theorem 2.7.

(a) *[Strauch(1966), Th. 4.1] Suppose the loss function is either non-negative, or bounded and non-positive. Then, for any strategy π, there exists a semi-Markov strategy $\hat{\pi}$ such that $v_x^\pi = v_x^{\hat{\pi}}$ for all $x \in \mathbf{X}$.*

(b) *[Strauch(1966), Th. 4.3] Suppose the loss function is non-negative. Then, for any strategy π, there exists a semi-Markov non-randomized strategy $\hat{\pi}$ such that $v_x^{\hat{\pi}} \leq v_x^\pi$ for all $x \in \mathbf{X}$.*

Remark 2.5. In this example, $\lim_{n\to\infty} v^n(x) = v^\infty(x) = v_x^*$ because the model is negative: see (2.10). Another MDP with a non-measurable Bellman function v_x^*, but with a measurable function $v^\infty(x)$ is described in [Bertsekas and Shreve(1978), Section 9.5, Ex. 2].

2.2.18 A randomized strategy is better than any selector (finite action space)

Theorem 2.8. *[Strauch(1966), Th. 8.3] If $c(x,a) \geq 0$ and there exists an optimal strategy, then there exists an optimal stationary selector.*

The following example, first published in [Bertsekas and Shreve(1978), Chapter 9, Ex. 3] shows that this assertion does not hold for negative models.

Let $\mathbf{X} = \{0, 1, 2, \ldots\}$; $\mathbf{A} = \{1, 2\}$; $p(0|0, a) \equiv 1$, $p(0|x, 1) \equiv 1$ for all $x \geq 0$, $p(x+1|x, 2) \equiv 1$ for all $x > 0$; all the other transition probabilities are zero; $c(0, a) \equiv 0$, $c(x, 1) = -2^x$ and $c(x, 2) \equiv 0$ for $x > 0$ (see Fig. 2.23).

Obviously, $v_0^* = 0$ and, for $x > 0$, $v_x^* = -\infty$ because $v_x^* \leq -2^x$; hence (if one applies action 2 in state x) $v_x^* \leq -2^{x+1}$; hence (if one applies action 2 in state $x+1$) $v_x^* \leq -2^{x+2}$, and so on. The only conserving strategy is $\varphi^*(x) \equiv 2$, but this is not equalizing and not optimal, since $v_x^{\varphi^*} \equiv 0$.

If $\varphi_t(h_{t-1})$ is an arbitrary selector and $x_0 > 0$ is the initial state then

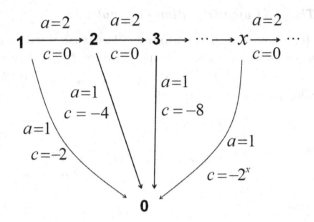

Fig. 2.23 Example 2.2.18: only a randomized strategy is (uniformly) optimal.

- either $\varphi_t(x_0, a_1, \ldots, x_{t-1}) \equiv 2$ for all $t > 0$, so that $v_{x_0}^\varphi = 0$,
- or $\varphi_t(x_0, a_1, \ldots, x_{t-1}) = 1$ for some (first) value of $t \geq 1$, in which case
$$v_{x_0}^\varphi = -2^{-(x_0+t-1)} > -\infty.$$

Thus, no one selector is (uniformly) optimal.

In the case where $P_0(x) = \left(\frac{1}{2}\right)^x$ for $x > 0$, the stationary selector $\varphi(x) \equiv 1$ is optimal (but certainly not uniformly optimal):
$$v^\varphi = \sum_{x=1}^\infty P_0(x)(-2^x) = -\infty.$$

Similarly, the randomized stationary strategy $\pi^*(1|x) = \pi^*(2|x) = \frac{1}{2}$ provides, for any $x > 0$,
$$v_x^{\pi^*} = -2^{x-1} + \frac{1}{2} v_{x+1}^{\pi^*} = \cdots = -k 2^{x-1} + \left(\frac{1}{2}\right)^k v_{x+k}^{\pi^*}, \quad k = 1, 2, \ldots$$

Moving to the limit as $k \to \infty$, we see that $v_x^{\pi^*} = -\infty$ (note that $v_y^{\pi^*} \leq 0$). Therefore, the π^* strategy is uniformly optimal (and also optimal for any initial distribution).

At the end of Section 2.2.4, one can find another example of a stationary (randomized) strategy π^s such that, for any stationary selector φ, there is an initial state $\hat{x}$ for which $v_{\hat{x}}^{\pi^s} < v_{\hat{x}}^\varphi$.

2.2.19 The fluid approximation does not work

Let $\mathbf{X} = \{0, 1, 2, \ldots\}$, $\mathbf{A}$ be an arbitrary Borel space, and suppose real functions $q^+(y, a) > 0$, $q^-(y, a) > 0$, and $\rho(y, a)$ on $\mathbb{R}^+ \times \mathbf{A}$ are given such that $q^+(y, a) + q^-(y, a) \leq 1$; $q^+(0, a) = q^-(0, a) = \rho(0, a) \stackrel{\triangle}{=} 0$. For a fixed $n \in \mathbb{N}$ (scaling parameter), we consider a random walk defined by

$$^n p(y|x, a) = \begin{cases} q^+(x/n, a), & \text{if } y = x + 1; \\ q^-(x/n, a), & \text{if } y = x - 1; \\ 1 - q^+(x/n, a) - q^-(x/n, a), & \text{if } y = x; \\ 0 & \text{otherwise} \end{cases}$$

$$^n c(x, a) = \frac{\rho(x/n, a)}{n}.$$

Below, we give the conditions under which this random walk is absorbing.

Let a piece-wise continuous function $\psi(y) : \mathbb{R}^+ \to \mathbf{A}$ be fixed, and introduce the continuous-time stochastic process

$$^n Y(\tau) = I\left\{\tau \in \left[\frac{t}{n}, \frac{t+1}{n}\right)\right\} {}^n X_t/n, \quad t = 0, 1, 2, \ldots, \qquad (2.19)$$

where the discrete-time Markov chain $^n X_t$ is governed by control strategy $\varphi(x) \stackrel{\triangle}{=} \psi(x/n)$. Under rather general conditions, if $\lim_{n \to \infty} {}^n X_0/n = y_0$ then, for any τ, $\lim_{n \to \infty} {}^n Y(\tau) = y(\tau)$ almost surely, where the deterministic function $y(\tau)$ satisfies

$$y(0) = y_0; \quad \frac{dy}{d\tau} = q^+(y, \psi(y)) - q^-(y, \psi(y)). \qquad (2.20)$$

(See, e.g., [Gairat and Hordijk(2000)]; the proof is based on the law of large numbers.) Hence, it is not surprising that in the absorbing case (if $q^-(y, a) > q^+(y, a)$) the objective

$$^n v^\varphi_{{}^n X_0} = E^\varphi_{{}^n X_0}\left[\sum_{t=1}^\infty {}^n c(X_{t-1}, A_t)\right]$$

converges to

$$\tilde{v}^\psi(y_0) = \int_0^\infty \rho(y(\tau), \psi(y(\tau))) d\tau \qquad (2.21)$$

as $n \to \infty$. (To be more rigorous, one has to keep the index n at the expectation $^n E$.) More precisely, the following statement was proved in [Piunovskiy(2009b), Th. 1].

Theorem 2.9. *Suppose all the functions $q^+(y,\psi(y))$, $q^-(y,\psi(y))$, $\rho(y,\psi(y))$ are piece-wise continuously differentiable;*

$$q^-(y,\psi(y)) > \underline{q} > 0, \quad \inf_{y>0} \frac{q^-(y,\psi(y))}{q^+(y,\psi(y))} = \tilde{\eta} > 1; \quad \sup_{y>0} \frac{|\rho(y,\psi(y))|}{\eta^y} < \infty,$$

where $\eta \in (1,\tilde{\eta})$.
Then, for an arbitrary fixed $\hat{y} \geq 0$,

$$\lim_{n\to\infty} \sup_{0\leq x\leq \hat{y}n} |\,^n v_x^\varphi - \tilde{v}^\psi(x/n)| = 0.$$

As a corollary, if one solves a rather simple optimization problem $\tilde{v}^\psi(y) \to \inf_\psi$, then the control strategy $\varphi^*(x) = \psi^*(x/n)$, derived from the optimal (or nearly optimal) feedback strategy ψ^*, will be nearly optimal in the underlying MDP, if n is large enough. More details, including an estimate of the rate of convergence, can be found in [Piunovskiy and Zhang(2011)]. Although that article is about controlled continuous-time chains, the statements can be reformulated for the discrete-time case using, e.g., the uniformization technique [Puterman(1994), Section 11.5.1]. The fluid approximation to an absorbing (discrete-time) uncontrolled random walk was discussed in [Piunovskiy(2009b)]. Example 3 in that article shows that the condition $\sup_{y>0}\frac{|\rho(y,\psi(y))|}{\eta^y} < \infty$ in Theorem 2.9 is important. Below, we present a slight modification of this example.

Let $\mathbf{A} = [1,2]$, $q^+(y,a) = ad^+$, $q^-(y,a) = ad^-$ for $y > 0$, where $d^- > d^+ > 0$ are fixed numbers such that $2(d^+ + d^-) \leq 1$. Put $\rho(y,a) = a^2\gamma^{y^2}$, where $\gamma > 1$ is a constant.

To solve the fluid model $\tilde{v}^\psi(y) \to \inf_\psi$, we use the dynamic programming approach. One can see that the Bellman function $\tilde{v}^*(y) \stackrel{\triangle}{=} \inf_\psi \tilde{v}^\psi(y)$ has the form

$$\tilde{v}^*(y) = \int_0^y \inf_{a\in\mathbf{A}} \left\{\frac{\rho(u,a)}{q^-(u,a) - q^+(u,a)}\right\} du$$

and satisfies the Bellman equation

$$\inf_{a\in\mathbf{A}} \left\{\frac{d\tilde{v}^*(y)}{dy}\left[q^+(y,a) - q^-(y,a)\right] + \rho(y,a)\right\} = 0, \quad \tilde{v}^*(0) = 0.$$

Technical details can be found in the proof of Lemma 2 in [Piunovskiy and Zhang(2011)]. Hence, the function

$$\tilde{v}^*(y) = \tilde{v}^{\psi^*}(y) = \int_0^y \frac{\gamma^{u^2}}{d^- - d^+} du$$

is well defined, and $\psi^*(y) \equiv 1$ is the optimal strategy.

Conversely, for any control strategy π in the underlying MDP, ${}^n v_x^\pi = \infty$ for all $x > 0$, $n \in \mathbb{N}$. Indeed, starting from any state $x > 0$, the probability of reaching state $x + k$ is not smaller than $(d^+)^k$, so that

$${}^n v_x^\pi \geq (d^+)^k \inf_{a \in A} {}^n c(x+k, a) = (d^+)^k \frac{\gamma^{\left(\frac{x+k}{n}\right)^2}}{n}$$

for all $k = 1, 2, \ldots$. Hence

$${}^n v_x^\pi \geq \lim_{k \to \infty} (d^+)^k \frac{\gamma^{\left(\frac{x+k}{n}\right)^2}}{n} = \infty.$$

Equation (2.2) cannot have finite non-negative solutions. To prove this for an arbitrary fixed value of n, suppose ${}^n v(x)$ is such a solution to the equation

$$v(0) = 0; \qquad \text{for } x > 0 \ \ v(x) =$$

$$\inf_{a \in A} \left\{ \frac{a^2}{n} \gamma^{(x/n)^2} + a d^- v(x-1) + a d^+ v(x+1) + (1 - a d^- - a d^+) v(x) \right\},$$

that is

$$\inf_{a \in A} \left\{ \frac{a}{n} \gamma^{(x/n)^2} + d^- \, {}^n v(x-1) + d^+ \, {}^n v(x+1) - (d^- + d^+) \, {}^n v(x) \right\}$$

$$= \frac{1}{n} \gamma^{(x/n)^2} + d^- \, {}^n v(x-1) + d^+ \, {}^n v(x+1) - (d^- + d^+) \, {}^n v(x) = 0.$$

If ${}^n v(0) = 0$ and ${}^n v(1) = b \geq 0$, then

$${}^n v(x) = b \frac{\tilde{\eta}^x - 1}{\tilde{\eta} - 1} - \frac{1}{nd^+(\tilde{\eta} - 1)} \sum_{j=1}^{x-1} \gamma^{(j/n)^2} (\tilde{\eta}^{x-j} - 1)$$

$$\leq \frac{1}{nd^+(\tilde{\eta} - 1)} \left[nd^+ b(\tilde{\eta}^x - 1) - \gamma^{[(x-1)/n]^2} (\tilde{\eta} - 1) \right],$$

where, as before, $\tilde{\eta} = d^- / d^+ > 1$. Hence, for sufficiently large x, ${}^n v(x) < 0$, which is a contradiction.

2.2.20 The fluid approximation: refined model

Consider the same situation as in Example 2.2.19 and assume that all conditions of Theorem 2.9 are satisfied, except for $q^-(y,\psi(y)) > \underline{q} > 0$. Since the control strategies ψ and φ are fixed, we omit them in the formulae below. Since $q^-(y)$ can approach zero and $q^+(y) < q^-(y)$, the stochastic process ${}^nX_t$ can spend too much time around the (nearly absorbing) state $x > 0$ for which $q^-(x/n) \approx q^+(x/n) \approx 0$, so that ${}^nv_{{}^nX_0}$ becomes big and can even approach infinity as $n \to \infty$. The situation becomes good again if, instead of inequalities

$$q^-(y) > \underline{q} > 0 \qquad \sup_{y>0} \frac{|\rho(y)|}{\eta^y} < \infty,$$

we impose the condition

$$\sup_{y>0} \frac{|\rho(y)|}{[q^+(y) + q^-(y)]\eta^y} < \infty.$$

Now we can make a (random) change of the time scale, and take into account only the original time moments when the state of the process ${}^nX_t$ actually changes. As a result, the one-step loss for $x > 0$ becomes

$$\hat{c}(x) = \frac{{}^nc(x)}{{}^np(x+1|x) + {}^np(x-1|x)},$$

because the time spent in the current state x has a geometric distribution with parameter $1 - {}^np(x+1|x) - {}^np(x-1|x)$. Hence, ${}^nv_{{}^nX_0} = {}^n\hat{v}_{{}^nX_0}$, where the hat corresponds to a new model with parameters

$$^n\hat{c}(x,a) = \frac{{}^nc(x,a)}{{}^np(x+1|x,a) + {}^np(x-1|x,a)},$$

$$^n\hat{p}(y|x,a) = \frac{{}^np(y|x,a)}{{}^np(x+1|x,a) + {}^np(x-1|x,a)} \quad (y = x \pm 1).$$

Formally, notice that for a fixed control strategy, the dynamic programming equations in the initial and transformed models have coincident solutions:

$$^nv_x = {}^nc(x) + {}^np(x+1|x)\,{}^nv_{x+1} + {}^np(x-1|x)\,{}^nv_{x-1} + {}^np(x|x)\,{}^nv_x, \quad x \geq 1;$$

$$^n\hat{v}_x = {}^n\hat{c}(x) + {}^n\hat{p}(x+1|x)\,{}^n\hat{v}_{x+1} + {}^n\hat{p}(x-1|x)\,{}^n\hat{v}_{x-1} \quad x \geq 1,$$

because the second equation coincides with the first one, divided by ${}^np(x+1|x) + {}^np(x-1|x)$.

Now we apply Theorem 2.9 to the transformed functions

$$\hat{q}^+(y) \triangleq \frac{q^+(y)}{q^+(y) + q^-(y)}; \quad \hat{q}^-(y) \triangleq \frac{q^-(y)}{q^+(y) + q^-(y)}; \quad \hat{\rho}(y) \triangleq \frac{\rho(y)}{q^+(y) + q^-(y)}.$$

(note that $\hat{q}^-(y) > \frac{\tilde{\eta}}{1+\tilde{\eta}} > 0$): for any $\hat{y} \geq 0$,

$$\lim_{n\to\infty} \sup_{0\leq x\leq \hat{y}n} |{}^n\hat{v}_x - \tilde{v}(x/n)| = \lim_{n\to\infty} \sup_{0\leq x\leq \hat{y}n} |{}^nv_x - \tilde{v}(x/n)| = 0. \quad (2.22)$$

We call the "hat" deterministic model

$$y(0) = y_0; \quad \frac{dy}{du} = \hat{q}^+(y) - \hat{q}^-(y); \quad \tilde{v}(y_0) = \int_0^\infty \hat{\rho}(y(u))du,$$

similar to (2.20) and (2.21), the refined fluid model. It corresponds to the change of time

$$\frac{du}{d\tau} = q^+(y(\tau)) + q^-(y(\tau)).$$

It is interesting to compare the initial fluid model (2.20), (2.21) with the initial stochastic process (2.19). Although the trajectories still converge almost surely (i.e. $\lim_{n\to\infty} {}^nY(\tau) = y(\tau)$, even uniformly on finite intervals), it can easily happen that $\lim_{n\to\infty} |{}^nv_x^\varphi - \tilde{v}^\psi(x/n)| > 0$. Since the derivative $\frac{dy}{d\tau} = q^+(y) - q^-(y)$, although negative for positive y, is not separated from zero, the limit $\lim_{\tau\to\infty} y(\tau)$ can be strictly positive, i.e. the process $y(\tau)$ decreases, but never reaches zero.

As an example, suppose that

$$q^-(y) = 0.1\ I\{y \in (0,1]\} + 0.125\ (y-1)^2 I\{y \in (1,3]\} + 0.5\ I\{y > 3\};$$

$$q^+(y) = 0.2\ q^-(y); \quad \rho(y) = 8\ q^-(y)$$

(see Fig. 2.24).

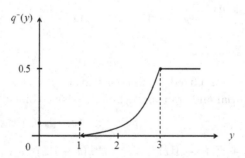

Fig. 2.24   Example 2.21: transition probability q^-.

Equation (2.20) takes the form $\frac{dy}{d\tau} = -0.1\ (y-1)^2$, and, if the initial state $y_0 = 2$, then $y(\tau) = 1 + \frac{10}{\tau+10}$, so that $\lim_{\tau\to\infty} y(\tau) = 1$. Conversely, since $q^-, q^+ > 0$ for $y > 0$, and there is a negative trend, the process ${}^nY(\tau)$ starting from ${}^nX_0/n = y_0 = 2$ will be absorbed at zero, but the moment of absorption is postponed until later and later as $n \to \infty$, because the process spends more and more time in the neighbourhood of 1. See Figs 2.25 and 2.26, where typical trajectories of ${}^nY(\tau)$ are shown along with the continuous curve $y(\tau)$.

On any finite interval $[0, T]$, we have

$$\lim_{n \to \infty} E_{2n} \left[\sum_{t=1}^{\infty} I\{t/n \le T\} \, {}^nc(X_{t-1}, A_t) \right]$$

$$= \lim_{n \to \infty} E \left[\int_0^T \rho({}^nY(\tau)) d\tau \right] = \int_0^T \rho(y(\tau)) d\tau = 10 - \frac{100}{T+10}.$$

Therefore,

$$\lim_{T \to \infty} \lim_{n \to \infty} E_{2n} \left[\sum_{t=1}^{\infty} I\{t/n \le T\} \, {}^nc(X_{t-1}, A_t) \right] = \lim_{T \to \infty} \int_0^T \rho(y(\tau)) d\tau = 10.$$

However, we are interested in the expected total cost at large values of n, as in the following limit:

$$\lim_{n \to \infty} \lim_{T \to \infty} E_{2n} \left[\sum_{t=1}^{\infty} I\{t/n \le T\} \, {}^nc(X_{t-1}, A_t) \right] = \lim_{n \to \infty} {}^nv_{2n},$$

a quantity which is far different from 10. Indeed, according to Theorem 2.9 applied to the refined model,

$$\lim_{n \to \infty} {}^nv_{2n} = \lim_{n \to \infty} {}^n\hat{v}_{2n} = \tilde{\hat{v}}(2) = \int_0^{\infty} \hat{\rho}(y(u)) du = \int_0^3 \frac{8}{1.2} du = 20,$$

because $\hat{\rho}(y) = \frac{8}{1.2}$ and, in the time scale u, the y process equals $y(u) = 2 - \frac{2}{3}u$ and hence is absorbed at zero at $u = 3$.

If one has an optimal control strategy $\psi^*(y)$ in the original model of (2.20) and (2.21), in the time scale τ, the corresponding strategy $\varphi^*(x) = \psi^*(x/n)$ can be far from optimal in the underlying MDP even for large values of n, simply because the values $\psi^*(y)$ for $y < \kappa \stackrel{\triangle}{=} \lim_{\tau \to \infty} y(\tau)$ play no role when $\lim_{\tau \to \infty} y(\tau) > 0$ under a control strategy ψ^*. On the other hand, the refined model (time scale u) is helpful for calculating a nearly optimal strategy φ^*. The example presented is a discrete-time version of Example 1 from [Piunovskiy and Zhang(2011)].

Fluid scaling is widely used in queueing theory. See, e.g., [Gairat and Hordijk(2000)], although more often continuous-time chains are studied

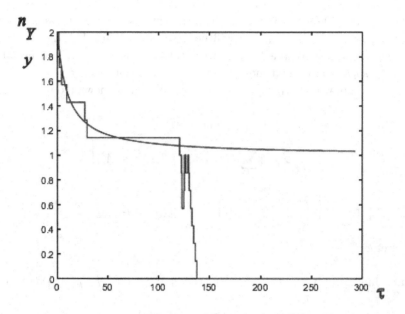

Fig. 2.25 Example 2.21: a stochastic process and its fluid approximation, $n = 7$.

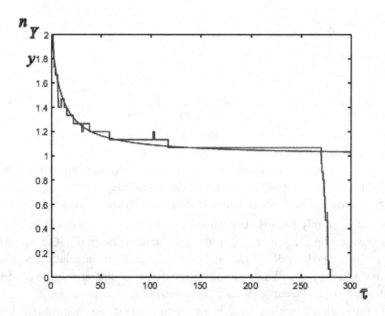

Fig. 2.26 Example 2.21: a stochastic process and its fluid approximation, $n = 15$.

[Pang and Day(2007); Piunovskiy and Zhang(2011)]. The scaling parameter n corresponds both to the size of one job and to the time unit, both being proportional to $1/n$. The arrival probability of one job and the probability of the service completion during one time unit can both depend on the current amount of work in the system x/n, where x is the integer number of jobs. The same is true for the one-step loss $\rho(x/n,a)/n$ which is divided by n because one step (the time unit) is $1/n$.

2.2.21 *Occupation measures: phantom solutions*

Definition 2.2. For a fixed control strategy π, the *occupation measure* η^π is the measure on $\mathbf{X} \times \mathbf{A}$ given by the formula

$$\eta^\pi(\Gamma^\mathbf{X} \times \Gamma^\mathbf{A}) \triangleq \sum_{t=0}^{\infty} P_{P_0}^\pi \{X_t \in \Gamma^\mathbf{X}, A_{t+1} \in \Gamma^\mathbf{A}\}, \quad \Gamma^\mathbf{X} \in \mathcal{B}(\mathbf{X}), \quad \Gamma^\mathbf{A} \in \mathcal{B}(\mathbf{A}).$$

For any π, the occupation measure η^π satisfies the equation

$$\eta(\Gamma \times \mathbf{A}) = P_0(\Gamma) + \int_{\mathbf{X} \times \mathbf{A}} p(\Gamma|y,a) d\eta(y,a) \qquad (2.23)$$

[Hernandez-Lerma and Lasserre(1999), Lemma 9.4.3].

Usually (e.g. in positive and negative models),

$$v^\pi = \int_{\mathbf{X} \times \mathbf{A}} c(x,a) d\eta^\pi(x,a),$$

and investigation of MDP in terms of occupation measures (the so-called *convex analytic approach*) is fruitful, especially in constrained problems.

Recall that MDP is called *absorbing* at 0 if $p(0|0,a) \equiv 1$, $c(0,a) \equiv 0$: state 0 is absorbing and there is no future loss after the absorption. Moreover, we require that

$$\forall \pi \quad E_{P_0}^\pi[T_0] < \infty, \qquad (2.24)$$

where $T_0 \triangleq \min\{t \geq 0: X_t = 0\}$ is the time to absorption [Altman(1999), Section 7.1]. Sometimes, the absorbing state is denoted as Δ.

In the absorbing case, for each strategy π, the occupation measure η^π satisfies the following equations

$$\eta^\pi((X \setminus \{0\}) \times \mathbf{A}) = E_{P_0}^\pi[T_0] < \infty, \quad \eta^\pi(\{0\} \times \mathbf{A}) = \infty$$

and

$$\eta((\Gamma \setminus \{0\}) \times \mathbf{A}) = P_0(\Gamma \setminus \{0\}) + \int_{(\mathbf{X} \setminus \{0\}) \times \mathbf{A}} p(\Gamma \setminus \{0\}|y,a) d\eta(y,a). \quad (2.25)$$

Equation (2.25) also holds for transient models: see Section 2.2.22.

If $\pi^s(\Gamma^A|y)$ is a stationary strategy in an absorbing MDP with a countable state space $\mathbf{X}$, then the measure $\hat{\eta}^{\pi^s}(x) \stackrel{\triangle}{=} \eta^{\pi^s}(x \times \mathbf{A})$ on $\mathbf{X} \setminus \{0\}$ satisfies the equation

$$\hat{\eta}^{\pi^s}(x) = P_0(x) + \sum_{y \in \mathbf{X} \setminus \{0\}} \int_{\mathbf{A}} p(x|y,a) \pi^s(da|y) \hat{\eta}^{\pi^s}(y). \qquad (2.26)$$

For given P_0, p and π^s, equation (2.26) w.r.t. $\hat{\eta}^{\pi^s}$ can have many solutions, but only the *minimal* non-negative solution gives the occupation measure [Altman(1999), Lemma 7.1]; non-minimal solutions are usually phantom and do not correspond to any control strategy.

The following example shows that equation (2.26) can indeed have many non-minimal (phantom) solutions. Let $\mathbf{X} = \{0, 1, 2, \ldots\}$, $\mathbf{A} = \{0\}$ (a dummy action). In reality, the model under consideration is uncontrolled, as there exists only one control strategy. We put

$$p(0|0,a) \equiv 1, \quad \forall x > 0 \ p(y|x,a) = \begin{cases} p_+, & \text{if } y = x+1; \\ p_-, & \text{if } y = x-1; \\ 0 & \text{otherwise,} \end{cases}$$

where $p_+ + p_- = 1$, $p_+ < p_-$, $p_+, p_- \in (0,1)$ are arbitrary numbers. The loss function does not play any role. See Fig. 2.27.

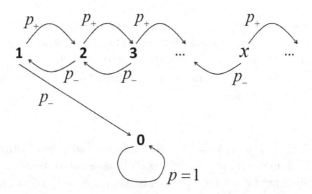

Fig. 2.27 Example 2.2.21: phantom solutions to equation (2.25).

Equation (2.25) is expressed as follows:
$$\eta(x \times \mathbf{A}) = P_0(x) + p_+ \eta((x-1) \times \mathbf{A}) + p_- \eta((x+1) \times \mathbf{A}), \quad x > 1;$$
$$\eta(1 \times \mathbf{A}) = P_0(1) + p_- \eta(2 \times \mathbf{A}).$$

Suppose $P_0(x) = \begin{cases} 1; & \text{if } x = 1, \\ 0, & \text{if } x \neq 1, \end{cases}$ then any function η_x of the form

$$\eta_x = d\left[1 - \left(\frac{p_+}{p_-}\right)^x\right] + \frac{1}{p_+}\left(\frac{p_+}{p_-}\right)^x, \quad x \geq 1$$

provides a solution, and the minimal non-negative solution corresponds to $d = 0$, negative values of d resulting in $\eta_x < 0$ for large values of x.

Putting $p_+ = 0$; $p_- = 1$, then equation (2.25) takes the form

$$\eta(x \times \mathbf{A}) = P_0(x) + \eta((x+1) \times \mathbf{A}), \quad x \geq 1.$$

The minimal non-negative solution is given by

$$\eta(x \times \mathbf{A}) = \sum_{i=x}^{\infty} P_0(i), \quad x \geq 1,$$

which is the unique finite solution: $\mu(\mathbf{X} \setminus \{0\} \times \mathbf{A}) < \infty$. At the same time, one can obviously add any constant to this solution, and equation (2.25) remains satisfied. A similar example was discussed in [Dufour and Piunovskiy(2010)].

Definition 2.3. [Altman(1999), Def. 7.4] Let the state space $\mathbf{X}$ be countable and let 0 be the absorbing state. A function $\mu : \mathbf{X} \to [1, \infty)$ is said to be a *uniform Lyapunov function* if

(i) $1 + \sum_{y \in \mathbf{X} \setminus \{0\}} p(y|x, a)\mu(y) \leq \mu(x)$;

(ii) $\forall x \in \mathbf{X}$, the mapping $a \to \sum_{y \in \mathbf{X} \setminus \{0\}} p(y|x, a)\mu(y)$ is continuous;

(iii) for any stationary selector φ, $\forall x \in \mathbf{X}$

$$\lim_{t \to \infty} E_x^\varphi[\mu(X_t)I\{X_t \neq 0\}] = 0.$$

If a uniform Lyapunov function exists, then the MDP is absorbing, i.e. equation (2.24) holds [Altman(1999), Lemma 7.2].

For an MDP with a uniform Lyapunov function, a solution η to equation (2.25) corresponds to some policy ($\eta = \eta^\pi$) if and only if $\eta(\mathbf{X}\setminus\{0\} \times \mathbf{A}) < \infty$ [Altman(1999), Th. 8.2]. In this case, $\eta = \eta^\pi = \eta^{\pi^s}$, where the stationary strategy π^s satisfies the equation

$$\eta^\pi(y \times \Gamma^\mathbf{A}) = \pi^s(\Gamma^\mathbf{A}|y)\eta^\pi(y \times \mathbf{A}), \quad \Gamma^\mathbf{A} \in \mathcal{B}(\mathbf{A}) \qquad (2.27)$$

[Altman(1999), Lemma 7.1, Th. 8.1].

In the convex analytic framework, the optimal control problem (2.1) is reformulated as

$$\int_{\mathbf{X}\times\mathbf{A}} c(y,a)d\eta(y,a) \to \inf_{\eta} \qquad (2.28)$$

subject to (2.25) and $\eta \geq 0$.

In the general case, when the MDP is not necessarily absorbing or transient, one has to consider equation (2.23) instead of (2.25). This is the so-called *Primal Linear Program*. To be successful in finding its solution η^*, one must be sure that it is not phantom; in that case, an optimal control strategy π^s is given by decomposition (2.27), e.g. if a uniform Lyapunov function exists.

2.2.22 Occupation measures in transient models

Suppose the state space $\mathbf{X}$ is countable.

Definition 2.4. [Altman(1999), Section 7.1]. A control strategy π is called *transient* if

$$\hat{\eta}^\pi(x) \stackrel{\Delta}{=} \eta^\pi(x \times \mathbf{A}) = \sum_{t=0}^{\infty} P_{P_0}^\pi(X_t = x) < \infty \quad \text{for any } x \in \mathbf{X}. \qquad (2.29)$$

In the case where state 0 is absorbing, we consider only $x \in \mathbf{X} \setminus \{0\}$ in (2.29). Sometimes the absorbing state is denoted as Δ.

An MDP is called *transient* if all its strategies are transient. Any absorbing MDP is also transient, but not *vice versa*.

In transient models, occupation measures η^π are finite on singletons but can only be σ-finite. They satisfy equations (2.23) or (2.25), but those equations can have phantom σ-finite solutions (see Section 2.2.21).

The following example [Feinberg and Sonin(1996), Ex. 4.3] shows that if π^s is a stationary strategy defined by (2.27) then it can happen that $\eta^{\pi^s} \neq \eta^\pi$. One can only claim that $\hat{\eta}^{\pi^s} \leq \hat{\eta}^\pi(x)$, where, as usual, $\hat{\eta}^\pi(x) \stackrel{\Delta}{=} \eta^\pi(x \times \mathbf{A})$ is the marginal (see [Altman(1999), Th. 8.1]).

Let $\mathbf{X} = \{0, 1, 2, \ldots\}$, $\mathbf{A} = \{f, b\}$. State 0 is absorbing: $p(0|0, a) \equiv 1$, $c(0, a) \equiv 0$. But the model will be transient, not absorbing, i.e. for each control strategy, formula (2.29) holds, but (2.24) is not guaranteed. We put

$$p(x - 1|x, b) = \gamma_x \quad \text{and} \quad p(0|x, b) = 1 - \gamma_x \quad \text{for } x \geq 2;$$

$$p(2|1, b) = \gamma_2; \quad p(0|1, b) = 1 - \gamma_2;$$

$$p(x+1|x,f) = \gamma_{x+1} \quad \text{and} \quad p(0|x,f) = 1 - \gamma_{x+1} \quad \text{for } x \geq 1.$$

Here $0 < \gamma_x \leq \gamma_{x+1} < 1$, $x = 1, 2, 3, \ldots$ are numbers such that

$$\gamma_1 \prod_{j=2}^{\infty} \gamma_j^{2^j+1} > 0.9. \tag{2.30}$$

Other transition probabilities are zero (see Fig. 2.28). The loss function does not play any role. We assume that $P_0(x) = I\{x = 1\}$.

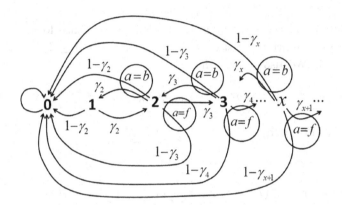

Fig. 2.28 Example 2.2.22: $\eta^{\pi^s} \neq \eta^\pi$ for π^s given by equation (2.27).

First of all, this MDP is transient, since for each strategy and for each state $x \in \mathbf{X} \setminus \{0\}$, the probability of returning to state $x = 2, 3, \ldots$ is bounded above by γ_{x+1}, so that

$$\hat{\eta}^\pi(x) = \eta^\pi(x \times \mathbf{A}) \leq 1 + \gamma_{x+1} + \gamma_{x+1}^2 + \cdots = \frac{1}{1 - \gamma_{x+1}} < \infty;$$

for $x = 1$, we observe that

$$\hat{\eta}^\pi(1) \leq 1 + \hat{\eta}^\pi(2) < \infty.$$

We consider the following control strategy π:

$$\pi_t(b|x_0, a_1, \ldots, x_{t-1}) = \begin{cases} 1, & \text{if } m = \sum_{n=0}^{t-1} I\{x_n = x_{t-1}\} \leq 2^{x_{t-1}-1}; \\ 0 & \text{otherwise;} \end{cases}$$

$$\pi_t(f|x_0, a_1, \ldots, x_{t-1}) = 1 - \pi_t(b|x_0, a_1, \ldots, x_{t-1}).$$

In fact, if the process visits state $j \geq 1$ for the mth time, then the strategy π selects action b if $m \leq 2^{j-1}$, and action f otherwise. For this strategy, the process will never be absorbed into 0 with probability $\prod_{j=2}^{\infty} \gamma_j^{2^j+1}$: starting from $X_0 = 1$, the process will visit state 2 a total of 2^{2-1} times and state 1 ($2^{2-1} + 1$) times, so that the absorption at zero with probability $1 - \gamma_2$ must be avoided ($2^2 + 1$) times. Similar reasoning applies to states 2 and 3, 3 and 4, and so on. Therefore,

$$P_1^\pi(T_0 = \infty) = \prod_{j=2}^{\infty} \gamma_j^{2^j+1} > 0.9,$$

where $T_0 \stackrel{\triangle}{=} \min\{t \geq 0 : X_t = 0\}$. We have proved that the model is not absorbing, because the requirement (2.24) is not satisfied for π.

Consider an arbitrary state $x \geq 2$. Clearly, $\hat{\eta}^\pi(x) \leq 2^{x-1} + 2^x + 1 = 3 \cdot 2^{x-1} + 1$. We also observe that

$$\hat{\eta}^\pi(x) = P_1^\pi(T_0 = \infty)[3 \cdot 2^{x-1} + 1] + P_1^\pi(T_0 < \infty)E_1^\pi\left[\sum_{t=0}^{\infty} I\{X_t = x\} | T_0 < \infty\right]$$
$$\geq P_1^\pi(T_0 = \infty)[3 \cdot 2^{x-1} + 1] > 0.9[3 \cdot 2^{x-1} + 1] \qquad (2.31)$$

and

$$\eta^\pi(x \times f) = P_1^\pi(T_0 = \infty)E_1^\pi\left[\sum_{t=0}^{\infty} I\{X_t = x, A_{t+1} = f\} | T_0 = \infty\right]$$
$$+ P_1^\pi(T_0 < \infty)E_1^\pi\left[\sum_{t=0}^{\infty} I\{X_t = x, A_{t+1} = f\} | T_0 < \infty\right]$$
$$\geq P_1^\pi(T_0 = \infty)[2^x + 1].$$

Therefore, the stationary control strategy π^s from (2.27) satisfies

$$\pi^s(f|x) = \frac{\eta^\pi(x \times f)}{\hat{\eta}^\pi(x)} \geq \frac{2^x + 1}{3 \cdot 2^{x-1} + 1} P_1^\pi(T_0 = \infty), \quad x \geq 2.$$

Below, we use the notation $\lambda_x \stackrel{\triangle}{=} \gamma_{x+1} \pi^s(f|x)$, $\mu_x \stackrel{\triangle}{=} \gamma_x \pi^s(b|x)$, $x \geq 2$, and $\lambda_1 \stackrel{\triangle}{=} \gamma_2$ for brevity.

Now, according to [Altman(1999), Lemma 7.1], the occupation measure $\hat{\eta}^{\pi^s}(x)$ is the minimal non-negative solution to equation (2.26), which takes the form

$$\hat{\eta}^{\pi^s}(1) = 1 + \mu_2 \hat{\eta}^{\pi^s}(2);$$
$$\hat{\eta}^{\pi^s}(x) = \lambda_{x-1} \hat{\eta}^{\pi^s}(x-1) + \mu_{x+1} \hat{\eta}^{\pi^s}(x+1), \qquad x \geq 2.$$

According to Lemma B.2,

$$\hat{\eta}^{\pi^s}(1) \leq 1 + \sum_{j=2}^{\infty}\left(\frac{\mu_2\cdots\mu_j}{\lambda_2\cdots\lambda_j}\right);$$

$$\hat{\eta}^{\pi^s}(x) \leq \sum_{j=x}^{\infty}\left(\frac{\mu_2\cdots\mu_j}{\lambda_2\cdots\lambda_j}\right)\Big/\left(\frac{\mu_2\cdots\mu_x}{\lambda_2\cdots\lambda_{x-1}}\right), \quad x \geq 2.$$

In fact, all μ_j appear in the numerators and λ_j stay in the denominators. We know that, for $x \geq 2$,

$$\lambda_x \geq \gamma_1 \pi^s(f|x) > \frac{2^x+1}{3\cdot 2^{x-1}+1}\cdot 0.9 > \frac{2}{3}\cdot 0.9 = \frac{3}{5}$$

and

$$\mu_x \leq 1 - \lambda_x < \frac{2}{5},$$

so that, for $x \geq 2$,

$$\hat{\eta}^{\pi^s}(x) \leq \left(\frac{2}{3}\right)^{x-1}\frac{1}{1-2/3}\Big/\left[\left(\frac{2}{3}\right)^{x-2}\cdot\frac{2}{5}\right] = 5,$$

but, according to (2.31), $\hat{\eta}^{\pi}(x) > 0.9 \cdot 7 = 6.3 > \hat{\eta}^{\pi^s}(x)$.

2.2.23 Occupation measures and duality

If a countable MDP is transient and *positive* then the value of the primal linear program (2.28) coincides with $v^* \stackrel{\triangle}{=} \inf_\pi v^\pi$ [Altman(1999), Th. 8.5]. Moreover, this statement holds also for a general MDP if the state space **X** is arbitrary Borel, action space **A** is finite, and the optimal value of program (2.28) is finite [Dufour and Piunovskiy(submitted), Th. 4.10].

The following example shows that the imposed conditions are important. Let $\mathbf{X} = \{\ldots,-2,-1,0,1,2,\ldots\}$, $\mathbf{A} = \{a\}$: the model is actually uncontrolled. We put $p(x+1|x,a) \equiv 1$ for all $x \in \mathbf{X}$, $c(x,a) = -I\{x=0\}$, $P_0(0) = 1$. This model is transient (but not absorbing), see Fig. 2.29.

Since the model is uncontrolled, we omit the argument a in the program (2.28):

$$-\eta(0) \to \inf_\eta \qquad (2.32)$$

subject to $\eta(x) = I\{x=0\} + \eta(x-1)$ for $x \in \mathbf{X}$, $\eta \geq 0$.

The optimal value equals $-\infty$, but $v^* = -1$.

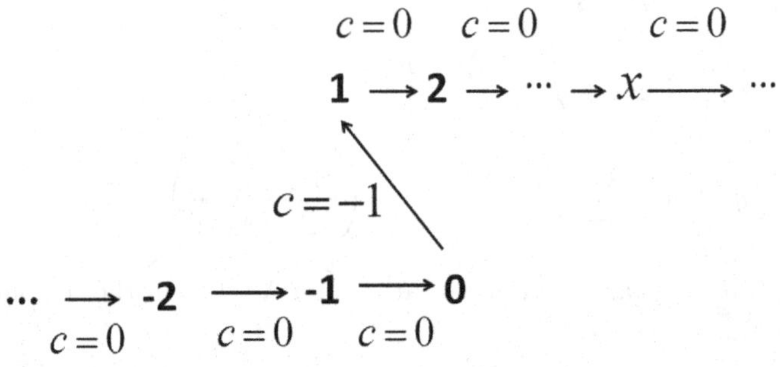

$$c=0 \quad c=0 \qquad c=0$$
$$1 \longrightarrow 2 \longrightarrow \cdots \longrightarrow X \longrightarrow \cdots$$
$$c=-1$$
$$\cdots \longrightarrow -2 \longrightarrow -1 \longrightarrow 0$$
$$c=0 \qquad c=0 \quad c=0$$

Fig. 2.29 Example 2.2.23: phantom solutions to the linear programs in duality.

In a general negative MDP with a countable state space, the *Dual Linear Program* looks as follows [Altman(1999), p. 123]:

$$\sum_{x \in \mathbf{X}} P_0(x)\tilde{v}(x) \to \sup_{\tilde{v}} \qquad (2.33)$$

$$\text{subject to } \tilde{v}(x) \le c(x,a) + \sum_{y \in \mathbf{X}} p(y|x,a)\tilde{v}(y).$$

In arbitrary negative Borel models, if $\tilde{v}(x) \le 0$ and $\tilde{v}(x) \le c(x,a) + \int_{\mathbf{X}} \tilde{v}(y)p(dy|x,a)$, then $\tilde{v}(x) \le v_x^*$ [Bertsekas and Shreve(1978), Prop. 9.10]. Thus, it is not surprising that the optimal value of the program (2.33), with the additional requirement $\tilde{v}(x) \le 0$, coincides with v^*, if the model is negative. We recall that the Bellman function v_x^* is feasible for the program (2.33) because it satisfies the optimality equation (2.2).

For the example presented above (see Fig. 2.29) the Dual Linear Program looks as follows:

$$\tilde{v}(0) \to \sup_{\tilde{v}} \qquad (2.34)$$

$$\text{subject to } \tilde{v}(x) \le -I\{x=0\} + \tilde{v}(x+1) \text{ for all } x \in \mathbf{X}$$

and has the optimal value $+\infty$, if we do not require that $\tilde{v} \le 0$.

One can rewrite programs (2.32) and (2.34) in the following form:
Primal Linear Program: $\sup_{\tilde{v}} L(\eta, \tilde{v}) \to \inf_{\eta \ge 0}$,

where $L(\eta, \tilde{v}) \stackrel{\triangle}{=} -\eta(0) + \sum_{x \in \mathbf{X}} [I\{x=0\} + \eta(x-1) - \eta(x)]\tilde{v}(x);$

$$(2.35)$$

Dual Linear Program: $\inf_{\eta \geq 0} \tilde{L}(\eta, \tilde{v}) \to \sup_{\tilde{v}},$

where $\tilde{L}(\eta, \tilde{v}) \stackrel{\triangle}{=} \tilde{v}(0) + \sum_{x \in \mathbf{X}} \eta(x)[-I\{x = 0\} + \tilde{v}(x+1) - \tilde{v}(x)]$

correspondingly. In spite of their titles, linear programs (2.35) do not yet make a dual pair because $L(\eta, \tilde{v}) \neq \tilde{L}(\eta, \tilde{v})$. That is why the primal optimal value is $-\infty$, and the dual optimal value is $+\infty$. To make the Lagrangeans $L = \tilde{L}$ coincident, we must impose conditions making the series $\sum_{i=-\infty}^{\infty} \eta(i)|v(i)|$ and $\sum_{i=-\infty}^{\infty} \eta(i)|v(i+1)|$ convergent. For example, restrict ourselves with absolutely summable functions $\tilde{v} \in l^1$ and σ-finite measures η, uniformly bounded on singletons. Then the primal optimal value is $-\infty$, and the dual linear program (2.34) has no feasible solutions leading to the optimal value $-\infty$. If we consider uniformly bounded functions $\tilde{v}$ and finite measures η, then the primal linear program (2.32) has no feasible solutions leading to the optimal value $+\infty$, and the dual optimal value is $+\infty$. Finally, let us consider such functions $\tilde{v}$, that $\sum_{x>0} |\tilde{v}(x)| < \infty$ and $\sup_{x \leq 0} |\tilde{v}(x)| < \infty$, and such measures η, that $\sup_{x \geq 0} |\eta(x)| < \infty$ and $\sum_{x \leq 0} |\eta(x)| < \infty$. Then the both programs (2.32) and (2.34) are feasible and have the coincident optimal value -1. Only this last case makes sense.

In specific examples it is often unclear, what class of measures and functions should be considered in the primal and dual programmes (2.35).

2.2.24 Occupation measures: compactness

We again consider the MDP described in Section 2.2.13. Since the model is semi-continuous (see Conditions 2.3), the space of all strategic measures $\mathcal{D} = \{P_{P_0}^\pi, \pi \in \Delta^{\text{All}}\}$ is compact in so-called ws^∞-topology [Schäl(1975a), Th. 6.6], which is the coarsest topology rendering the mapping $P \to \int_{\mathbf{H}} f(h_T) dP(h_T)$ continuous for each function $f(h_T) = f(x_0, a_1, x_1, \ldots, a_T, x_T)$, $0 \leq T < \infty$. (Those functions must be continuous w.r.t. $(a_1, a_2, \ldots, a_T)$ under arbitrarily fixed $x_0, x_1, \ldots, x_T$, although this requirement can be ignored since the spaces $\mathbf{X}$ and $\mathbf{A}$ are discrete with the discrete topology.) Now the mapping $P \to \int_{\mathbf{H}} s(h_T) P(dh_T)$ is lower semi-continuous for any function s on $\mathbf{H}_T$, meaning that, in the finite-horizon version of this MDP, there is an optimal strategy. But, as we already know, there are no optimal strategies in the entire infinite-horizon model. Note that Condition (C) from [Schäl(1975a)] is violated for the given loss function c:

$$\inf_{T\geq n}\inf_{\pi\in\Delta^{\text{All}}}\sum_{t=n}^{T}E_x^\pi[c(X_{t-1},A_t)] \leq \inf_{\pi\in\Delta^{\text{All}}}E_x^\pi[c(X_{n-1},A_n)]$$

$$\leq E_x^{\varphi^n}[c(X_{n-1},A_n)] \leq \left(\frac{1}{2}\right)^{n-1}(1-2^{x+n-1}) \leq 1-2^x$$

does not go to zero as $n\to\infty$. Here φ^n is the following Markov selector:

$$\varphi_t^n(x)=\begin{cases}2, & \text{if } t<n;\\ 1, & \text{if } t\geq n.\end{cases}$$

On the other hand, one can use Theorem 8.2 from [Altman(1999)]. Below, we assume that the initial distribution is concentrated at state 1: $P_0(1)=1$. If $\nu(x,1)=2^x$ is the weight function then, according to Remark 2.4, the general uniform Lyapunov function does not exist. But, for example, for $\nu(x,a)\equiv 1$, inequality (2.16) holds for $\mu(x)=4$. Now the space of all occupation measures $\{\eta^\pi,\ \pi\in\Delta^{\text{All}}\}$ on $\mathbf{X}\setminus\{0\}\times\mathbf{A}$ is convex compact [Altman(1999), Th. 8.2(ii)], but the mapping

$$\eta \to \sum_{(x,a)\in\mathbf{X}\times\mathbf{A}}c(x,a)\eta(x,a) \tag{2.36}$$

is *not* lower semi-continuous. To show this, consider the sequence of stationary selectors

$$\varphi^n(x)=\begin{cases}1, & \text{if } x\geq n,\\ 2, & \text{if } x<n;\end{cases}\quad n=1,2,\ldots.$$

It is easy to see that

$$\eta^{\varphi^n}(x,a)=\begin{cases}\left(\frac{1}{2}\right)^{x-1}, & \text{if } x<n,\ a=2;\\ \left(\frac{1}{2}\right)^{n-1}, & \text{if } x=n,\ a=1;\\ 0 & \text{otherwise.}\end{cases}$$

Therefore,

$$\sum_{(x,a)\in\mathbf{X}\times\mathbf{A}}c(x,a)\eta^{\varphi^n}(x,a)=(1-2^n)\left(\frac{1}{2}\right)^{n-1}$$

$$=\left(\frac{1}{2}\right)^{n-1}-2\to -2\quad\text{as } n\to\infty,$$

but $\lim_{n\to\infty}\eta^{\varphi^n}=\eta^{\varphi^\infty}$, where $\varphi^\infty(x)\equiv 2$, $\eta^{\varphi^\infty}(x,a)=\left(\frac{1}{2}\right)^{x-1}I\{a=2\}$. Convergence in the space of occupation measures is standard: for any

bounded continuous function $s(x,a)$ (for an arbitrary bounded function in the discrete case)
$$\sum_{(x,a)\in \mathbf{X}\times\mathbf{A}} s(x,a)\eta^{\varphi^n}(x,a) \to \sum_{(x,a)\in \mathbf{X}\times\mathbf{A}} s(x,a)\eta^{\varphi^\infty}(x,a).$$
Now
$$\sum_{(x,a)\in \mathbf{X}\times\mathbf{A}} c(x,a)\eta^{\varphi^\infty}(x,a) = 0 > -2,$$
and mapping (2.36) is not lower semi-continuous. Note that mapping (2.36) would have been lower semi-continuous if function $c(x,a)$ were lower semi-continuous and *bounded below* [Bertsekas and Shreve(1978), Prop. 7.31]; see also Theorem A.13, where $q(x,a|\eta) = \eta(x,a)$; $f(x,a,\eta) = c(x,a)$.

Finally, we slightly modify the model in such a way that the one-step loss $\hat{c}(x,a)$ becomes bounded (and continuous). A similar trick was demonstrated in [Altman(1999), Section 7.3]. As a result, the mapping (2.36) will be continuous (see Theorem A.13).

We introduce artificial states $1', 2'\ldots$, so that the loss $c(i,1) = 1 - 2^i$ is accumulated owing to the long stay in state i'. In other words, $\mathbf{X}\{0,1,1',2,2',\ldots\}$, $\mathbf{A} = \{1,2\}$, $p(0|0,a) \equiv 1$, for $i > 0$ $p(i'|i,1) = 1$, $p(0|i,2) = p(i+1|i,2) = 1/2$, $p(i'|i',a) = \frac{2^i-2}{2^i-1}$, $p(0|i',a) = \frac{1}{2^i-1}$; all the other transition probabilities are zero. We put $\hat{c}(i,a) \equiv 0$ for $i \geq 0$, $\hat{c}(i',a) \equiv -1$ for all $i \geq 1$. See Fig. 2.30.

Only actions in states $1,2,\ldots$ play a role; as soon as $A_{t+1} = 1$ and $X_t = i$, the process jumps to state $X_{t+1} = i'$ and remains there for T_i time units, where T_i is geometrically distributed with parameter $p_i = \frac{1}{2^i-1}$. Since $\hat{c}(i',a) = -1$, the total expected loss, up to absorption at zero from state i', equals $\frac{-1}{p_i} = 1 - 2^i$, meaning that this modified model essentially coincides with the MDP from Section 2.2.13. Function $\hat{c}$ is bounded.

Remark 2.6. This trick can be applied to any MDP; as a result, the loss function $|\hat{c}|$ can be always made smaller than 1.

Now the mapping (2.36) is continuous. At the same time, the space $\{\eta^\pi,\ \pi \in \Delta^{\text{All}}\}$ is not compact. Although inequality (2.16) holds for $\nu = 0$, $\mu(i) = 2^i + 2$, $\mu(i') = 2^i - 1$, for $\varphi^2(x) \equiv 2$, the mathematical expectation
$$E_i^{\varphi^2}[\mu(X_t)] = \left(\frac{1}{2}\right)^t (2^{i+t} + 2) = 2^i + 2^{1-t}$$
does not approach zero as $t \to \infty$, and the latter is one of the conditions on the Lyapunov function [Altman(1999), Def. 7.4]. Therefore, Theorem 8.2 from [Altman(1999)] is not applicable.

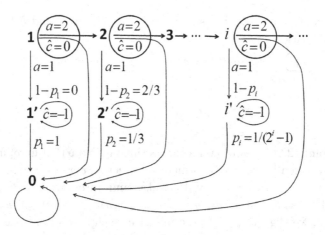

Fig. 2.30 Example 2.2.24: how to make the loss function bounded.

2.2.25 The bold strategy in gambling is not optimal (house limit)

Suppose a gambler wishes to obtain at least a certain fortune, say 100, in a primitive casino. He can stake any amount of fortune in his possession but no more than he possesses, and he gains his stake with probability w and and loses his stake with the complementary probability $\bar{w} = 1 - w$. How much should the gambler stake every time so as to maximize his chance of eventually obtaining 100?

It is known that the gambler should play boldly in case the casino is unfair ($w < 1/2$): $a = \varphi(x) = \min\{x, (100 - x)\}$, where $x \in (0, 100)$ is the current value of the fortune [Dubins and Savage(1965)],[Bertsekas(1987), Section 6.6].

Suppose now that there is a house limit $h \in (0, 100)$. It is known that the gambler should still play boldly; that is, $a = \varphi^b(x) = \min\{x, 100 - x, h\}$, when $h = \frac{100}{n}$ for some $n = 1, 2, \ldots$ But as is shown in [Heath et al.(1972)], if $h \in \left(\frac{100}{n+1}, \frac{100}{n}\right)$ for some integer $n \geq 3$, then the bold strategy is not optimal for all w sufficiently close to zero.

One can model this game as an MDP in the following way. $\mathbf{X} = [0, 200)$, $\mathbf{A} = (0, 100)$, and state 0 is absorbing with no future costs. For $0 < x < 100$,

$$p(\Gamma^\mathbf{X}|x, a) = wI\left\{\Gamma^\mathbf{X} \ni (x + a)\right\} + \bar{w}I\left\{\Gamma^\mathbf{X} \ni \max\{0, (x - a)\}\right\};$$

$$c(x,a) = \begin{cases} 0, & \text{if } a \le \min\{x,h\}; \\ +\infty, & \text{if } a > \min\{x,h\}. \end{cases}$$

For $x \ge 100$,

$$p(\Gamma^X | x, a) = I\{\Gamma^X \ni 0\}, \qquad c(x,a) = -1.$$

See Fig. 2.31.

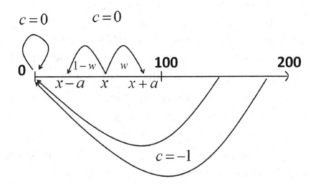

Fig. 2.31   Example 2.2.25: gambling.

Below, we assume that $h = 22$ is an integer, and we consider the finite model $\mathbf{X} = \{0, 1, 2, \ldots, 198\}$, $\mathbf{A} = \{1, 2, \ldots, 99\}$. Following [Heath et al.(1972)], we shall prove that the bold strategy is not optimal for small values of w.

Consider the initial state $\hat{x} = 37$ and action $a = 20$. After the first decision, starting from the new states 57 or 17, the strategy is bold. We call this strategy "timid" and denote it as $\hat{\varphi}$. We intend to prove that $v_{37}^{\hat{\varphi}} < v_{37}^{\varphi^b}$ for small enough w. (Remember that the performance functional to be minimized equals minus the probability of success.) The main point is that, starting from $x = 17$, it is possible to reach 100 in 4 steps, but it is impossible to do so starting from $x = 15$, i.e. after losing the bold stake $a = 22$.

When playing boldly, the gambler can win in no more than four plays in only the following three cases:

fortune	37	59	81	100	
result of the game		win	win	win	

fortune	37	59	37	59	81	100
result of the game		win	loss	win	win	win

fortune	37	59	81	59	81	100
result of the game		win	win	loss	win	win

When using the timid strategy $\hat{\varphi}$, the value of the fortune will change (decrease by 2), but there will still be no other ways to reach 100 in four or fewer successful plays, apart from the aforementioned path
$$37 \to 17 \to 34 \to 56 \to 78 \to 100.$$
We can estimate the number of plays $M(k)$ such that at the end, the bold gambler reaches 0 or 100 experiencing at most k winning bets.
$$M(0) \leq 9 \text{ (if starting from } x = 99);$$
$$M(1) \leq 8 + 1 + M(0) = 18;$$
$$M(2) \leq 8 + 1 + M(1) = 27;$$
$$M(3) \leq 36; \quad M(4) \leq 45.$$
Thus, after 45 plays, either the game is over, or the gambler wins at least five times, and there are no more than 2^{45} such paths.

Summing up, we conclude that
$$v_{37}^{\varphi^b} \geq -(w^3 + 2w^4\bar{w} + 2^{45}w^5).$$
But
$$v_{37}^{\hat{\varphi}} \leq -(w^3 + 3w^4\bar{w}),$$
and $v_{37}^{\hat{\varphi}} < v_{37}^{\varphi^b}$ if w is sufficiently small.

Detailed calculations with $w = 0.01$ give the following values (φ^* is the optimal stationary selector and $v_x^{\varphi^*} = v_x^*$ is the Bellman function).

x	36	37	38	39
$v_x^{\varphi^b}$	-102041×10^{-11}	-102061×10^{-11}	-102071×10^{-11}	-103070×10^{-11}
$v_x^{\varphi^*}$	-102060×10^{-11}	-103031×10^{-11}	-103050×10^{-11}	-103070×10^{-11}
$\varphi^b(x)$	22	22	22	22
$\varphi^*(x)$	20	19	18	22

The graph of $\varphi^*(x)$ is presented in Fig. 2.32.

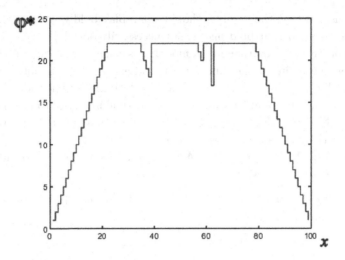

Fig. 2.32 Example 2.2.25: the optimal strategy in gambling with a house limit.

2.2.26 The bold strategy in gambling is not optimal (inflation)

The problem under study is due to [Chen et al.(2004)], where the situation is described as follows.

A gambler with a (non-random) initial fortune X_0 less than 1, wants to buy a house which sells today for 1. Owing to inflation, the price of the house tomorrow will be $\frac{1}{\beta} \geq 1$, and will continue to go up at this rate each day, so as to become $\left(\frac{1}{\beta}\right)^n$ on the nth day. Once each day, the gambler can stake any amount a of the fortune in his possession, but no more than he possesses, in a primitive casino. If he makes a bet, he gains r times his stake with probability $w \in (0, 1)$ and loses his stake with the complementary probability $\bar{w} = 1 - w$. How much should the gambler stake each day so as to maximize his chance of eventually catching up with inflation and being able to buy the house?

It is known that, if there is no inflation ($\beta = 1$), the gambler should play boldly in case the casino is unfair (that is if $w < \frac{1}{1+r}$): $a = \varphi(x) = \max\{x, (1-x)/r\}$, where x is the current value of fortune, since there is no other strategy that offers a higher probability of reaching the goal [Dubins and Savage(1965), Chapter 6]. The presence of inflation would intuitively motivate the gambler to recognize the time value of his fortune, and we would suspect that the gambler should again play boldly. However, in this example we show that bold play is not necessarily optimal.

To construct the mathematical model, it is convenient to accept that the house price remains at the same level 1, and express the current fortune in terms of the actual house price. In other words, if the fortune today is x, the stake is a, and the gambler wins, then his fortune increases to $(x+ra)\beta$. If he loses, the value becomes $(x-a)\beta$. We assume that $(r+1)\beta > 1$, because otherwise the gambler can never reach his goal.

Therefore, $\mathbf{X} = [0, (r+1)\beta)$, $\mathbf{A} = [0,1)$, and state 0 is absorbing with no future costs. For $0 < x < 1$,

$$p(\Gamma^{\mathbf{X}}|x,a) = wI\{\Gamma^{\mathbf{X}} \ni (x+ra)\beta\} + \bar{w}I\{\Gamma^{\mathbf{X}} \ni \max\{0,(x-a)\beta\}\};$$

$$c(x,a) = \begin{cases} 0, & \text{if } a \leq x; \\ +\infty, & \text{if } a > x. \end{cases}$$

For $x \geq 1$,

$$p(\Gamma^{\mathbf{X}}|x,a) = I\{\Gamma^{\mathbf{X}} \ni 0\}, \quad c(x,a) = -1$$

(see Fig. 2.33). Recall that we deal with minimization problems; the value $c(x,a) = +\infty$ prevents the stakes bigger than the current fortune.

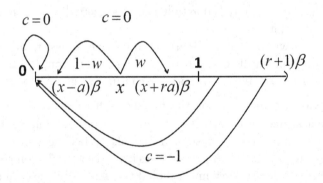

Fig. 2.33 Example 2.2.26: gambling in the presence of inflation.

Now we are in the framework of problem (2.1), and the Bellman equation (2.2) is expressed as follows:

$$v(x) = \inf_{a\in[0,x]} \{wv\left((x+ra)\beta\right) + \bar{w}v\left((x-a)\beta\right)\} \text{ if } x < 1;$$

$$v(x) = -1 \text{ if } x \geq 1; \qquad v(0) = 0.$$

Because of inflation, the bold strategy looks slightly different:
$$\varphi(x) = \min\left\{x, \frac{1-\beta x}{r\beta}\right\}.$$

Function v_x^φ satisfies the equation
$$v_x^\varphi = wv_{(x+r\varphi(x))\beta}^\varphi + \bar{w}v_{(x-\varphi(x))\beta}^\varphi, \text{ if } x < 1;$$
$$v_x^\varphi = -1 \text{ if } x \geq 1; \qquad v_0^\varphi = 0.$$

Therefore,
$$v_x^\varphi = -1, \text{ if } x \geq 1,$$
$$v_x^\varphi = -w + \bar{w}v_{\frac{\beta(r+1)x-1}{r}}^\varphi, \text{ if } \frac{1}{(r+1)\beta} \leq x < 1,$$
$$v_x^\varphi = wv_{(r+1)x\beta}^\varphi, \text{ if } 0 < x < \frac{1}{(r+1)\beta}, \qquad (2.37)$$
$$v_x^\varphi = 0, \text{ if } x = 0.$$

In what follows, we put $\gamma \triangleq \frac{1}{(1+r)\beta}$ and assume that $r > 1$, $\gamma < 1$, and β is not too big, so that $\gamma \in [B, 1)$, where

$$B = \begin{cases} \max\{\frac{1}{1+r}, r^{-(1+1/K)}\}, & \text{if } K \geq 1; \\ \frac{1}{1+r}, & \text{if } K = 0, \end{cases}$$

$K \triangleq \left\lfloor \frac{\ln(w)}{\ln(\bar{w})} \right\rfloor$ being the integer part. Finally, we fix a positive integer m such that $r\gamma^m < 1 - \gamma$. Note that

$$\gamma \sum_{i=0}^{\infty} (r\gamma^m)^i = \frac{\gamma}{1 - r\gamma^m} < 1.$$

If $x = \gamma$ then the second equation (2.37) implies that $v_\gamma^\varphi = -w$ and, as a result of the third equation, for all $i = 1, 2, \ldots$,
$$v_{\gamma^i}^\varphi = -w^i.$$

If $x = \gamma + r\gamma^{i+1} < 1$, then the second equation (2.37) implies that
$$v_{\gamma+r\gamma^{i+1}}^\varphi = -w + \bar{w}v_{\gamma^i}^\varphi = -w - \bar{w}w^i,$$
and, because of the third equation, for all $j = 1, 2, \ldots$,
$$v_{\gamma^{j+1}+r\gamma^{j+i+1}}^\varphi = -w^j[w + \bar{w}w^i] = -w^{j+1} - \bar{w}w^{i+j}.$$

If we continue in the same way, we see that, in a rather general case,
$$v_x^\varphi = -\sum_{l=0}^k \bar{w}^l w^{n_l - l}, \text{ if } x = \sum_{l=0}^k r^l \gamma^{n_l}$$
(see [Chen et al.(2004), Th. 2]). In particular, for
$$\hat{x} = \gamma^2[1 + r\gamma^m + r^2\gamma^{2m} + \cdots + r^{K+2}\gamma^{(K+2)m}]$$
we have
$$v_{\hat{x}}^\varphi = -w^2[1 + \bar{w}w^{m-1} + \bar{w}^2 w^{2(m-1)} + \cdots + \bar{w}^{K+2}w^{(K+2)(m-1)}].$$

We intend to show that
$$\inf_{a \in [0,\hat{x}]} \left\{ w v_{(\hat{x}+ra)\beta}^\varphi + \bar{w} v_{(\hat{x}-a)\beta}^\varphi - v_{\hat{x}}^\varphi \right\} < 0. \tag{2.38}$$

To do so, take
$$\hat{a} \stackrel{\triangle}{=} \hat{x} - \gamma\{\gamma^m + r\gamma^{2m} + \cdots + r^K \gamma^{(K+1)m}\}/\beta.$$

Now
$$(\hat{x} - \hat{a})\beta = \gamma^{m+1} + r\gamma^{2m+1} + \cdots + r^K \gamma^{(K+1)m+1},$$

so that
$$v_{(\hat{x}-\hat{a})\beta}^\varphi = -w^2[w^{m-1} + \bar{w}w^{2(m-1)} + \cdots + \bar{w}^K w^{(K+1)(m-1)}].$$

Since
$$\gamma < (\hat{x} + r\hat{a})\beta = \gamma + r^{K+2}\gamma^{(K+2)m+1} < 1,$$

then, according to (2.37),
$$v_{(\hat{x}+r\hat{a})\beta}^\varphi = -w + \bar{w} v_{r^{K+1}\gamma^{(K+2)m}}^\varphi \leq -w + \bar{w} v_{\gamma^{(K+2)m-K}}^\varphi,$$

because the function v_x^φ decreases with x [Chen et al.(2004), Th. 1] and because $r^{K+1}\gamma^K \geq 1$ (recall that $\gamma \geq r^{-(1+1/K)}$ if $K \geq 1$).

Therefore,
$$w v_{(\hat{x}+r\hat{a})\beta}^\varphi + \bar{w} v_{(\hat{x}-\hat{a})\beta}^\varphi - v_{\hat{x}}^\varphi \leq -w^2 + w\bar{w}[-w^{(K+2)m-K}]$$
$$- \bar{w}w^2[w^{m-1} + \bar{w}w^{2(m-1)} + \cdots + \bar{w}^K w^{(K+1)(m-1)}]$$
$$+ w^2[1 + \bar{w}w^{m-1} + \bar{w}^2 w^{2(m-1)} + \cdots + \bar{w}^{K+2}w^{(K+2)(m-1)}]$$
$$= \bar{w}w^{(K+2)(m-1)+2}[\bar{w}^{K+1} - w] < 0,$$

since $0 < w < 1$ and $w > \bar{w}^{K+1}$ according to the definition of K. Inequality (2.38) is proved. Thus, having $\hat{x}$ in hand, making stake $\hat{a}$ and playing boldly afterwards is better than just playing boldly. For more about this model see [Chen et al.(2004)].

2.2.27 Search strategy for a moving target

Suppose that a moving object is located in one of two possible locations: 1 or 2. Suppose it moves according to a Markov chain with the transition matrix $\begin{bmatrix} 1-q_1 & q_1 \\ q_2 & 1-q_2 \end{bmatrix}$, where $q_1, q_2 \in (0,1)$. The initial probability of being in location 1, x_0, is given. The current position is unknown, but on each step the decision maker can calculate the *posteriori* probability x of the object to be in location 1. Based on this information, the objective is to discover the object in the minimum expected time. Similar problems were studied in [Ross(1983), Chapter III, Section 5].

In fact, we have a model with imperfect state information, and the *posteriori* probability x is the so-called sufficient statistic which plays the role of the state in the MDP under study: see [Bertsekas and Shreve(1978), Chapter 10]. Obviously, x can take values in the segment $[0,1]$, so that the state space is $\mathbf{X} = [0,1] \cup \{\Delta\}$, where Δ means that the object is discovered and the search process stopped. We put $\mathbf{A} = \{1,2\}$. The action a means looking at the corresponding location. The transition probability (for the sufficient statistic x) and loss function are equal to

$$p(dy|x,a) = \begin{cases} (1-x)\delta_{q_2}(dy), & \text{if } x \in [0,1] \text{ and } a = 1; \\ x\delta_{1-q_1}(dy), & \text{if } x \in [0,1] \text{ and } a = 2; \end{cases}$$

$$p(\Delta|x,a) = \begin{cases} x, & \text{if } x \in [0,1] \text{ and } a = 1; \\ 1-x, & \text{if } x \in [0,1] \text{ and } a = 2; \end{cases}$$

$$p(\Delta|\Delta,a) \equiv 1; \qquad c(x,a) = \begin{cases} 1, & \text{if } x \in [0,1]; \\ 0, & \text{if } x = \Delta. \end{cases}$$

See Fig. 2.34.

It seems plausible that the optimal strategy is simply to search the location that gives the highest probability of finding the object:

$$\varphi(x) = \begin{cases} 1, & \text{if } x > 1/2; \\ 2, & \text{if } x \leq 1/2. \end{cases} \qquad (2.39)$$

We shall show that this strategy is not optimal unless $q_1 = q_2$ or $q_1 + q_2 = 1$.

The optimality equation (2.2) can be written as follows

$$v(x) = 1 + \min\{(1-x)v(q_2),\ xv(1-q_1)\}; \qquad v(\Delta) = 0. \qquad (2.40)$$

The model under consideration is positive, and, starting from $t = 1$, $X_t \in \{q_2, 1-q_1, \Delta\}$. Thus, we can solve equation (2.40) for $x \in \{q_2, 1-q_1\}$. The following assertions can be easily verified; see Fig. 2.35.

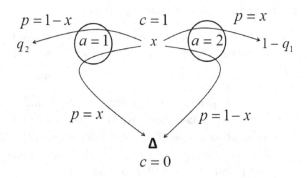

Fig. 2.34 Example 2.2.27: optimal search.

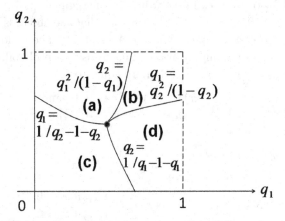

Fig. 2.35 Example 2.2.27: structure of the Bellman function.

(a) If
$$q_1 \geq \frac{1}{q_2} - 1 - q_2 \text{ and } q_2 \geq \frac{q_1^2}{1 - q_1}$$

then
$$v(q_2) = \frac{1}{q_2}; \quad v(1 - q_1) = 1 + \frac{q_1}{q_2},$$

and the stationary selector $\varphi^*(q_2) = \varphi^*(1 - q_1) = 1$ provides the minimum in (2.40).

(b) If
$$q_1 \leq \frac{q_2^2}{1-q_2} \text{ and } q_2 \leq \frac{q_1^2}{1-q_1}$$
then
$$v(q_2) = \frac{1}{q_2}; \quad v(1-q_1) = \frac{1}{q_1},$$
and the stationary selector $\varphi^*(q_2) = 1$, $\varphi^*(1-q_1) = 2$ provides the minimum in (2.40).

(c) If
$$q_1 \leq \frac{1}{q_2} - 1 - q_2 \text{ and } q_2 \leq \frac{1}{q_1} - 1 - q_1$$
then
$$v(q_2) = \frac{1+q_2}{1-q_1q_2}; \quad v(1-q_1) = \frac{1+q_1}{1-q_1q_2},$$
and the stationary selector $\varphi^*(q_2) = 2$, $\varphi^*(1-q_1) = 1$ provides the minimum in (2.40).

(d) If
$$q_1 \geq \frac{q_2^2}{1-q_2} \text{ and } q_2 \geq \frac{1}{q_1} - 1 - q_1$$
then
$$v(q_2) = 1 + \frac{q_2}{q_1}; \quad v(1-q_1) = \frac{1}{q_1},$$
and the stationary selector $\varphi^*(q_2) = \varphi^*(1-q_1) = 2$ provides the minimum in (2.40).

For small (large) values of x, the minimum in (2.40) is provided by the second (first) term: $\varphi^*(x) = 2$ ($\varphi^*(x) = 1$). Therefore, the stationary selector (2.39) is optimal if and only if $v(q_2) = v(1-q_1)$. In cases (a) and (d), we conclude that $q_1 + q_2 = 1$, and in cases (b) and (c) we see that $q_1 = q_2$.

A good explanation of why the strategy (2.39) is not optimal in the asymmetric case is given in [Ross(1983), Chapter III, Section 5.3]. Suppose $q_1 = 0.99$ and $q_2 = 0.5$. If $x = 0.51$ then an immediate search of location 1 will discover the object with probability 0.51, whereas a search of location 2 discovers it with probability 0.49. However, an unsuccessful search of location 2 leads to a near-certain discovery at the next stage (because $q_1 = 0.99$ is the probability of the object to be in the second location), whereas an unsuccessful search of location 1 results in complete uncertainty as to where it will be at the time of the next search. Here, we have the situation in case (d) with $v(q_2) \approx 1.505$; $v(1-q_1) \approx 1.010$ and
$$0.49v(q_2) \approx 0.737 > 0.51v(1-q_1) \approx 0.515,$$
and $\varphi^*(0.51) = 2$.

2.2.28 The three-way duel ("Truel")

The sequential truel is a game that generalizes the simple duel. Three marksmen, A, B, and C, have accuracies $0 < \alpha < \beta < \gamma \le 1$. Because of this disparity the marksmen agree that A shall shoot first, followed by B, followed by C, this sequential rotation continuing until only one man (the winner) remains standing. When all three men are standing, the active marksman must decide who to shoot at. Every marksman wants to maximize his probability of winning the game. We assume that every player behaves in the same way under identical circumstances. Obviously, nobody will intentionally miss if only two men are standing (otherwise, the probability of winning is zero for the player who decides not to shoot). The following notation will be used, just for brevity: $p_B(ABC)$ is the probability that B wins the game if all three men are standing and A shoots first; $p_C(BC)$ is the probability that C wins the game if B and C are standing and B shoots first, and so on. In what follows, $\bar{\alpha} = 1 - \alpha$, $\bar{\beta} = 1 - \beta$ and $\bar{\gamma} = 1 - \gamma$.

Suppose for the moment that it is not allowed to miss intentionally, and consider the behaviour of marksmen B and C if all three men are standing. The probability $p_C(AC)$ satisfies the equation

$$p_C(AC) = \bar{\alpha}[\gamma + \bar{\gamma}p_C(AC)] :$$

C wins with the probability γ if A does not hit him; the probability of reaching the same state AC equals $\bar{\alpha}\bar{\gamma}$. Thus

$$p_C(AC) = \frac{\bar{\alpha}\gamma}{\alpha + \bar{\alpha}\gamma}.$$

Similarly,

$$p_C(BC) = \frac{\bar{\beta}\gamma}{\beta + \bar{\beta}\gamma}.$$

Since $\bar{\alpha} > \bar{\beta}$, $p_C(AC) > p_C(BC)$. Now it is clear that in state CAB, the marksman C will shoot B as the situation does not change if he misses, but if he hits the target, it is better to face A afterwards rather than B who is stronger.

By a similar argument,

$$p_B(AB) = \frac{\bar{\alpha}\beta}{\alpha + \bar{\alpha}\beta}; \quad p_B(CB) = \frac{\bar{\beta}\gamma}{\beta + \bar{\beta}\gamma},$$

and, in state BCA, marksman B will shoot C.

Below, we assume that marksmen B and C behave as described above, and we allow A to miss intentionally. Now, in the case of marksman A (when he is first to shoot), the game can be modelled by the following MDP. $\mathbf{X} = \{A, AB, AC, ABC, \Delta\}$, where Δ means the game is over (for A). $\mathbf{A} = \{0, \hat{b}, \hat{c}\}$, where 0 means "intentionally miss", $\hat{b}(\hat{c})$ means "shoot B(C)". In states AB and AC the both actions $\hat{b}$ and $\hat{c}$ mean "shoot the standing partner" (do not miss intentionally).

$$p(ABC|ABC, a) = \begin{cases} \bar{\beta}\bar{\gamma}, & \text{if } a = 0; \\ \bar{\alpha}\bar{\beta}\bar{\gamma}, & \text{if } a \neq 0 \end{cases}$$

$$p(AB|ABC, a) = \begin{cases} \beta, & \text{if } a = 0; \\ \bar{\alpha}\beta, & \text{if } a = \hat{b}; \\ \alpha\bar{\beta} + \bar{\alpha}\beta, & \text{if } a = \hat{c} \end{cases}$$

$$p(AC|ABC, a) = \begin{cases} \bar{\beta}\gamma, & \text{if } a = 0; \\ \alpha\bar{\gamma} + \bar{\alpha}\bar{\beta}\gamma, & \text{if } a = \hat{b}; \\ \bar{\alpha}\bar{\beta}\gamma, & \text{if } a = \hat{c} \end{cases}$$

$$p(A|ABC, a) \equiv 0; \quad p(\Delta|ABC, a) = \begin{cases} 0, & \text{if } a = 0; \\ \alpha\gamma, & \text{if } a = \hat{b}; \\ \alpha\beta, & \text{if } a = \hat{c} \end{cases}$$

$$p(AB|AB, a) = \begin{cases} \bar{\beta}, & \text{if } a = 0; \\ \bar{\alpha}\bar{\beta}, & \text{if } a \neq 0 \end{cases}$$

$$p(A|AB, a) = \begin{cases} 0, & \text{if } a = 0; \\ \alpha, & \text{if } a \neq 0 \end{cases} \quad p(\Delta|AB, a) = \begin{cases} \beta, & \text{if } a = 0; \\ \bar{\alpha}\beta, & \text{if } a \neq 0 \end{cases}$$

$$p(AC|AC, a) = \begin{cases} \bar{\gamma}, & \text{if } a = 0; \\ \bar{\alpha}\bar{\gamma}, & \text{if } a \neq 0 \end{cases}$$

$$p(A|AC, a) = \begin{cases} 0, & \text{if } a = 0; \\ \alpha, & \text{if } a \neq 0 \end{cases} \quad p(\Delta|AC, a) = \begin{cases} \gamma, & \text{if } a = 0; \\ \bar{\alpha}\gamma, & \text{if } a \neq 0 \end{cases}$$

$$p(\Delta|\Delta, a) \equiv 1 \quad c(x, a) = \begin{cases} -1, & \text{if } x = A; \\ 0 & \text{otherwise} \end{cases} \quad P_0(ABC) = 1.$$

All other transition probabilities are zero. See Fig. 2.36.

After we define the MDP in this way, the modulus of the performance functional $|v^\pi|$ coincides with the probability for A to win the game, and

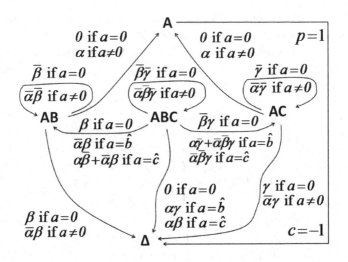

Fig. 2.36 Example 2.2.28: truel. The arrows are marked with the transition probabilities.

the minimization $v^\pi \to \inf_\pi$ means the maximization of that probability. The optimality equation (2.2) is given by

$$v(\Delta) = 0;$$
$$v(A) = -1 + v(\Delta) = -1;$$
$$v(AB) = \min\{\bar{\beta}\, v(AB) + \beta\, v(\Delta);\quad \bar{\alpha}\bar{\beta}\, v(AB) + \alpha\, v(A) + \bar{\alpha}\beta\, v(\Delta)\};$$
$$v(AC) = \min\{\bar{\gamma}\, v(AC) + \gamma\, v(\Delta);\quad \bar{\alpha}\bar{\gamma}\, v(AC) + \alpha\, v(A) + \bar{\alpha}\gamma\, v(\Delta)\};$$
$$v(ABC) = \min\{\bar{\beta}\bar{\gamma}\, v(ABC) + \beta\, v(AB) + \bar{\beta}\gamma\, v(AC);$$
$$\bar{\alpha}\bar{\beta}\bar{\gamma}\, v(ABC) + \bar{\alpha}\beta\, v(AB) + (\alpha\bar{\gamma} + \bar{\alpha}\bar{\beta}\gamma)v(AC) + \alpha\gamma\, v(\Delta);$$
$$\bar{\alpha}\bar{\beta}\bar{\gamma}\, v(ABC) + (\alpha\bar{\beta} + \bar{\alpha}\beta)v(AB) + \bar{\alpha}\bar{\beta}\gamma\, v(AC) + \alpha\beta\, v(\Delta)\}.$$

Therefore, $v(\Delta) = 0$, $v(A) = -1$, $v(AB) = -\frac{\alpha}{\beta + \alpha\bar{\beta}}$, $v(AC) = -\frac{\alpha}{\alpha + \bar{\alpha}\gamma}$, and actions $\varphi^*(AB) = b$, $\varphi^*(AC) = c$ are optimal (for marksman A).

Lemma 2.1.

(a) In state ABC, action $\hat{b}$ is not optimal.

(b) If $\alpha \geq \frac{(\bar{\beta})^2\gamma^2 - \beta\gamma}{(\bar{\beta})^2\gamma^2 + \beta^2\bar{\gamma}}$ then $\varphi^*(ABC) = 0$ is optimal and

$$v(ABC) = -\alpha \frac{\beta\gamma + \alpha\beta\bar{\gamma} + \beta\bar{\beta}\gamma + \alpha(\bar{\beta})^2\gamma}{(\beta + \alpha\bar{\beta})(\alpha + \bar{\alpha}\gamma)(\beta + \bar{\beta}\gamma)}.$$

(c) If $\alpha \leq \frac{(\bar\beta)^2\gamma^2 - \beta\gamma}{(\bar\beta)^2\gamma^2 + \beta^2\bar\gamma}$ then $\varphi^*(ABC) = \hat{c}$ is optimal and

$$v(ABC) = -\alpha \frac{\alpha^2\bar\beta + \alpha\bar\alpha\beta + \alpha\bar\alpha\bar\beta\gamma + (\bar\alpha)^2\beta\gamma + \bar\alpha\beta\bar\beta\gamma + \alpha\bar\alpha(\bar\beta)^2\gamma}{(\beta + \alpha\bar\beta)(\alpha + \bar\alpha\gamma)(1 - \bar\alpha\bar\beta\bar\gamma)}.$$

The proof is presented in Appendix B.

Below, we assume that $\alpha \geq \frac{(\bar\beta)^2\gamma^2 - \beta\gamma}{(\bar\beta)^2\gamma^2 + \beta^2\bar\gamma}$.

Along with the probability $p_A(ABC)$ that marksman A wins the truel (which equals $-v(ABC)$), it is not hard to calculate the similar probabilities for B and C:

$$p_B(ABC) = \bar\beta\bar\gamma \, p_B(ABC) + \beta \, p_B(AB),$$

so that $p_B(ABC) = \frac{\bar\alpha\beta^2}{(\beta + \alpha\bar\beta)(\beta + \bar\beta\gamma)}$, and

$$p_C(ABC) = \bar\beta\bar\gamma \, p_C(ABC) + \bar\beta\gamma \, p_C(AC),$$

so that $p_C(ABC) = \frac{\bar\alpha\bar\beta\gamma^2}{(\alpha + \bar\alpha\gamma)(\beta + \bar\beta\gamma)}$.

Take $\alpha = 0.3$, $\beta = 0.5$ and $\gamma = 0.6$. Then $\varphi^*(ABC) = 0$. Marksman A intentionally misses his first shot and wins the game with probability $p_A(ABC) \approx 0.445$; the other marksmen B and C win the game with probabilities $p_B(ABC) \approx 0.337$ and $p_C(ABC) \approx 0.219$ correspondingly. The order of shots is more important than their accuracies. The decision to miss intentionally allows A to wait until the end of the duel between B and C; after that he has the advantage of shooting first. In the case where $\alpha < \gamma \frac{\gamma(\bar\beta)^2 - \beta}{(\bar\beta)^2\gamma^2 + \beta^2\bar\gamma}$, marksman C is too dangerous for A, and B has too few chances to hit him. Now the best decision for A is to help B to hit C (Lemma 2.1(c)).

If marksmen B and C are allowed to miss intentionally then, generally speaking, the situation changes if A decides not to shoot at the very beginning. For $\alpha = 0.3$, $\beta = 0.5$ and $\gamma = 0.6$, the scenario will be the same as described above: one can check that neither B nor C will miss intentionally and the first phase is just their duel. Suppose now that α increases to $\alpha = 0.4$. According to Lemma 2.1, marksman A will intentionally miss if all three marksmen are standing. But now, assuming that A behaves like this all the time, it is better for B to shoot A, and marksman C will wait (intentionally miss). In the end, there will be a duel between B and C, when C shoots first. Of course, in a more realistic model, the marksmen adjust their behaviour accordingly. In the second round, A (if he is still alive) will probably shoot B, and marksman B will switch to C, who will respond. After A observes the duel between B and C, he will miss intentionally and this unstable process will repeat, as is typical for proper games with complete information.

The reader may find more about truels in [Kilgour(1975)].

Chapter 3

Homogeneous Infinite-Horizon Models: Discounted Loss

3.1 Preliminaries

This chapter is about the following problem:

$$v^\pi = E_{P_0}^\pi \left[\sum_{t=1}^\infty \beta^{t-1} c(X_{t-1}, A_t) \right] \to \inf_\pi, \qquad (3.1)$$

where $\beta \in (0,1)$ is the *discount factor*. As usual, v^π is called the *performance functional*. All notation is the same as in Chapter 2. Moreover, the discounted model is a particular case of an MDP with total expected loss. To see this, modify the transition probability $p(dy|x,a) \to \beta p(dy|x,a)$ and introduce an absorbing state Δ: $p(\{\Delta\}|\Delta, a) \equiv 1$, $p(\{\Delta\}|x,a) \stackrel{\triangle}{=} (1-\beta)$. Investigation of problem (3.1) is now equivalent to the investigation of the modified (absorbing) model, with finite, totally bounded expected absorption time. Nevertheless, discounted models traditionally constitute a special area in MDP.

The optimality equation takes the form

$$v(x) = \inf_{a \in \mathbf{A}} \left\{ c(x,a) + \beta \int_{\mathbf{X}} v(y) p(dy|x,a) \right\}. \qquad (3.2)$$

All definitions and notation are similar to those introduced in Chapter 2 and earlier, and all the main theorems from Chapter 2 apply. Incidentally, a stationary selector φ is called *equalizing* if

$$\forall x \in \mathbf{X} \quad \lim_{t \to \infty} E_x^\varphi \left[\beta^t \, v_{X_t}^* \right] \geq 0.$$

Sometimes we use notation $v^{\pi,\beta}$, $v_x^{*,\beta}$ etc. if it is important to underline the dependence on the discount factor β.

If $\sup_{(x,a) \in \mathbf{X} \times \mathbf{A}} |c(x,a)| < \infty$ (and in many other cases when the investigation is performed, e.g. in weighted normed spaces; see [Hernandez-Lerma and Lasserre(1999), Chapter 8]), the solution to equation (3.2) is unique in the space of bounded functions, coincides with the Bellman function $v_x^* \triangleq \inf_\pi v_x^\pi$, and can be built using the *value iteration algorithm*:

$$v^0(x) \equiv 0;$$

$$v^{n+1}(x) = \inf_{a \in \mathbf{A}} \left\{ c(x,a) + \beta \int_{\mathbf{X}} v^n(y) p(dy|x,a) \right\}, \quad n = 0, 1, 2, \ldots$$

(we leave aside the question of the measurability of v^{n+1}). In many cases, e.g. if the model is positive or negative, or $\sup_{(x,a) \in \mathbf{X} \times \mathbf{A}} |c(x,a)| < \infty$, there exists the limit $v^\infty(x) \triangleq \lim_{n \to \infty} v^n(x)$.

Note also Remark 2.1 about Markov and semi-Markov strategies.

3.2 Examples

3.2.1 Phantom solutions of the optimality equation

Let $\mathbf{X} = \mathbb{R}^1$, $\mathbf{A} = \mathbb{R}^1$, $p(\Gamma|x,a) = \int_\Gamma h(y - bx) dy$, where h is a fixed density function with $\int_{-\infty}^\infty y h(y) dy = 0$ and $\int_{-\infty}^\infty y^2 h(y) dy = 1$.

Remark 3.1. One can represent the evolution of the process as the following *system equation*:

$$X_t = b X_{t-1} + \zeta_t,$$

where $\{\zeta_t\}$ is a sequence of independent random variables with the probability density h (see Fig. 3.1).

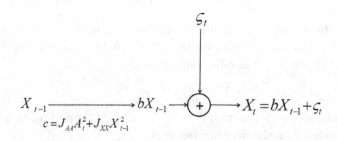

Fig. 3.1 Example 3.2.1: phantom solutions of the optimality equation.

We put $c(x,a) = J_{AA}a^2 + J_{XX}x^2$, where $J_{AA}, J_{XX} > 0$. The described model is a special case of linear–quadratic systems [Piunovskiy(1997), Chapter 4].

The optimality equation (3.2) takes the form

$$v(x) = \inf_{a \in \mathbf{A}} \left\{ J_{AA}a^2 + J_{XX}x^2 + \beta \int_{-\infty}^{\infty} h(y - bx)v(y)dy \right\}$$

and, if $\beta b^2 \neq 1$, has a solution which does not depend on h:

$$v(x) = \frac{J_{XX}}{1 - \beta b^2}x^2 + \frac{\beta J_{XX}}{(1-\beta)(1-\beta b^2)}. \tag{3.3}$$

The stationary selector $\varphi^*(x) \equiv 0$ provides the infimum in the optimality equation.

Value iterations give the following:

$$v^n(x) = f_n x^2 + q_n,$$

where

$$f_n = \begin{cases} J_{XX}\frac{1-(\beta b^2)^n}{1-\beta b^2}, & \text{if } \beta b^2 \neq 1; \\ nJ_{XX}, & \text{if } \beta b^2 = 1 \end{cases}$$

$$q_n = \beta f_{n-1} + \beta q_{n-1}, \qquad q_0 = 0.$$

In the case where $\beta b^2 < 1$ we really have the point-wise convergence $\lim_{n \to \infty} v^n(x) = v^\infty(x) = v(x)$ and $v(x) = v_x^*$. But if $\beta b^2 \geq 1$ then $v^\infty(x) = \infty$, and one can prove that $v_x^* = \infty \neq v(x)$.

Note that the X_t process is not stable if $|b| > 1$ (i.e. $\lim_{t \to \infty} |X_t| = \infty$ if $X_0 \neq 0$ and $\zeta_t \equiv 0$; see Definition 3.2), and nevertheless the MDP is well defined if the discount factor $\beta < \frac{1}{b^2}$ is small enough. This example first appeared in [Piunovskiy(1997), Section 1.2.2.3].

Another example is similar to Section 2.2.2, see Fig. 2.2.

Let $\mathbf{X} = \{0, 1, 2, \ldots\}$, $\mathbf{A} = \{0\}$ (a dummy action), $p(0|0,a) = 1$,

$$\forall x > 0 \quad p(y|x,a) = \begin{cases} \lambda, & \text{if } y = x+1; \\ \mu, & \text{if } y = x-1; \\ 1 - \lambda - \mu, & \text{if } y = x; \\ 0 & \text{otherwise;} \end{cases} \quad c(x,a) = I\{x > 0\}.$$

The process is absorbing at zero, and the one-step loss equals 1 for all positive states. For simplicity, we take $\lambda + \mu = 1$, $\beta \in (0,1)$ is arbitrary. Now the optimality equation (3.2) takes the form

$$v(x) = 1 + \beta\mu v(x-1) + \beta\lambda v(x+1), \qquad x > 0,$$

and has the following general solution, satisfying the obvious condition $v(0) = 0$:

$$v(x) = \frac{1}{1-\beta} - \left[\frac{1}{1-\beta} + C\right]\gamma_1^x + C\gamma_2^x,$$

where

$$\gamma_{1,2} = \frac{1 \pm \sqrt{1 - 4\beta^2\lambda\mu}}{2\beta\lambda},$$

and one can show that $0 < \gamma_1 < 1$, $\gamma_2 > 1$.

The Bellman function v_x^* coincides with the minimal positive solution corresponding to $C = 0$ and, in fact, is the only bounded solution.

Another beautiful example was presented in [Bertsekas(1987), Section 5.4, Ex. 2]: let $\mathbf{X} = [0, \infty)$, $\mathbf{A} = \{a\}$ (a dummy action), $p(x/\beta|x, a) = 1$ with all the other transition probabilities zero. Put $c(x, a) \equiv 0$. Now the optimality equation (3.2) takes the form

$$v(x) = \beta v(x/\beta),$$

and is satisfied by any linear function $v(x) = kx$. But the Bellman function $v_x^* \equiv 0$ coincides with the unique bounded solution corresponding to $k = 0$.

Other simple examples, in which the loss function c is bounded and the optimality equation has unbounded phantom solutions, can be found in [Hernandez-Lerma and Lasserre(1996a), p. 51] and [Feinberg(2002), Ex. 6.4]. See also Example 3.2.3.

3.2.2 When value iteration is not successful: positive model

It is known that, if $c(x, a) \leq 0$, or $\sup_{(x,a) \in \mathbf{X} \times \mathbf{A}} |c(x, a)| < \infty$, or the action space $\mathbf{A}$ is finite and $c(x, a) \geq 0$, then $v^\infty(x) = v_x^*$ [Bertsekas and Shreve(1978), Prop. 9.14, Corollary 9.17.1], so that the MDP is stable. A positive MDP is stable if and only if

$$v^\infty(x) = \inf_{a \in \mathbf{A}}\left\{c(x, a) + \beta \int_{\mathbf{X}} v^\infty(y)p(dy|x, a)\right\}$$

[Bertsekas and Shreve(1978), Prop. 9.16]. Below, we present a positive discounted MDP which is not stable; the idea is similar to Example 2.2.7.

Let $\mathbf{X} = \{\Delta, 0, (n, i) : n = 1, 2, \ldots; i = 1, 2, \ldots, n\}$; $\mathbf{A} = \{1, 2, \ldots\}$; $p(\Delta|\Delta, a) \equiv 1$, $p(\Delta|(n, 1), a) \equiv 1$, $p((n, i-1)|(n, i), a) \equiv 1$ for all $i = 2, 3, \ldots, n$, and $p((a, a)|0, a) \equiv 1$, with all the other transition probabilities zero. We put $\beta = 1/2$, $c((n, 1), a) \equiv 2^n$, all other losses equal zero (see Fig. 3.2).

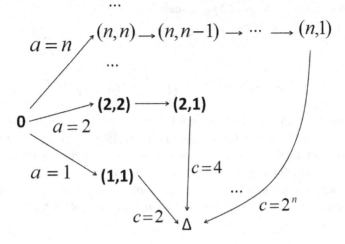

Fig. 3.2   Example 3.2.2: value iteration does not converge to the Bellman function.

The optimality equation (3.2) takes the form

$$v(\Delta) = \frac{1}{2}v(\Delta); \quad v((n,1)) = 2^n + \frac{1}{2}v(\Delta);$$

$$v((n,i)) = \frac{1}{2}v((n,i-1)), \quad i = 2,3,\ldots,n; \quad v(0) = \inf_{a \in A}\left\{\frac{1}{2}v((a,a))\right\},$$

and has a single finite solution $v(\Delta) = 0$,

$$v((n,n)) = 2, \quad v((n,i)) = 2^{n-i+1}, \quad i = 1, 2, \ldots, n, \quad n = 1, 2, \ldots;$$

$$v(0) = 2.$$

Value iteration results in the following sequence:

$$v^0(0) = v^0(\Delta) = 0, \quad v^0((i,j)) \equiv 0;$$

$$v^1((1,1)) = 2, \quad v^1((2,1)) = 4, \quad v^1((3,1)) = 8, \ldots, \quad v^1((n,1)) = 2^n, \ldots,$$

and these values remain unchanged in the further calculations. The remainder values of $v^1(\cdot)$ are zero. $v^1(0) = 0$ because $v^0((1,1)) = 0$.

$$v^2((2,2)) = 2, \quad v^2((3,2)) = 4, \ldots, \quad v^2((n,2)) = 2^{n-1}, \ldots,$$

and these values remain unchanged in the further calculations. $v^2(0) = 0$ because $v^1((2,2)) = 0$.

And so on.

We see that $v^\infty(\Delta) = v(\Delta) = 0$ and

$$v^\infty((n,i)) = v((n,i)) = 2^{n-i+1} \quad \text{for all } i = 1, 2, \ldots, n,$$

but

$$v^\infty(0) = 0 < 2 = \inf_{a \in \mathbf{A}} \left\{ \frac{1}{2} v^\infty((a,a)) \right\}.$$

Sufficient conditions for the equality $v^\infty = v_x^*$ in positive models are also given in [Hernandez-Lerma and Lasserre(1996a), Lemma 4.2.8]. In the current example, all of these conditions are satisfied apart from the assumption that, for any $x \in \mathbf{X}$, the loss function c is inf-compact.

Another example in which value iteration does not converge to the Bellman function in a positive model can be found in [Bertsekas(2001), Exercise 3.1].

3.2.3 A non-optimal strategy $\hat{\pi}$ for which $v_x^{\hat{\pi}}$ solves the optimality equation

Theorem 3.1.

(a) *[Bertsekas and Shreve(1978), Prop. 9.13]*
If $\sup_{(x,a) \in \mathbf{X} \times \mathbf{A}} |c(x,a)| < \infty$ then a stationary control strategy $\hat{\pi}$ is uniformly optimal if and only if

$$v_x^{\hat{\pi}} = \inf_{a \in \mathbf{A}} \left\{ c(x,a) + \beta \int_\mathbf{X} v_y^{\hat{\pi}} p(dy|x,a) \right\}. \tag{3.4}$$

In negative models, this statement holds even for $\beta = 1$ and without any restrictions on the loss growth.

(b) *[Bertsekas and Shreve(1978), Prop. 9.12]*
If $\sup_{(x,a) \in \mathbf{X} \times \mathbf{A}} |c(x,a)| < \infty$ then a stationary control strategy $\hat{\pi}$ is uniformly optimal if and only if it is conserving. In positive models, this statement holds even for $\beta = 1$ and without any restrictions on the loss growth.

Now, consider an arbitrary positive discounted model, where $v_x^\pi < \infty$ for all $x \in \mathbf{X}$, for some strategy π. If the loss function c is not uniformly bounded, a stationary control strategy $\hat{\pi}$ is uniformly optimal if, in addition to (3.4),

$$v_x^{\hat{\pi}} < \infty \text{ and } \lim_{T \to \infty} \beta^T E_x^\pi \left[v_{X_T}^{\hat{\pi}} \right] = 0 \tag{3.5}$$

for all $x \in \mathbf{X}$ and for any strategy π satisfying inequality $v_y^\pi < \infty$ for all $y \in \mathbf{X}$ (there is no reason to consider other strategies).

[Bertsekas(2001), Ex. 3.1.4] presents an example of a non-optimal strategy $\hat\pi$ in a positive model satisfying equation (3.4), for which $v_x^{\hat\pi} = \infty$ and $\lim_{T\to\infty} \beta^T E_x^\pi [v_{X_T}^{\hat\pi}] = \infty$ for some states x. Below, we present another illustrative example where all functions are finite.

Let $\mathbf{X} = \{0,1,2\ldots\}$, $\mathbf{A} = \{1,2\}$, $p(0|0,a) \equiv 1$, $c(0,a) \equiv 0$. For $x > 0$ we put $p(x+1|x,1) = p(0|x,1) \equiv 1/2$, $c(x,1) = 2^x$, $p(x+1|x,2) \equiv 1$, $c(x,2) \equiv 1$. The discount factor $\beta = 1/2$ (see Fig. 3.3).

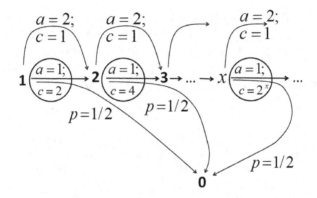

Fig. 3.3 Example 3.2.3: the selector $\hat\varphi(x) \equiv 1$ is not optimal.

The optimality equation (3.2) is given by
$$v(0) = \frac{1}{2} v(0);$$
for $x > 0$ $v(x) = \min\{2^x + \frac{1}{4}v(x+1) + \frac{1}{4}v(0);\ 1 + \frac{1}{2}v(x+1)\},$
and has the minimal non-negative solution
$$v(x) = v_x^* \equiv 2 \quad \text{for } x > 0; \qquad v(0) = 0,$$
coincident with the Bellman function. The stationary selector $\varphi^*(x) \equiv 2$ is conserving and equalizing and hence uniformly optimal; see Theorem 3.1(b).

Now look at the stationary selector $\hat\varphi(x) \equiv 1$. The performance functional $v_x^{\hat\varphi}$ is given by $v_x^{\hat\varphi} = 2^{x+1}$ (for $x > 0$) and satisfies equation (3.4):
$$1 + \frac{1}{2} v_{x+1}^{\hat\varphi} = 1 + 2^{x+1} > 2^{x+1} = v_x^{\hat\varphi} = 2^x + \frac{1}{4} v_{x+1}^{\hat\varphi}.$$

The second equation (3.5) is violated for selector φ^*:

$$\beta^T E_x^{\varphi^*}\left[v_{X_T}^{\hat\varphi}\right]=\beta^T\,2^{x+T+1}=2^{x+1},$$

and the selector $\hat\varphi$ is certainly non-optimal.

We see that the both functions $v_x^*=v_x^{\varphi^*}$ and $v_x^{\hat\varphi}$ solve the optimality equation (3.2), but only v_x^* is the minimal non-negative solution. Note that equation (3.2) has many other solutions, e.g. $v(x)=2+k\,2^x$ with $k\in[0,2]$.

3.2.4 The single conserving strategy is not equalizing and not optimal

This example is based on the same ideas as Example 2.2.4, and is very similar to the example described in [Hordijk and Tijms(1972)].

Let $\mathbf{X}=\{0,1,2,\ldots\}$, $\mathbf{A}=\{1,2\}$, $p(0|0,a)\equiv 1$, $c(0,a)\equiv 0$, $\forall x>0\ p(0|x,1)\equiv 1$, $p(x+1|x,2)\equiv 1$, with all other transition probabilities zero; $c(x,1)=1-2^x$, $c(x,2)\equiv 0$, $\beta=1/2$ (see Fig. 3.4).

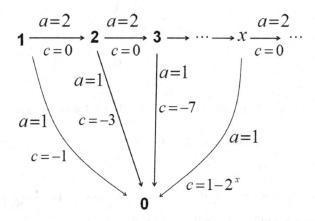

Fig. 3.4 Example 3.2.4: no optimal strategies.

Like any other negative MDP, this model is stable [Bertsekas and Shreve(1978), Prop. 9.14]; the optimality equation (3.2) is given by

$$v(0)=\frac{1}{2}v(0);$$

$$\text{for } x>0\quad v(x)=\min\left\{1-2^x+\frac{1}{2}v(0);\ \frac{1}{2}v(x+1)\right\},$$

and has the maximal non-positive solution $v(0) = 0$; $v(x) = -2^x = v_x^*$ for $x > 0$, which coinsides with the Bellman function.

The stationary selector $\varphi^*(x) \equiv 2$ is the single conserving strategy at $x > 0$, but

$$\lim_{t \to \infty} E_x^{\varphi^*} \left[\beta^t v_{X_t}^* \right] = -2^x < 0,$$

so that it is not equalizing and not optimal. Note that equality (3.4) is violated for φ^* because $v_x^{\varphi^*} \equiv 0$.

There exist no optimal strategies in this model, but the selector

$$\varphi_t^\varepsilon(x) = \begin{cases} 2, & \text{if } t < 1 - \frac{\ln \varepsilon}{\ln 2}; \\ 1 & \text{otherwise} \end{cases}$$

is (uniformly) ε-optimal; $\forall \varepsilon > 0$ $v_x^{\varphi^\varepsilon} < \varepsilon - 2^x$.

Another trivial example of a negative MDP where a conserving strategy is not optimal can be found in [Bertsekas(2001), Ex. 3.1.3].

3.2.5 Value iteration and convergence of strategies

Suppose the state and action spaces $\mathbf{X}$ and $\mathbf{A}$ are finite. Then the optimality equation (3.2) has a single bounded solution $v(x) = v_x^*$ coincident with the Bellman function, and any stationary selector from the set of *conserving* selectors

$$\Phi^* \triangleq \left\{ \varphi : \mathbf{X} \to \mathbf{A} : c(x, \varphi(x)) + \beta \sum_{y \in \mathbf{X}} v(y) p(y|x, \varphi(x)) = v(x) \right\} \quad (3.6)$$

is optimal.

We introduce the sets

$$\Phi_n^* \triangleq \left\{ \varphi : \mathbf{X} \to \mathbf{A} : c(x, \varphi(x)) + \beta \sum_{y \in \mathbf{X}} v^n(y) p(y|x, \varphi(x)) = v^{n+1}(x) \right\},$$

$$n = 0, 1, 2, \ldots \quad (3.7)$$

It is known that for all sufficiently large n, $\Phi_n^* \subseteq \Phi^*$ [Puterman(1994), Th. 6.8.1]. The following example from [Puterman(1994), Ex. 6.8.1] illustrates that, even if Φ^* contains all stationary selectors, the inclusion $\Phi_n^* \subset \Phi^*$ can be proper for all $n \geq 0$.

Let $\mathbf{X} = \{1, 2, 3, 4, 5\}$, $\mathbf{A} = \{1, 2\}$, $\beta = 3/4$; $p(2|1, 1) = p(4|1, 2) = 1$, $p(3|2, a) = p(2|3, a) = p(5|4, a) = p(4|5, a) \equiv 1$, with other transition

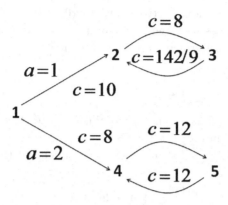

Fig. 3.5 Example 3.2.5: erratic value iterations.

probabilities zero; $c(1,1) = 10$, $c(1,2) = 8$, $c(2,a) \equiv 8$, $c(3,a) \equiv 142/9$, $c(4,a) = c(5,a) \equiv 12$ (see Fig. 3.5).

The value iterations can be written as follows:
$$v^{n+1}(1) = \min\{10 + 3/4 \cdot v^n(2);\quad 8 + 3/4 \cdot v^n(4)\};$$
$$v^{n+1}(2) = 8 + 3/4 \cdot v^n(3);$$
$$v^{n+1}(3) = 142/9 + 3/4 \cdot v^n(2);$$
$$v^{n+1}(4) = 12 + 3/4 \cdot v^n(5);$$
$$v^{n+1}(5) = 12 + 3/4 \cdot v^n(4).$$

Obviously,
$$v^n(4) = v^n(5) = 48\left[1 - \left(\frac{3}{4}\right)^n\right];$$
$$v^{n+1}(2) = \frac{119}{16} + \frac{9}{16}v^{n-1}(2).$$

Since $v^0(2) = 0$ and $v^1(2) = \frac{71}{6}$, we conclude that
$$v^n(2) = \begin{cases} \frac{136}{3} - 34\left(\frac{3}{4}\right)^{n-1}, & \text{if } n \text{ is even;} \\ \frac{136}{3} - \frac{112}{3} \cdot \left(\frac{3}{4}\right)^{n-1}, & \text{if } n \text{ is odd.} \end{cases}$$

The rigorous proof can be done by induction.

Now, Φ^* coincides with the set of all stationary selectors: it is sufficient to look at state 1 only:
$$v(2) = \lim_{n\to\infty} v^n(2) = \frac{136}{3}; \quad v(4) = \lim_{n\to\infty} v^n(4) = 48.$$

If $a = 1$ then
$$c(1,a) + \beta \sum_{y \in \mathbf{X}} v(y) p(y|1,a) = 10 + \frac{3}{4} \cdot v(2) = 44 = v(1);$$
if $a = 2$ then
$$c(1,a) + \beta \sum_{y \in \mathbf{X}} v(y) p(y|1,a) = 8 + \frac{3}{4} \cdot v(4) = 44 = v(1).$$

As for Φ_n^*, we see that, for fixed n, $\varphi \in \Phi_n^*$ iff
$$\varphi(1) = \begin{cases} 1, & \text{if } \Delta^n \leq 0; \\ 2, & \text{if } \Delta^n \geq 0, \end{cases}$$
where
$$\Delta^n = 2 + \frac{3}{4}[v^n(2) - v^n(4)] = \begin{cases} 2 \cdot \left(\frac{3}{4}\right)^n, & \text{if } n \text{ is even}; \\ -\left(\frac{3}{4}\right)^{n-1}, & \text{if } n \text{ is odd}. \end{cases}$$

Therefore, the selector $\varphi^1(x) \equiv 1$ belongs to Φ_n^* only if n is odd, and the selector $\varphi^2(x) \equiv 2$ belongs to Φ_n^* only if n is even.

If either of the spaces $\mathbf{X}$ or $\mathbf{A}$ is not finite, it can easily happen that $\Phi_n^* \cap \Phi^* = \emptyset$ for all $n = 0, 1, 2, \ldots$. The next example illustrates this point.

3.2.6 Value iteration in countable models

Let $\mathbf{X} = \{1, 2, 3\}$, $\mathbf{A} = \{0, 1, 2, \ldots\}$, $\beta = 1/2$,
$$p(2|1,a) = \begin{cases} \frac{1}{2} - \left(\frac{1}{2}\right)^a, & \text{if } a > 0; \\ \frac{1}{2}, & \text{if } a = 0, \end{cases} \quad p(3|1,a) = 1 - p(2|1,a),$$
$p(2|2,a) = p(3|3,a) \equiv 1$, with other transition probabilities zero; $c(1,a) = \left(\frac{1}{2} - \left(\frac{1}{2}\right)^a\right)^2$, $c(2,a) \equiv 1$, $c(3,a) \equiv 2$ (see Fig. 3.6).

For $x = 2$ and $x = 3$, value iteration gives
$$v^n(2) = 2 - \left(\frac{1}{2}\right)^{n-1}; \quad v^n(3) = 4 - \left(\frac{1}{2}\right)^{n-2}.$$

We consider state $x = 1$ and calculate firstly the following infimum
$$\inf_{a>0} \left\{ \left(\frac{1}{2} - \left(\frac{1}{2}\right)^a\right)^2 + \frac{1}{2}\left[\left(\frac{1}{2} - \left(\frac{1}{2}\right)^a\right)\left(2 - \left(\frac{1}{2}\right)^{n-1}\right) \right.\right.$$
$$\left.\left. + \left(\frac{1}{2} + \left(\frac{1}{2}\right)^a\right)\left(4 - \left(\frac{1}{2}\right)^{n-2}\right)\right]\right\}.$$

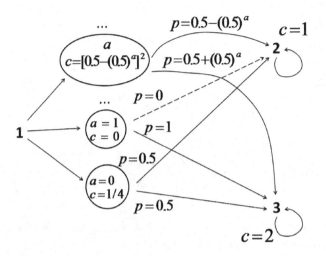

Fig. 3.6 Example 3.2.6: value iteration, countable action space.

Writing $\gamma \triangleq \left(\frac{1}{2}\right)^a$, we see that the infimum w.r.t. γ in the expression

$$\left(\frac{1}{2}-\gamma\right)^2 + \frac{1}{2}\left[\left(\frac{1}{2}-\gamma\right)\left(2-\left(\frac{1}{2}\right)^{n-1}\right) + \left(\frac{1}{2}+\gamma\right)\left(4-\left(\frac{1}{2}\right)^{n-2}\right)\right]$$

is attained at $\gamma = \left(\frac{1}{2}\right)^{n+1}$ and equals $\frac{7}{4} - \frac{3}{2}\cdot\left(\frac{1}{2}\right)^n - \left(\frac{1}{2}\right)^{2n+2}$. Since

$$c(1,0) + \beta p(2|1,0)v^n(2) + \beta p(3|1,0)v^n(3) = \frac{7}{4} - \frac{3}{2}\cdot\left(\frac{1}{2}\right)^n,$$

we conclude that, for each $n = 0, 1, 2, \ldots$, the action $a^* = n+1$ provides the infimum in the formula

$$v^{n+1}(1) = \inf_{a \in \mathbf{A}}\{c(1,a) + \beta p(2|1,a)v^n(2) + \beta p(3|1,a)v^n(3)\}$$

$$= \frac{7}{4} - \frac{3}{2}\cdot\left(\frac{1}{2}\right)^n - \left(\frac{1}{2}\right)^{2n+2}.$$

Following (3.7),

$$\Phi_n^* = \{\varphi: \quad \varphi(1) = n+1\} \text{ for all } n = 0, 1, 2, \ldots.$$

But

$$\inf_{a \in \mathbf{A}}\{c(1,a) + \beta p(2|1,a)v(2) + \beta p(3|1,a)v(3)\}$$

$$= \inf_{a \in A} \left\{ \left(\frac{1}{2} - \left(\frac{1}{2}\right)^a\right)^2 + p(2|1,a) + 2p(3|1,a) \right\} = \frac{7}{4},$$

and the infimum is attained at $a^* = 0$. To see this, using the previous notation $\gamma \stackrel{\triangle}{=} \left(\frac{1}{2}\right)^a$ for $a = 1, 2, \ldots$, one can compute

$$\inf_{\gamma > 0} \left\{ \left(\frac{1}{2} - \gamma\right)^2 + \frac{1}{2} - \gamma + 2\left(\frac{1}{2} + \gamma\right) \right\} = \frac{7}{4};$$

this infimum is attained at $\gamma = 0$, corresponding to no one action $a > 0$.

Therefore, following (3.6),

$$\Phi^* = \{\varphi : \varphi(1) = 0\} \cap \Phi_n^* = \emptyset$$

for all $n = 0, 1, 2, \ldots$

Now take $\mathbf{X} = \{0, 1, 2, \ldots\}$, $\mathbf{A} = \{1, 2\}$, $\beta = 1/2$, $p(0|0,a) \equiv 1$, for $x > 0$ put $p(x|x,1) \equiv 1$ and $p(x-1|x,2) \equiv 1$, with other transition probabilities zero; $c(0,a) \equiv -1$, for $x > 0$ put $c(x,1) \equiv 0$ and $c(x,2) = \left(\frac{1}{4}\right)^x$ (see Fig. 3.7).

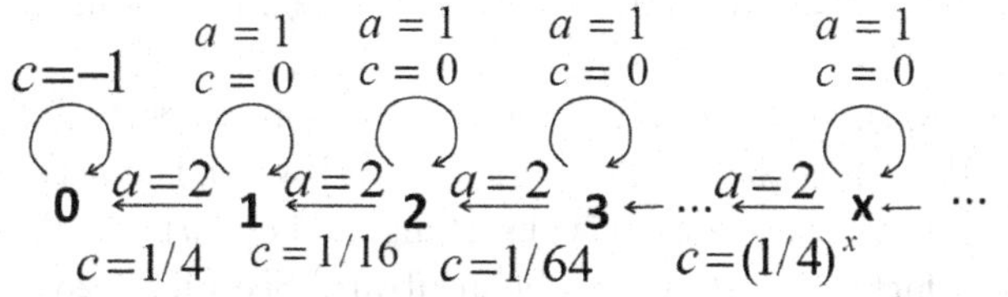

Fig. 3.7 Example 3.2.6: value iteration, countable state space.

Value iteration gives the following table:

x	0	1	2	3	4 ...
$v^0(x)$	0	0	0	0	0
$v^1(x)$	-1	0	0	0	0
$v^2(x)$	$-3/2$	$-1/4$	0	0	0
$v^3(x)$	$-7/4$	$-1/2$	$-1/16$	0	0
$v^4(x)$	$-15/8$	$-5/8$	$-3/16$	$-1/64$	0
...					...

In general,

$$v^n(0) = -2 + \left(\frac{1}{2}\right)^{n-1},$$

and for $x > 0$, $n > 0$ we have

$$v^n(x) = \min\left\{\frac{1}{2}v^{n-1}(x);\quad \left(\frac{1}{4}\right)^x + \frac{1}{2}v^{n-1}(x-1)\right\}$$

$$= \begin{cases} 0, & \text{if } n \leq x; \\ -\left(\frac{1}{2}\right)^x - \left(\frac{1}{4}\right)^x + \left(\frac{1}{2}\right)^{n-1}, & \text{if } n \geq x+1. \end{cases}$$

Therefore,

$$\Phi_n^* = \{\varphi:\quad \text{for } x > 0,\quad \varphi(x) = 1 \text{ if } x \geq n, \text{ and } \varphi(x) = 2 \text{ if } x < n\},$$

but $\Phi^* = \{\varphi:\quad \text{for } x > 0,\quad \varphi(x) \equiv 2\}$.

3.2.7 The Bellman function is non-measurable and no one strategy is uniformly ε-optimal

This example, described in [Blackwell(1965), Ex. 2] and in [Puterman(1994), Ex. 6.2.3], is similar to the examples in Sections 1.4.14, 1.4.15 and 2.2.17.

Let $\mathbf{X} = [0,1]$, $\mathbf{A} = [0,1]$ and let B be a Borel subset of $\mathbf{X} \times \mathbf{A}$ with projection $B^1 = \{x \in \mathbf{X}: \exists a \in \mathbf{A}: (x,a) \in B\}$ which is not Borel. We put $p(x|x,a) \equiv 1$, so that each state is absorbing, and $c(x,a) = -I\{(x,a) \in B\}$. The discount factor $\beta \in (0,1)$ can be arbitrary. See Fig. 1.23.

For any $x \in \mathbf{X} \setminus B^1$, $v_x^* \equiv 0$ and for any fixed $\hat{x} \in B^1$, $v_{\hat{x}}^* \equiv \frac{-1}{1-\beta}$: it is sufficient to take the stationary selector $\hat{\varphi}(x) \equiv \hat{a}$ with $\hat{a}$ such that $(\hat{x}, \hat{a}) \in B$. Thus, the Bellman function v_x^* is not measurable.

Now, consider an arbitrary control strategy π. Since $\pi_1(da|x)$ is measurable w.r.t. x, $\int_\mathbf{A} c(x,a)\pi(da|x)$ is also measurable and

$$\left\{x \in \mathbf{X}:\quad \int_\mathbf{A} c(x,a)\pi_1(da|x) < 0\right\} \subseteq B^1$$

is a Borel subset of $\mathbf{X}$; hence there is $y \in B^1$ such that $\int_\mathbf{A} c(x,a)\pi_1(da|x) = 0$, meaning that

$$v_y^\pi \geq -\beta - \beta^2 - \cdots = -\frac{\beta}{1-\beta} = \frac{-1}{1-\beta} + 1 = v_y^* + 1,$$

i.e. the strategy π is not uniformly ε-optimal for any $\varepsilon < 1$.

3.2.8 No one selector is uniformly ε-optimal

In this example, first published in [Dynkin and Yushkevich(1979), Chapter 6, Section 7], the Bellman function $v_x^* \equiv 2$ is well defined.

Let $\mathbf{X} = [0, 1]$, $\mathbf{A} = [0, 1]$ and let $Q \subset \mathbf{X} \times \mathbf{A}$ be a Borel subset such that $\forall x \in \mathbf{X}\ \exists a : (x, a) \in Q$ and Q does not contain graphs of measurable maps $\mathbf{X} \to \mathbf{A}$. (Such a subset was constructed in [Dynkin and Yushkevich(1979), App. 3, Section 3].) We put $c(x, a) = -I\{(x, a) \in Q\}$, $p(\{0\}|x, a) \equiv 1$, and $\beta = 1/2$. See Fig. 3.8.

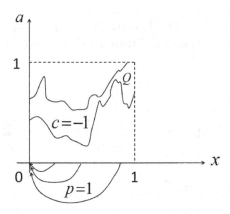

Fig. 3.8 Example 3.2.8: no uniformly ε-optimal selectors.

For any $x \in \mathbf{X}$ $v_x^* \equiv -2$. But for any selector φ_t,

$$v_x^\varphi \geq c(x, \varphi_1(x)) - 1$$

(the second term, equal to the total discounted loss starting from state $X_1 = 0$, cannot be smaller than -1). Since $\varphi_1(x)$ is a measurable map $\mathbf{X} \to \mathbf{A}$, there is a point $\hat{x} \in \mathbf{X}$ such that $(\hat{x}, \varphi_1(\hat{x})) \notin Q$ and $v_{\hat{x}}^\varphi \geq -1$. Thus the selector φ is not uniformly ε-optimal if $\varepsilon < 1$.

3.2.9 Myopic strategies

Definition 3.1. A stationary strategy $\pi(da|x)$, uniformly optimal in the homogeneous one-step model ($T = 1$) with $C(x) \equiv 0$, is called *myopic*. In other words, $\int_\mathbf{A} c(x, a)\pi(da|x) \equiv \inf_{a \in \mathbf{A}} \{c(x, a)\}$.

When the loss function $c(x)$ does not depend on the action a, a stationary strategy π is called *myopic* if $\int_A \int_X p(dy|x,a) c(y) \pi(da|x) \equiv \inf_{a \in A} \int_X p(dy|x,a) c(y)$.

It is known that a myopic strategy is uniformly optimal in the discounted MDP with an arbitrary discount factor $\beta \in (0,1)$, if it is uniformly optimal in the finite-horizon case with any time horizon $T < \infty$, without final loss ($C(x) \equiv 0$), and without discounting ($\beta = 1$). See [Piunovskiy(2006), Lemma 2.1]. The converse assertion is not always true, as the following example, published in the above article, shows.

Let $\mathbf{X} = \{1,2,3\}$, $\mathbf{A} = \{1,2\}$, $p(1|1,1) = 1$, $p(2|1,2) = 1$, $p(3|2,a) = p(3|3,a) \equiv 1$, with other transition probabilities zero; $c(1,1) = -2$, $c(1,2) = -3$, $c(2,a) \equiv 0$, $c(3,a) \equiv -3$ (see Fig. 3.9). The discount factor $\beta \in (0,1)$ is arbitrarily fixed.

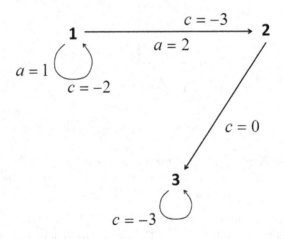

Fig. 3.9 Example 3.2.9: optimal myopic strategy.

From the optimality equation, we obtain

$$v(3) = -\frac{3}{1-\beta}; \quad v(2) = -\frac{3\beta}{1-\beta}; \quad v(1) = -3 - \frac{3\beta^2}{1-\beta};$$

the stationary selector $\varphi^*(x) \equiv 2$ is (uniformly) optimal independently of the value of β, and $v_x^* \equiv v(x)$. The selector φ^* is myopic.

At the same time, if $T = 2$ then, in this non-discounted finite-horizon model,
$$v_2(x) = C(x) \equiv 0, \quad v_1(x) = \begin{cases} -3, & \text{if } x = 1; \\ 0, & \text{if } x = 2; \\ -3, & \text{if } x = 3, \end{cases} \quad v_0(x) = \begin{cases} -5, & \text{if } x = 1; \\ -3, & \text{if } x = 2; \\ -3, & \text{if } x = 3, \end{cases}$$
the selector
$$\varphi_t(1) = \begin{cases} 1, & \text{if } t = 1; \\ 2, & \text{if } t = 2, \end{cases} \quad \varphi_t(2) = \varphi_t(3) \equiv 2$$
is optimal, and the myopic selector φ^* is *not* optimal for the initial state $X_0 = 1$.

3.2.10 Stable and unstable controllers for linear systems

MDPs with finite-dimensional Euclidean spaces $\mathbf{X}$ and $\mathbf{A}$ are often defined by *system equations* of the form
$$X_t = bX_{t-1} + cA_t + d\zeta_t, \tag{3.8}$$
where $\{\zeta_t\}_{t=1}^\infty$ is a sequence of *disturbances*, i.e. independent random vectors with $E[\zeta_t] = 0$ and $E[\zeta_t \zeta_t'] = I$ (the identity matrix). In what follows, all vectors are columns; $b, c,$ and d are the given matrices of appropriate dimensionalities, and the dash means transposition. The transition probability can be defined through the density function of ζ_t; see Example 3.2.1. In the framework of dynamical systems (3.8), stationary selectors are called (feedback) *controllers/controls*.

Definition 3.2. A system is called *stable* if, in the absence of disturbances ζ_t, the state $X_t = bX_{t-1} + cA_t$ tends to zero as $t \to \infty$ (for all or several initial states X_0); see [Bertsekas(2005), p. 153]. Likewise, the controller is called *stable* if $A_t \to 0$ in the absence of disturbances.

The following example, similar to [Bertsekas(2005), Ex. 5.3.1], shows that the stabilizing (and even minimum-variance) control can itself be unstable.

Let $\mathbf{X} = \mathbb{R}^2$, $\mathbf{A} = \mathbb{R}^1$, $b = \begin{bmatrix} 1 & -2 \\ 0 & 0 \end{bmatrix}$, $c = \begin{bmatrix} 1 \\ 1 \end{bmatrix}$, $d = \begin{bmatrix} 1 \\ 0 \end{bmatrix}$, $\zeta_t \in \mathbb{R}^1$. The system equation for $X = \begin{pmatrix} X^1 \\ X^2 \end{pmatrix}$,
$$\begin{cases} X_t^1 = X_{t-1}^1 - 2X_{t-1}^2 + A_t + \zeta_t \\ X_t^2 = A_t \end{cases}$$

can be rewritten as
$$X_t^1 = X_{t-1}^1 - 2A_{t-1} + A_T + \zeta_t.$$
We put $c(x,a) = (x^1)^2$; that is, the goal is to minimize the total (discounted) variance of the main component X^1. The discount factor $\beta \in (0,1)$ is arbitrarily fixed. See Fig. 3.10.

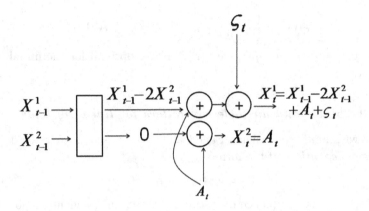

Fig. 3.10 Example 3.2.10: an unstable controller.

The optimality equation takes the form
$$v(x^1, x^2) = \inf_{a \in \mathbf{A}} \left\{ (x^1)^2 + \beta \, E[v(x^1 - 2x^2 + a + \zeta, a)] \right\},$$
where the expectation E corresponds to the random variable ζ. Using the standard convention $E[\zeta] = 0$, $E[\zeta^2] = 1$, one can calculate the minimal non-negative solution
$$v(x^1, x^2) = (x^1)^2 + \frac{\beta}{1-\beta},$$
which coincides with the Bellman function v_x^* because the model is positive. The optimal selector is given by the formula
$$\varphi^*(x) = 2x^2 - x^1,$$
so that $X_t^1 = \zeta_t$ for all $t = 1, 2, \ldots$; the system is stable.

At the same time, the feedback control φ^* results in the sequence
$$A_1 = 2X_0^2 - X_0^1,$$
$$A_2 = 2A_1 - \zeta_1,$$
$$A_3 = 4A_1 - 2\zeta_1 - \zeta_2,$$

and, in general,
$$A_t = 2^{t-1}A_1 - \sum_{i=1}^{t-1} 2^{t-i-1}\zeta_i,$$

meaning that the optimal controller (as well as the second component X_t^2) is unstable. Moreover, it is even "discounted"-unstable: in the absence of disturbances, $\lim_{t\to\infty} \beta^{t-1}A_t \neq 0$ for $\beta \geq 1/2$, if $A_1 \neq 0$. Note that the selector φ^* is myopic.

One can find an acceptable solution to this control problem by taking into account all the variables; that is, we put

$$c(x,a) = k_1(x^1)^2 + k_2(x^2)^2 + k_3 a^2 \qquad k_1, k_2, k_3 > 0.$$

Now we can use the well developed theory of linear–quadratic control, see for example [Piunovskiy(1997), Section 1.2.2.5].

The maximal eigenvalue of matrix b equals 1. Moreover, for the selector

$$\varphi_t(x) = \begin{cases} x^1, & \text{if } t \text{ is even;} \\ 0, & \text{if } t \text{ is odd} \end{cases}$$

we have

$$X_t^1 = \begin{cases} 2(\zeta_{t-2} + \zeta_{t-1}) + \zeta_t, & \text{if } t \text{ is even;} \\ \zeta_{t-1} + \zeta_t, & \text{if } t \text{ is odd} \end{cases} \quad X_t^2 = \begin{cases} \zeta_{t-2} + \zeta_{t-1}, & \text{if } t \text{ is even;} \\ 0, & \text{if } t \text{ is odd} \end{cases}$$

for all $t \geq 3$, so that all the processes are stable. Therefore, for an arbitrary fixed discount factor $\beta \in (0,1)$, the optimal stationary selector and the Bellman function are given by the formulae

$$\varphi^*(x) = -\frac{\beta c' f b}{k_3 + \beta c' f c} x; \qquad v_x^* = x' f x + q,$$

where $q = \frac{\beta f_{11}}{1-\beta}$ and $f = \begin{bmatrix} f_{11} & f_{12} \\ f_{21} & f_{22} \end{bmatrix}$ (a symmetric matrix) is the unique positive-definite solution to the equation

$$f = \beta b' f b + \begin{bmatrix} k_1 & 0 \\ 0 & k_2 \end{bmatrix} - \frac{\beta^2 b' f c c' f b}{k_3 + \beta c' f c}.$$

Moreover,
$$\lim_{T\to\infty} E^{\varphi^*}[\beta^T X_T' f X_T] = \lim_{T\to\infty} E^{\varphi^*}[\beta^T X_T] = 0.$$

The last equalities also hold in the case of no disturbances, i.e. the system is "discounted"-stable. If there are no disturbances, then all the formulae and statements survive (apart from $q = 0$) for any $\beta \in [0,1]$.

It can happen that system (3.8) is stable only for some initial states X_0 [Bertsekas(2001), Ex. 3.1.1]. Let $\mathbf{X} = \mathbb{R}^1$, $\mathbf{A} = [-3, 3]$, $b = 3$, $c = 1$ and suppose there are no disturbances ($\zeta_t \equiv 0$):
$$X_t = 3X_{t-1} + A_t.$$
If $|X_0| < 3/2$ then we can put $A_t = -3\,\text{sign}(X_{t-1})$ for $t = 1, 2, \ldots$, up to the moment when $|X_\tau| \leq 1$; we finish with $X_t \equiv 0$ afterwards, so that the system is stable. The system is unstable if $|X_0| \geq 3/2$.

Now let
$$\zeta_t = \begin{cases} +1 & \text{with probability } 1/2 \\ -1 & \text{with probability } 1/2 \end{cases}$$
and consider the performance functional
$$v_x^\pi = E_x^\pi \left[\sum_{t=1}^\infty \beta^{t-1} |X_{t-1}| \right]$$
with $\beta = 1/2$.

Firstly, it can be shown that, for any control strategy π, $v_x^\pi = \infty$ if $|x| > 1$. In the case where $x > 1$, there is a positive probability of having $\zeta_1 = \zeta_2 = \cdots = \zeta_\tau = 1$, up to the moment when $X_\tau > 4$: the sequence
$$X_1 \geq 3x - 3 + 1; \quad X_2 \geq 3X_1 - 3 + 1, \quad \ldots$$
approaches $+\infty$. Thereafter,
$$X_{\tau+1} \geq 3X_\tau - 3 - 1 = 2X_\tau + (X_\tau - 4) > 2X_\tau$$
and, for all $i = 0, 1, 2, \ldots$, $X_{\tau+i+1} > 2X_{\tau+i}$, meaning that $v_x^\pi = \infty$. The reasoning for $x < -1$ is similar. Hence $v_x^* = \infty$ for $|x| > 1$.

If $|x| \leq 1$ then $v_x^* = |x| + \frac{\beta}{1-\beta} = |x| + 1$, and the control strategy $\varphi^*(x) = -3x$ is optimal (note that $\varphi^*(X_t) \in [-3, 3]$ $P_x^{\varphi^*}$-almost surely). The reader can study the optimality equation independently.

3.2.11 Incorrect optimal actions in the model with partial information

Suppose the transition probabilities depend on an unknown parameter $\theta \in \{\theta_1, \theta_2\}$, and the decision maker knows only the initial probability $q_0 = P\{\theta = \theta_1\}$. We assume that the loss $k(x, y)$ is associated with the transitions of the observable process, where function k is θ-independent.

It is known that the *posteriori* probability $q_t = P\{\theta = \theta_1 | X_0, X_1, \ldots, X_t\}$ is a sufficient statistic for one to investigate the model with complete information, where the pair $(X_t, q_t) \in \mathbf{X} \times [0, 1]$ plays the role of the

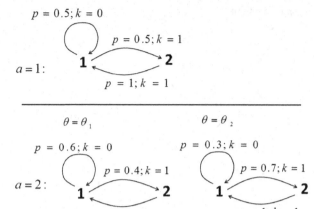

Fig. 3.11 Example 3.2.11: MDP with partial information.

state. The initial value q_0 is assumed to be X_0-independent. More about MDPs with partial information is given in [Bertsekas and Shreve(1978), Chapter 10] and [Dynkin and Yushkevich(1979), Chapter 8]. The following example appeared in [Bertsekas(2005), Ex. 6.1.6].

Let $\mathbf{X} = \{1,2\}$, $\mathbf{A} = \{1,2\}$, $k(1,1) = 0$, $k(1,2) = 1$, $k(2,1) = 1$, where transitions $2 \to 2$ never occur. Independently of θ,

$$p(1|1,1) = p(2|1,1) = 0.5 \qquad p(1|2,a) \equiv 1.$$

The transition probability $p^\theta(1|1,2)$ depends on the value of θ:

$$p^\theta(1|1,2) = \begin{cases} 0.6, & \text{if } \theta = \theta_1; \\ 0.3, & \text{if } \theta = \theta_2 \end{cases} \qquad p^\theta(2|1,2) = 1 - p^\theta(1|1,2)$$

(see Fig. 3.11). The discount factor $\beta \in (0,1)$ is arbitrarily fixed. Actions in state 2 play no role.

The *posteriori* probability q_t can change only if $X_{t-1} = 1$ and $A_t = 2$:

$$\text{if } X_t = 1 \text{ then } q_t = \frac{0.6 q_{t-1}}{0.6 q_{t-1} + 0.3(1 - q_{t-1})} = \frac{0.6 q_{t-1}}{0.3 + 0.3 q_{t-1}};$$

$$\text{if } X_t = 2 \text{ then } q_t = \frac{0.4 q_{t-1}}{0.4 q_{t-1} + 0.7(1 - q_{t-1})} = \frac{0.4 q_{t-1}}{0.7 - 0.3 q_{t-1}}.$$

The transition probabilities and the loss function for the model with complete information and state space $\mathbf{X} \times [0,1]$ are defined by the following formulae (see also Fig. 3.12):

$$\hat{p}((1,q)|(1,q),1) = \hat{p}((2,q)|(1,q),1) = 0.5; \quad \hat{p}((1,q)|(2,q),a) \equiv 1;$$

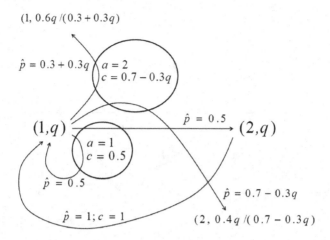

Fig. 3.12 Example 3.2.11: equivalent MDP with complete information.

$$\hat{p}\left(\left(1,\frac{0.6q}{0.3+0.3q}\right)\bigg|(1,q),2\right)=0.3+0.3q;$$

$$\hat{p}\left(\left(2,\frac{0.4q}{0.7-0.3q}\right)\bigg|(1,q),2\right)=0.7-0.3q;$$

(other transition probabilities are zero);

$$c((1,q),1)=0.5;\quad c((2,q),a)\equiv 1;\quad c((1,q),2)=0.7-0.3q.$$

The optimality equation (3.2) takes the form

$$v(1,q) = \min\Big\{0.5 + \beta[0.5\ v(1,q) + 0.5\ v(2,q)];$$

$$0.7 - 0.3q + \beta\left[(0.3+0.3q)v\left(1,\frac{0.6q}{0.3+0.3q}\right)\right.$$

$$\left.+(0.7-0.3q)v\left(2,\frac{0.4q}{0.7-0.3q}\right)\right]\Big\};$$

$$v(2,q) = 1 + \beta\ v(1,q),$$

and there exists an optimal stationary selector $\varphi^*((x,q))$ because the action space **A** is finite [Bertsekas and Shreve(1978), Corollary 9.17.1].

We show that there exist positive values of q for which $\varphi^*((1,q))=1$. Indeed, if $q\in(0,1)$ then $\frac{0.6q}{0.3+0.3q},\ \frac{0.4q}{0.7q-0.3q}\in(0,1)$, and we can concentrate on the model with complete information and with state space $\mathbf{X} \times (0,1)$ for which φ^* is also an optimal strategy. For the stationary selector $\varphi((x,q)) \equiv 2$, we have

$$v^{\varphi}_{(1,q)} = \frac{0.4(1+\beta)q}{1-0.6\beta-0.4\beta^2} + \frac{0.7(1+\beta)(1-q)}{1-0.3\beta-0.3\beta^2};$$

$$v^{\varphi}_{(2,q)} = 1 + \beta\, v(1,q).$$

But now

$$v^{\varphi}_{(1,q)} - \{0.5 + \beta[0.5 v^{\varphi}_{(1,q)} + 0.5 v^{\varphi}_{(2,q)}]\} =$$

$$(1+\beta)\,[1-0.5\beta(1+\beta)]\left[\frac{0.4q}{1-0.6\beta-0.4\beta^2} + \frac{0.7(1-q)}{1-0.3\beta-0.7\beta^2}\right] - 0.5(1+\beta).$$

This last function of q is continuous, and it approaches the positive value

$$\frac{0.7(1+\beta)(1-0.5\beta(1+\beta))}{1-0.3\beta-0.7\beta^2} - 0.5(1+\beta) = \frac{0.2(1-\beta^2)}{1-0.3\beta-0.7\beta^2} > 0$$

as $q \to 0$, even if $\beta \to 1-$. Therefore, in the model under investigation, the selector $\varphi((1,q)) \equiv 2$ is not optimal [Bertsekas and Shreve(1978), Proposition 9.13] meaning that, for some positive values of q, it is reasonable to apply action $a = 1$: $\varphi^*((1,q)) = 1$. But thereafter that, the value of q remains unchanged, and the decision maker applies action $a = 1$ every time. Therefore, for those values of q, with probability $q = P\{\theta = \theta_1\}$, in the original model (Fig. 3.11), the decision maker always applies action $a = 1$, although action $a = 2$ dominates action $a = 1$ in state 1 if $\theta = \theta_1$: the probability of remaining in state 1 (and of having zero loss) is higher if $a = 2$.

3.2.12 Occupation measures and stationary strategies

Definition 3.3. For a fixed control strategy π, the *occupation measure* η^π is the measure on $\mathbf{X} \times \mathbf{A}$ given by the formula

$$\eta^\pi(\Gamma^\mathbf{X} \times \Gamma^\mathbf{A}) \triangleq (1-\beta) \sum_{t=0}^{\infty} \beta^{t-1} P^\pi_{P_0}\{X_t \in \Gamma^\mathbf{X}, A_{t+1} \in \Gamma^\mathbf{A}\},$$

$$\Gamma^\mathbf{X} \in \mathcal{B}(\mathbf{X}), \quad \Gamma^\mathbf{A} \in \mathcal{B}(\mathbf{A}).$$

A finite measure η on $\mathbf{X} \times \mathbf{A}$ is an occupation measure if and only if it satisfies the equation

$$\eta(\Gamma \times \mathbf{A}) = (1-\beta)P_0(\Gamma) + \beta \int_{\mathbf{X} \times \mathbf{A}} p(\Gamma|y,a) d\eta(y,a)$$

for all $\Gamma \in \mathcal{B}(\mathbf{X})$ [Piunovskiy(1997), Lemma 25].

Usually (e.g. in positive and negative models and in the case where $\sup_{x \in \mathbf{X},\ a \in \mathbf{A}} |c(x,a)| < \infty$),

$$v^\pi = \frac{1}{1-\beta} \int_{\mathbf{X} \times \mathbf{A}} c(x,a)\ d\eta^\pi(x,a), \tag{3.9}$$

and investigation of the MDP in terms of occupation measures (the so-called *convex analytic approach*) is fruitful, especially in constrained problems.

The space of all occupation measures is convex and, for any strategy π, there exists a stationary strategy π^s such that $\eta^\pi = \eta^{\pi^s}$ [Piunovskiy(1997), Lemma 24 and p. 142]. According to (3.9), the performance functional v^π is linear on the space of occupation measures. Note that the space of stationary strategies is also convex, but the map $\pi^s \to \eta^{\pi^s}$ is not affine, and the function $v^{\pi^s} : \pi^s \to \mathbb{R}^1$ can be non-convex. The following example confirms this.

Let $\mathbf{X} = \{1,2\}$, $\mathbf{A} = \{0,1\}$, $p(1|1,0) = 1$, $p(2|1,1) = 1$, $p(1|2,a) \equiv 1$, with other transition probabilities zero; $P_0(1) = 1$, $c(2,a) \equiv 1$, $c(1,a) \equiv 0$ (see Fig. 3.13).

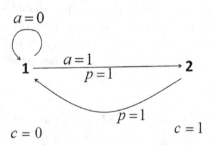

Fig. 3.13 Example 3.2.12: the performance functional is non-convex w.r.t. to strategies.

For any two stationary strategies π^{s^1} and π^{s^2}, strategy $\alpha \pi^{s^1}(a|x) + (1-\alpha)\pi^{s^2}(a|x) = \pi^{s^3}$ is of course stationary. Since the values of $\pi^s(a|2)$ play no role, we can describe any stationary strategy with the number $\gamma =$

$\pi^s(0|1)$. Now the strategy π^{s^3} corresponds to $\gamma^3 = \alpha\gamma^1 + (1-\alpha)\gamma^2$, where γ^1 and γ^2 describe strategies π^{s^1} and π^{s^2}, so that the convexity/linearity w.r.t. π coincides with the convexity/linearity in γ, and the convex space of stationary strategies is isomorphic to the segment $[0,1] \ni \gamma$.

For a fixed value of γ, the marginal $\hat{\eta}^\gamma(x) \stackrel{\triangle}{=} \eta^\gamma(x \times \mathbf{A})$ satisfies the equations

$$\hat{\eta}^\gamma(1) = (1-\beta) + \beta[\gamma\hat{\eta}^\gamma(1) + \hat{\eta}^\gamma(2)];$$
$$\hat{\eta}^\gamma(2) = \beta(1-\gamma)\hat{\eta}^\gamma(1)$$

(the index γ corresponds to the stationary strategy defined by γ). Therefore,

$$\hat{\eta}^\gamma(1) = \frac{1}{1+\beta-\beta\gamma}, \qquad \hat{\eta}^\gamma(2) = \frac{\beta-\beta\gamma}{1+\beta-\beta\gamma},$$

and the mapping $\pi^s \to \eta^{\pi^s}$ (or equivalently, the map $\gamma \to \eta^\gamma$) is not convex. Similarly, the function $v^{\pi^s} = \hat{\eta}^{\pi^s}(2) = \hat{\eta}^\gamma(2) = \frac{\beta-\beta\gamma}{1+\beta-\beta\gamma}$ is non-convex in γ (and in π^s).

One can encode occupation measures using the value of $\delta = \hat{\eta}(1)$: for any $\delta \in \left[\frac{1}{1+\beta}\right]$, the corresponding occupation measure is given by

$$\eta(1,0) = \frac{\delta}{\beta} + \delta - \frac{1}{\beta}; \quad \eta(1,1) = \frac{1}{\beta} - \frac{\delta}{\beta}; \quad \hat{\eta}(2) = 1 - \delta.$$

The separate values $\eta(2,0)$ and $\eta(2,1)$ are of no importance but, if needed, they can be defined by

$$\eta(2,0) = (1-\delta)\varepsilon, \quad \eta(2,1) = (1-\delta)(1-\varepsilon),$$

where $\varepsilon \in [0,1]$ is an arbitrary number corresponding to $\pi(0|2)$. Now the performance functional $v^\pi = \hat{\eta}^\pi(2) = 1 - \delta$ is affine in δ.

Lemma 3.1. *[Piunovskiy(1997), Lemma 2], [Dynkin and Yushkevich(1979), Chapter 3, Section 8]. For every control strategy π, there exists a Markov strategy π^m such that $\forall t = 1, 2, \ldots$*

$$P_{P_0}^\pi(X_{t-1} \in \Gamma^\mathbf{X}, A_t \in \Gamma^\mathbf{A}) = P_{P_0}^{\pi^m}(X_{t-1} \in \Gamma^\mathbf{X}, A_t \in \Gamma^\mathbf{A}) \qquad (3.10)$$

for any $\Gamma^\mathbf{X} \in \mathcal{B}(\mathbf{X})$ and $\Gamma^\mathbf{A} \in \mathcal{B}(\mathbf{A})$.

Clearly, (3.10) implies that $\eta^\pi = \eta^{\pi^m}$.

Suppose an occupation measure $\eta = \eta^\pi$ is not extreme: $\eta^\pi = \alpha\eta^{\pi^1} + (1-\alpha)\eta^{\pi^2}$, where $\eta^{\pi^1} \neq \eta^{\pi^2}$, $\alpha \in (0,1)$. On the one hand, as usual, η^π is generated by a stationary strategy π^s: $\eta^\pi = \eta^{\pi^s}$. On the other hand, according to Lemma 3.1, there exists a Markov strategy π^m for which equality (3.10) holds and hence $\eta^\pi = \eta^{\pi^m}$. In a typical situation, π^m is non-stationary: see Section 2.2.1, the reasoning after formula (2.5). Therefore, very often the same occupation measure can be generated by many different strategies.

3.2.13 *Constrained optimization and the Bellman principle*

Suppose we have two loss functions $^1c(x,a)$ and $^2c(x,a)$. Then every control strategy π results in two performance functionals $^1v^\pi$ and $^2v^\pi$ defined according to (3.1), the discount factor β being the same. The *constrained* problem looks like

$$^1v^\pi \to \inf_\pi \qquad ^2v^\pi \leq d, \qquad (3.11)$$

where d is a chosen number. Strategies satisfying the above inequality are called *admissible*. The following example, first published in [Altman and Shwartz(1991a), Ex. 5.3] shows that, in this framework, the optimal strategy depends on the initial distribution and, moreover, the Bellman principle fails to hold.

Let $\mathbf{X} = \{1,2\}$; $\mathbf{A} = \{1,2\}$; $p(1|x,a) \equiv 0.1$; $p(2|x,a) \equiv 0.9$; $\beta = 0.1$;

$$^1c(1,a) \equiv 0 \quad ^1c(2,a) = \begin{cases} 1, & \text{if } a = 1; \\ 0, & \text{if } a = 2 \end{cases}$$

$$^2c(1,a) \equiv 1 \quad ^2c(2,a) = \begin{cases} 0, & \text{if } a = 1; \\ 0.1, & \text{if } a = 2 \end{cases}$$

(see Fig. 3.14).

We take $d = \frac{91}{90}$, the minimal value of $^2v_1^\pi$, i.e. the value of the Bellman function $^2v_1^*$ associated with the loss ^2c. The actions in state 1 play no role. If the initial state is $X_0 = 1$ then, in state 2, one has to apply action 1 because otherwise the constraint in (3.11) is violated. Therefore, the stationary selector $\varphi^1(x) \equiv 1$ solves problem (3.11) if $X_0 = 1$ (a.s.).

Suppose now the initial state is $X_0 = 2$ (a.s.) and consider the stationary selector $\varphi^*(x) \equiv 2$. This solves the unconstrained problem $^1v_2^\pi \to \inf_\pi$. But in the constrained problem (3.11), $^2v_2^{\varphi^*} = \frac{109}{900} < d = \frac{910}{900}$ is also admissible. Therefore, the optimal strategy depends on the initial state. The Bellman principle fails to hold because the optimal actions in state 2 at the later decision epochs depend on the value of X_0. Another simple example, confirming that stationary strategies are not sufficient for solving constrained problems, can be found in [Frid(1972), Ex. 2].

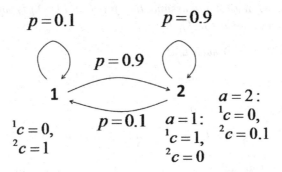

Fig. 3.14 Example 3.2.13: constrained problem.

Suppose $X_0 = 1$ (a.s.) and suppose we have the observed trajectory
$$X_0 = 1 \to X_1 = 2 \to X_2 = 2.$$
The optimal strategy φ^1 prescribes the actions $A_1 = A_2 = A_3 = \cdots = 1$. At the same time, at decision epoch 3, we know the current value of the accumulated realized second-type loss
$$^2c(X_0, 1) + \beta\ ^2c(X_1, 1) = 1 + 0 = 1, \qquad (3.12)$$
so that, in the case where we apply selector φ^* starting from decision epoch 3, then the future expected second-type loss equals
$$\beta^2 \cdot {}^2v_2^{\varphi^*} = 0.01 \cdot \frac{109}{900} \approx 0.0012$$
which, together with one unit from (3.12), makes $1.0012 < d = \frac{91}{90}$. Thus, why do we not use selector φ^* resulting in a smaller value for the main functional ^1v in this situation? The answer is that if we do that then, after many repeated experiments, the average value of $^2v_1$ will be greater than d.

3.2.14 Constrained optimization and Lagrange multipliers

When solving constrained problems like (3.11), the following statement is useful.

Proposition 3.1. *Suppose the performance functionals $^1v^\pi$ and $^2v^\pi$ are finite for any strategy π and $^2v^{\hat\pi} < d$ for some strategy $\hat\pi$. Let*
$$L(\pi, \lambda) = {}^1v^\pi + \lambda(\,{}^2v^\pi - d), \qquad \lambda \geq 0$$

be the Lagrange function and assume that function $L(\pi, 1)$ is bounded below. If π^* solves problem (3.11), then

(a) there is $\lambda^* \geq 0$ such that
$$\inf_\pi L(\pi, \lambda^*) = \sup_{\lambda \geq 0} \inf_\pi L(\pi, \lambda);$$

(b) strategy π^* is such that
$$L(\pi^*, \lambda^*) = \min_\pi L(\pi, \lambda^*), \quad {}^2v^{\pi^*} \leq d \quad \text{and} \quad \lambda^* \cdot ({}^2v^{\pi^*} - d) = 0.$$

More about the Lagrange approach can be found in [Piunovskiy(1997), Section 2.3].

The function $g(\lambda) \triangleq \inf_\pi L(\pi, \lambda)$ is called *dual functional*. It is obviously helpful for solving constrained problems, so its properties are of the great importance. It is known that g is concave. If, for example, the loss functions ${}^1c$ and ${}^2c$ are bounded below then the dual functional g is continuous for $\lambda > 0$, like any other concave function with the full domain ($\forall \lambda \geq 0 \; g(\lambda) > -\infty$); see [Rockafellar(1970), Th. 10.1]. Incidentally, the article [Frid(1972)] contains a minor mistake: on p. 189, the author claims that the dual functional g is continuous on $[0, \infty)$. The following example shows that functional g can be discontinuous at zero (see [Sennott(1991), Ex. 3.1]).

Let $\mathbf{X} = \{0, 1, 2, \ldots\}$; $\mathbf{A} = \{0, 1\}$; $p(0|0, 0) = 1$, $p(1|0, 1) = 1$, $p(i+1|i, a) \equiv 1$ for all $i = 1, 2, \ldots$, with other transition probabilities zero. We put

$${}^1c(x, a) = \begin{cases} 1, & \text{if } x = 0; \\ 0, & \text{if } x > 0 \end{cases} \qquad {}^2c(x, a) = 2^x;$$

$\beta = 1/2$; $P_0(0) = 1$ and $d = 1$ (see Fig. 3.15).

Only the actions in state $x = 0$ play any role.

Since ${}^2v^\pi = +\infty$ for any strategy except for those which apply action $a = 0$ in state $x = 0$, we conclude that

$$g(\lambda) = {}^1v^{\varphi^0} + \lambda({}^2v^{\varphi^0} - 1) = 2 + \lambda$$

if $\lambda > 0$. Here $\varphi^0(x) \equiv 0$. If $\lambda = 0$ then $\min_\pi {}^1v^\pi = 1$ is provided by the stationary selector $\varphi^1(x) \equiv 1$, and $g(0) = 1$. Hence, the dual functional is discontinuous at zero. The solution to the constrained problem (3.11) is provided by the selector φ^0. Proposition 3.1 is not valid because the performance functional ${}^2v^\pi$ is not finite, e.g. for the selector φ^1; λ^* maximizing the dual functional does not exist.

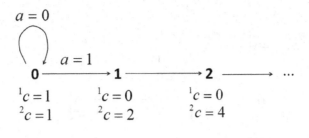
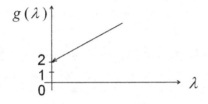

Fig. 3.15 Example 3.2.14: discontinuous dual functional.

When the loss functions ^1c and ^2c are bounded below and $^2v^{\hat\pi} < \infty$ for a strategy $\hat\pi$ providing $\min_\pi {}^1v^\pi$, the dual functional is also continuous at $\lambda = 0$: the proof of Lemma 3.5 [Sennott(1991)] also holds in the general situation (not only for countable $\mathbf{X}$).

The following example shows that the *Slater* condition $^2v^{\hat\pi} < d$ is also important in Proposition 3.1.

Let $\mathbf{X} = [1,\infty) \cup \{0\}$; $\mathbf{A} = [1,\infty) \cup \{0\}$; $p(\Gamma^{\mathbf{X}}|0,a) = I\{a \in \Gamma^{\mathbf{X}}\}$, $p(\Gamma^{\mathbf{X}}|x,a) \equiv I\{x \in \Gamma^{\mathbf{X}}\}$;

$$^1c(x,a) = \begin{cases} 2, & \text{if } x = 0,\ a = 0; \\ 0, & \text{if } x = 0,\ a \geq 1; \\ 1 - \frac{1}{x}, & \text{if } x \geq 1, \end{cases} \qquad ^2c(x,a) = \begin{cases} 0, & \text{if } x = 0; \\ \frac{1}{x^2}, & \text{if } x \geq 1, \end{cases}$$

$\beta = 1/2$, $P_0(\{0\}) = 1$ and $d = 0$ (see Fig. 3.16).

Only the actions in state $x = 0$ play any role, and only the strategies which apply action $a = 0$ in state $x = 0$ are admissible. When calculating the dual functional g, without loss of generality, we need consider only stationary selectors φ or, to be more precise, only the values $\varphi(0) = a$. Now

$$L(\varphi,\lambda) = {}^1v^\varphi + \lambda\, {}^2v^\varphi = \begin{cases} 4, & \text{if } a = 0; \\ 1 - \frac{1}{a} + \frac{\lambda}{a^2}, & \text{if } a \geq 1 \end{cases}$$

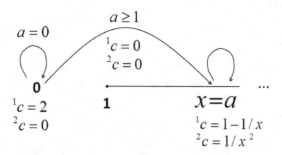

Fig. 3.16 Example 3.2.14: no optimal values of the Lagrange multiplier λ.

and

$$g(\lambda) = \inf_{\varphi} L(\varphi, \lambda) = \begin{cases} \lambda, & \text{if } \lambda < \frac{1}{2}; \\ 1 - \frac{1}{4\lambda}, & \text{if } \lambda \geq \frac{1}{2} \end{cases}$$

(see Fig. 3.17). The optimal value of $a = \varphi(0)$ providing the infimum equals $a^* = \begin{cases} 1, & \text{if } \lambda < \frac{1}{2}; \\ 2\lambda, & \text{if } \lambda \geq \frac{1}{2}. \end{cases}$ Obviously, $\sup_{\lambda \geq 0} g(\lambda) = 1$, but there is no one λ that provides this supremum. Proposition 3.1 fails to hold because the Slater condition is violated.

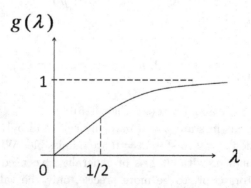

Fig. 3.17 Example 3.2.14: dual functional.

3.2.15 Constrained optimization: multiple solutions

Consider again the constrained problem (3.11). Section 3.2.13 demonstrated that, here, the Bellman principle can fail to hold. Below, we show that a solution to a simple constrained problem can be obtained by uncountably many different stationary strategies. Incidentally, in terms of occupation measures, the constrained problem

$$\int_{\mathbf{X}\times\mathbf{A}} {}^1c(x,a)d\eta^\pi(x,a) \to \inf_\pi \quad \int_{\mathbf{X}\times\mathbf{A}} {}^2c(x,a)d\eta^\pi(x,a) \leq (1-\beta)d$$

usually has a solution η^π which is a non-extreme occupation measure, and thus is generated by many different control strategies: see the end of Section 3.2.12.

Let $\mathbf{X} = \{0,1,2,\ldots\}$, $\mathbf{A} = \{0,1\}$, $p(x+1|x,a) \equiv 1$, with all other transition probabilities zero. We put ${}^1c(x,a) = a$, ${}^2c(x,a) = 1-a$, $\beta = 1/4$, $P_0(0) = 1$ (see Fig. 3.18). Clearly, ${}^1v^\pi = \frac{1}{1-\beta} - {}^2v^\pi$.

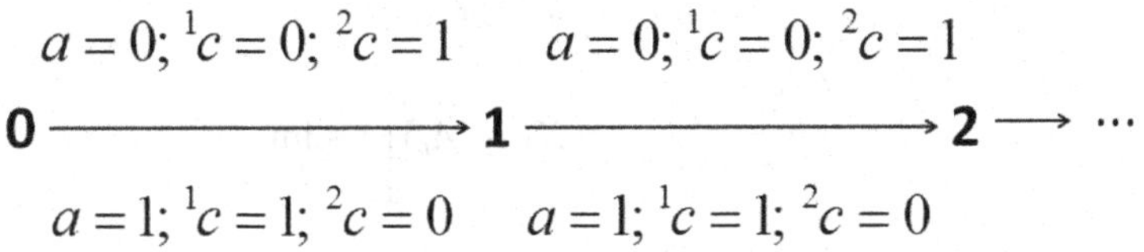

Fig. 3.18 Example 3.2.15: continuum constrained-optimal strategies.

Consider the constrained problem

$${}^1v^\pi \to \inf_\pi \quad {}^2v^\pi \leq 1/2.$$

Any strategy for which ${}^2v^\pi = 1/2$ is optimal, and we intend to build uncountably many stationary optimal strategies.

Let $M \subseteq \{1,2,\ldots\}$ be a non-empty set, put $q_M \triangleq \sum_{i\in M} \beta^i \leq 1/3$, and consider the following stationary strategy:

$$\pi^{(M)}(a|x) \triangleq \begin{cases} \frac{1}{2} - q_M, & \text{if } x=0,\ a=0; \\ \frac{1}{2} + q_M, & \text{if } x=0,\ a=1; \\ 1, & \text{if } x \in M \text{ and } a=0 \\ & \text{or if } x \notin M,\ x \neq 0 \text{ and } a=1. \end{cases}$$

Clearly,

$${}^2v^{\pi^{(M)}} = \frac{1}{2} - q_M + \sum_{t=2}^\infty \beta^{t-1} I\{X_{t-1} \in M\} = \frac{1}{2} - q_M + q_M = \frac{1}{2},$$

and all these strategies $\pi^{(M)}$ are optimal.

This example is similar to one developed by J.Gonzalez-Hernandez (unpublished communication).

3.2.16 Weighted discounted loss and (N,∞)-stationary selectors

Suppose there are two discount factors $1 > \beta_1 > \beta_2 > 0$ and two loss functions $^1c(x,a)$, $^2c(x,a)$, and consider the performance functional

$$v^\pi = E^\pi_{P_0}\left[\sum_{t=1}^\infty \{\beta_1^{t-1} \cdot {}^1c(X_{t-1}, A_t) + \beta_2^{t-1} \cdot {}^2c(X_{t-1}, A_t)\}\right] \to \inf_\pi. \tag{3.13}$$

If the model is finite then the following reasoning helps to solve problem (3.13).

Since $\beta_1 > \beta_2$, the main impact at big values of t comes from the first term ^1c, meaning that, starting from some $t = N \geq 1$, the optimal strategy (stationary selector ψ) is one that solves the problem

$$E^\pi_{P_0}\left[\sum_{t=1}^\infty \beta_1^{t-1} \cdot {}^1c(X_{t-1}, A_t)\right] \to \inf_\pi.$$

(If there are several different optimal selectors then one should choose the one that minimizes the loss of the second type.) As a result, we know the value

$$R(x) = E^\psi_{P_0}\left[\sum_{t=N}^\infty \{\beta_1^{t-1} \cdot {}^1c(X_{t-1}, A_t) + \beta_2^{t-1} \cdot {}^2c(X_{t-1}, A_t)\} \mid X_{N-1} = x\right].$$

We still need to solve the (non-homogeneous) finite-horizon problem with the one-step loss $c_t(x,a) = \beta_1^{t-1} \cdot {}^1c(x,a) + \beta_2^{t-1} \cdot {}^2c(x,a)$ and final loss $R(x)$, which leads to a Markov selector $\varphi_t(x)$ in the decision epochs $t \in \{1, 2, \ldots, N-1\}$. The resulting selector

$$\varphi_t^*(x) = \begin{cases} \varphi_t(x), & \text{if } t < N; \\ \psi(x), & \text{if } t \geq N \end{cases} \tag{3.14}$$

solves problem (3.13). More about the weighted discounted criteria can be found in [Feinberg and Shwartz(1994)].

Definition 3.4. Selectors looking like (3.14) with finite N are called (N,∞)-stationary.

The following examples, based on [Feinberg and Shwartz(1994)], show that this method does not work if the model is not finite.

Let $\mathbf{X} = \{0\}$, $\mathbf{A} = [0,1]$, $\beta_1 = 1/2$, $^1c(x,a) = a^2$, $\beta_2 = 1/4$, $^2c(x,a) = a^2 - a$. The only optimal strategy is

$$\varphi_t^*(x) = \frac{1}{2 + 2^t}$$

which minimizes the one-step loss $(1/2)^{t-1}a^2 + (1/4)^{t-1}(a^2 - a)$, so that an (N,∞)-stationary optimal selector does not exist.

Let $\mathbf{X} = \{0, 1, 2, \ldots\} \cup \{\Delta\}$, $\mathbf{A} = \{1, 2\}$, $P_0(0) = 1$,

$$p(i|0, a) = (1/2)^{i+1} \text{ for all } i \geq 0, \ a \in \mathbf{A},$$

$$p(\Delta|\Delta, a) = p(\Delta|i, a) \equiv 1 \text{ for all } i \geq 1,$$

with all other transition probabilities zero. We put $\beta_1 = 1/2$, $\beta_2 = 1/4$, $^1c(0,a) = {}^2c(0,a) = {}^1c(\Delta,a) = {}^2c(\Delta,a) \equiv 0$. For $i \geq 1$,

$$^1c(i,a) = \begin{cases} 0; & \text{if } a = 1; \\ 3 \cdot 2^{-i}, & \text{if } a = 2 \end{cases} \quad {}^2c(i,a) = \begin{cases} 1, & \text{if } a = 1; \\ 0, & \text{if } a = 2 \end{cases}$$

(see Fig. 3.19).

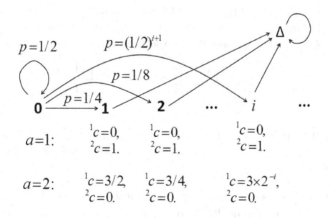

Fig. 3.19 Example 3.2.16: no (N, ∞)-stationary optimal selectors.

Note that the process can take any value $i \geq 1$ at any moment $t \geq 1$, and the loss equals

$$\beta_1^{t-1} \cdot {}^1c(i,a) + \beta_2^{t-1} \cdot {}^2c(i,a) = \begin{cases} (1/2)^{2(t-1)}, & \text{if } a = 1; \\ 3 \cdot (1/2)^{t+i-1}, & \text{if } a = 2. \end{cases}$$

Therefore, the optimal strategy (Markov selector) in states $x \geq 1$ is unique and is defined by
$$\varphi_t^* = \begin{cases} 1, & \text{if } x - t + 1 + \log_2 3 < 0; \\ 2, & \text{if } x - t + 1 + \log_2 3 > 0. \end{cases}$$
There is no $N < \infty$ such that there exists an optimal (N, ∞)-stationary selector.

3.2.17 *Non-constant discounting*

A natural generalization of the MDP with discounted loss (3.1) is as follows:
$$v^\pi = E_{P_0}^\pi \left[\sum_{t=1}^\infty f(t) c(X_{t-1}, A_t) \right] \to \inf_\pi, \qquad (3.15)$$
where function f decreases to zero rapidly enough.

Definition 3.5. A function $f : \mathbb{N}_0 \to \mathbb{R}$ is called *exponentially representable* if there exist sequences $\{d_k\}_{k=1}^\infty$ and $\{\gamma_k\}_{k=1}^\infty$ such that $\{\gamma_k\}_{k=1}^\infty$ is positive, strictly decreasing and $\gamma_1 < 1$; $f(t) = \sum_{k=1}^\infty d_k \gamma_k^t$, and the sum converges absolutely after some $N < \infty$.

The standard case, when $f(t) = \beta^{t-1}$, corresponds to $d_1 = 1/\beta$, $d_k = 0$ for $k \geq 1$, $\gamma_1 = \beta$, and $\{\gamma_k\}_{k=2}^\infty$ is a sufficiently arbitrary positive decreasing sequence. If function f is exponentially representable, then (3.15) is an infinite version of the weighted discounted loss considered in Section 3.2.16:
$$v^\pi = E_{P_0}^\pi \left[\sum_{t=1}^\infty \sum_{k=1}^\infty \beta_k^t d_k c(X_{t-1}, A_t) \right]$$
(here $\gamma_k = \beta_k$).

Theorem 3.2. *[Carmon and Shwartz(2009), Th. 1.5]. If the model is finite and function f is exponentially representable, then there exists an optimal (N, ∞)-stationary selector.*

If function f is not exponentially representable, then this statement can be false even if f monotonically decreases to zero, as the following example confirms.

Suppose $f(t) = \beta^{t-1} \cdot h(t)$ with $\beta = 1/4$, $h(t) = \begin{cases} 1, & \text{if } t \text{ is odd}; \\ 1/2, & \text{if } t \text{ is even}. \end{cases}$
This function $f(t)$ is not exponentially representable, because the necessary condition [Carmon and Shwartz(2009), Lemma 3.1.]
$$\exists \gamma \in (0,1): \quad \lim_{t \to \infty} \gamma^{-t} f(t) = c \neq 0 \text{ and } c < \infty$$

does not hold.

Let $\mathbf{X} = \{1,2,3\}$, $\mathbf{A} = \{0,1\}$, $p(1|1,a) = a$, $p(2|1,a) = 1-a$, $p(3|2,a) = p(1|3,a) \equiv 1$, with other transition probabilities zero. We put $c(1,a) \equiv 1$, $c(2,a) \equiv 5/4$, $c(3,a) \equiv 0$ (see Fig. 3.20). Note that the transitions are deterministic.

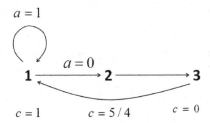

Fig. 3.20 Example 3.2.17: non-constant discounting.

To solve problem (3.15), we deal with a standard discounted MDP where the one-step loss depends on the value of $h(t)$; it is reasonable to extend the states by incorporating the values of $h(t) \in \{1, 1/2\}$. Thus, we put

$$\tilde{\mathbf{X}} = \{(1,1),\ (1,1/2),\ (2,1),\ (2,1/2),\ (3,1),\ (3,1/2)\},$$

the action space $\mathbf{A}$ remains the same,

$$\tilde{p}((1,1/2)|(1,1),a) = \tilde{p}((1,1)|(1,1/2),a) = a,$$
$$\tilde{p}((2,1/2)|(1,1),a) = \tilde{p}((2,1)|(1,1/2),a) = 1-a,$$
$$\tilde{p}((3,1/2)|(2,1),a) = \tilde{p}((3,1)|(2,1/2),a)$$
$$= \tilde{p}((1,1/2)|(3,1),a) = \tilde{p}((1,1)|(3,1/2),a) \equiv 1,$$

and other transition probabilities in the auxiliary tilde-model are zero. Finally,

$$\tilde{c}((1,1),a) \equiv 1, \quad \tilde{c}((1,1/2),a) \equiv 1/2, \quad \tilde{c}((2,1),a) \equiv 5/4,$$
$$\tilde{c}((2,1/2),a) \equiv 5/8, \quad \tilde{c}((3,1),a) = \tilde{c}((3,1/2).a) \equiv 0$$

(see Fig. 3.21).

The optimality equation (3.2) is given by

$$v(1,1) = \min\{1 + \beta v(2,1/2);\ 1 + \beta v(1,1/2)\};$$
$$v(1,1/2) = \min\{1/2 + \beta v(2,1);\ 1/2 + \beta v(1,1)\};$$
$$v(2,1) = 5/4 + \beta v(3,1/2); \quad v(2,1/2) = 5/8 + \beta v(3,1);$$
$$v(3,1) = \beta v(1,1/2); \quad v(3,1/2) = \beta v(1,1)$$

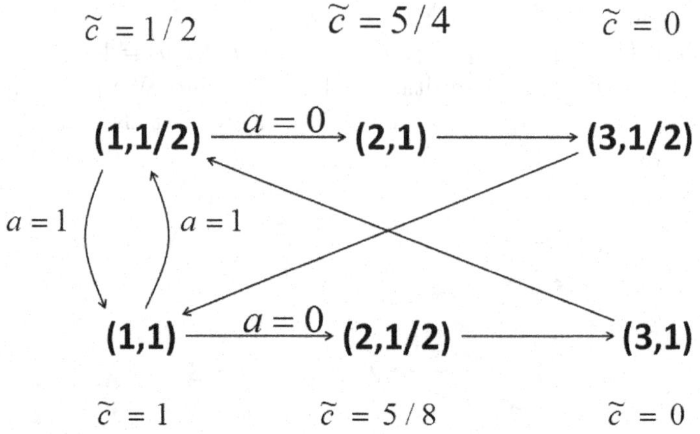

$$\tilde{c} = 1/2 \qquad \tilde{c} = 5/4 \qquad \tilde{c} = 0$$

$$(1,1/2) \xrightarrow{a=0} (2,1) \longrightarrow (3,1/2)$$

$a=1$ ⟲ $a=1$

$$(1,1) \xrightarrow{a=0} (2,1/2) \longrightarrow (3,1)$$

$$\tilde{c} = 1 \qquad \tilde{c} = 5/8 \qquad \tilde{c} = 0$$

Fig. 3.21 Example 3.2.17: non-constant discounting, auxiliary model.

and has the solution

$$v(1,1) = \frac{298}{255}; \qquad v(1,1/2) = \frac{202}{255};$$

$$v(2,1) = \frac{2699}{2040}; \qquad v(2,1/2) = \frac{172}{255};$$

$$v(3,1) = \frac{101}{510}; \qquad v(3,1/2) = \frac{149}{510}.$$

In state $(1,1)$, action $a = 0$ is optimal; in state $(1,1/2)$, action $a = 1$ is optimal. Therefore, if the initial state is $(1,1)$, corresponding to the initial state $x = 1$ in the original model, then only the following sequences of actions are optimal in the both models:

t	1	2	3	4	5	6
State of the auxiliary tilde-model	$(1,1)$	$(2,\frac{1}{2})$	$(3,1)$	$(1,\frac{1}{2})$	$(1,1)$	$(2,\frac{1}{2})$
State of the original model	1	2	3	1	1	2
action	0	0 or 1	0 or 1	1	0	0 or 1

In the original model, in state 1, the optimal actions always switch from 1 to 0, and there exists no (N, ∞)-stationary optimal selector.

3.2.18 *The nearly optimal strategy is not Blackwell optimal*

One of the approaches to problem (2.1) is the study of discounted problem (3.1), letting $\beta \to 1-$.

Definition 3.6. [Kallenberg(2010), Section 4.1], [Puterman(1994), Section 5.4.3]. A strategy π^* is called *Blackwell optimal* if $v_x^{\pi^*,\beta} = v_x^{*,\beta}$ for all $x \in \mathbf{X}$ and all $\beta \in [\beta_0, 1) \neq \emptyset$.

If the model is finite then a Blackwell optimal strategy does exist and can be found in the form of a stationary selector [Puterman(1994), Th. 10.1.4].

Definition 3.7. [Blackwell(1962), p. 721] A strategy π^* is called *nearly optimal* if
$$\lim_{\beta \to 1-} [v_x^{\pi^*,\beta} - v_x^{*,\beta}] = 0.$$

A nearly optimal strategy is also optimal in problem (2.1) (under Condition 2.1):
$$\forall x \in \mathbf{X}\ \forall \pi\ v_x^\pi = \lim_{\beta \to 1-} v_x^{\pi,\beta} \geq \lim_{\beta \to 1-} v_x^{*,\beta} = \lim_{\beta \to 1-} v_x^{\pi^*,\beta} = v_x^{\pi^*}.$$

Any Blackwell optimal strategy is obviously nearly optimal. The converse is not true, even in finite models, as the following example shows (see also [Blackwell(1962), Ex. 1]).

Let $\mathbf{X} = \{0, 1\}$, $\mathbf{A} = \{1, 2\}$, $p(1|1,1) = p(0|1,1) = 1/2$, $p(0|1,2) = 1$, $p(0|0,a) \equiv 1$, $c(1,a) = -a$, $c(0,a) \equiv 0$ (see Fig. 3.22).

The optimality equation (3.2) is given by
$$v^\beta(0) = \beta v^\beta(0),$$
$$v^\beta(1) = \min\{-1 + \frac{1}{2}\beta(v^\beta(0) + v^\beta(1));\ -2 + \beta v^\beta(0)\},$$

so that, for any $\beta \in (0,1)$, for stationary selector $\varphi^2(x) \equiv 2$ we have $v^\beta(0) = v_0^{*,\beta} = v_0^{\varphi^2,\beta} = 0$, $v^\beta(1) = v_1^{*,\beta} = v_1^{\varphi^2,\beta} = -2$, so that φ^2 is uniformly optimal and Blackwell optimal. At the same time, for selector $\varphi^1(x) \equiv 1$ we have
$$v_0^{\varphi^1,\beta} = 0,\quad v_1^{\varphi^1,\beta} = -\frac{2}{2-\beta},$$

so that φ^1 is nearly optimal, but certainly not Blackwell optimal. Both selectors φ^1 and φ^2 are uniformly optimal in problem (2.1).

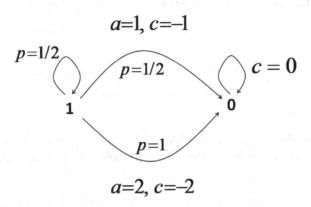

Fig. 3.22 Example 3.2.18: the nearly optimal strategy is not Blackwell optimal.

3.2.19 Blackwell optimal strategies and opportunity loss

If the time horizon T is finite, one can evaluate a control strategy π with the opportunity loss (or regret) $E_x^\pi \left[\sum_{t=1}^T c(X_{t-1}, A_t) \right] - V_x^T$, where as usual $V_x^T = \inf_\pi E_x^\pi \left[\sum_{t=1}^T c(X_{t-1}, A_t) \right]$. In the infinite-horizon models, the goal may be to find a strategy that solves the problem

$$\limsup_{T \to \infty} \left\{ E_x^\pi \left[\sum_{t=1}^T c(X_{t-1}, A_t) \right] - V_x^T \right\} \to \inf_\pi \text{ for all } x \in \mathbf{X}. \qquad (3.16)$$

The following example, based on [Flynn(1980), Ex. 3], shows that it can easily happen that a Blackwell optimal strategy does not solve problem (3.16) and, *vice versa*, a strategy minimizing the opportunity loss may not be Blackwell optimal.

Let $\mathbf{X} = \{0, 1, 2, 2', 3, 3'\}$, $\mathbf{A} = \{1, 2, 3\}$, $p(1|0, 1) = p(2|0, 2) = p(3|0, 3) = 1$, $p(1|1, a) \equiv 1$, $p(2'|2, a) = p(2|2', a) \equiv 1$, $p(3'|3, a) = p(3|3', a) \equiv 1$, $c(0, 1) = 1/2$, $c(0, 2) = -1$, $c(0, 3) = 1$, $c(2, a) \equiv 2$, $c(2', a) \equiv -2$, $c(3, a) \equiv -2$, $c(3', a) \equiv 2$ (see Fig. 3.23).

We only need study state 0, and only the stationary selectors $\varphi^1(x) \equiv 1$, $\varphi^2(x) \equiv 2$ and $\varphi^3(x) \equiv 3$ need be considered.

Since

$$v_0^{\varphi^1, \beta} = 1/2, \quad v_0^{\varphi^2, \beta} = -1 + \frac{2\beta}{1+\beta}, \quad v_0^{\varphi^3, \beta} = 1 - \frac{2\beta}{1+\beta},$$

only the stationary selector φ^2 is Blackwell optimal. At the same time, the

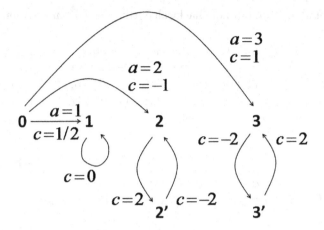

Fig. 3.23 Example 3.2.19: opportunity loss versus Blackwell optimality.

left-hand side of (3.16) equals 3/2, 2 and 2 at φ^1, φ^2 and φ^3 correspondingly, so that only the selector φ^1 solves problem (3.16).

Note that all three selectors φ^1, φ^2 and φ^3 are equally AC-optimal (see Chapter 4).

3.2.20 Blackwell optimal and n-discount optimal strategies

Definition 3.8. [Puterman(1994), Section 5.4.3] A strategy π^* is n-discount optimal for some $n \geq -1$ if

$$\limsup_{\beta \to 1-} (1-\beta)^{-n} [v_x^{\pi^*,\beta} - v_x^{\pi,\beta}] \leq 0$$

for all $x \in \mathbf{X}$, π.

Remark 3.2. Since

$$\limsup_{\beta \to 1-}[v_x^{\pi^*,\beta} - v_x^{\pi,\beta}] \leq \limsup_{\beta \to 1-}[v_x^{\pi^*,\beta} - v_x^{*,\beta}] + \limsup_{\beta \to 1-}[v_x^{*,\beta} - v_x^{\pi,\beta}],$$

we conclude that any nearly optimal strategy is 0-discount optimal. The converse is not true: see Section 4.2.27.

If π^* is a Blackwell optimal strategy then it is n-discount optimal for any $n \geq -1$ [Puterman(1994), Th. 10.1.5]. In finite models, the converse is also true: if a strategy is n-discount optimal for any $n \geq -1$ then it is Blackwell optimal [Puterman(1994), Section 5.4.3]. The following example, similar to [Puterman(1994), Ex. 10.1.1], shows that a strategy can be n-discount optimal for all $n < m$, but not Blackwell optimal and not n-discount optimal for $n \geq m$.

Let $\mathbf{X} = \{0, 1, 2, \ldots, m+1\}$, $\mathbf{A} = \{1, 2\}$, $p(1|0, 1) = 1$, $p(i+1|i, a) \equiv 1$ for all $i = 1, 2, \ldots, m$, $p(m+1|m+1, a) \equiv 1$, $p(m+1|0, 2) = 1$, with other transition probabilities zero. We put $c(0, 2) = 0$, $c(0, 1) = 1$, $c(i, a) \equiv (-1)^i \frac{m!}{i!(m-i)!}$ for $i = 1, 2, \ldots, m$; $c(m+1, a) = 0$. Fig. 3.24 illustrates the case $m = 4$.

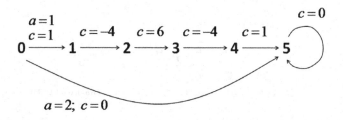

Fig. 3.24 Example 3.2.20: a 3-discount optimal strategy is not Blackwell optimal ($m = 4$).

In fact, here there are only two essentially different strategies (stationary selectors) $\varphi^1(x) \equiv 1$ and $\varphi^*(2) \equiv 2$ (actions in states $1, 2, \ldots, m+1$ play no role).

$$v_{m+1}^{\pi,\beta} \equiv v_{m+1}^{*,\beta} = 0, \quad v_m^{\pi,\beta} \equiv v_m^{*,\beta} = (-1)^m,$$
$$v_{m-1}^{\pi,\beta} \equiv v_{m-1}^{*,\beta} = (-1)^{m-1}m + (-1)^m \beta,$$
$$v_{m-2}^{\pi,\beta} \equiv v_{m-2}^{*,\beta} = (-1)^{m-2}\frac{m(m-1)}{2} + (-1)^{m-1}m\beta + (-1)^m \beta^2, \ldots,$$
$$v_1^{\pi,\beta} \equiv v_1^{*,\beta} = -m + \frac{m(m-1)}{2}\beta + \cdots + (-1)^{m-2}\frac{m(m-1)}{2}\beta^{m-3}$$
$$+ (-1)^{m-1}m\beta^{m-2} + (-1)^m \beta^{m-1},$$
$$v_0^{\varphi^1,\beta} = (1-\beta)^m, \quad v_0^{\varphi^2,\beta} = 0.$$

Therefore, φ^2 is Blackwell optimal, and φ^1 is not. At the same time, φ^1 is n-discount optimal for all $n = -1, 0, 1, 2, \ldots, m-1$:

$$\lim_{\beta \to 1-} (1-\beta)^{-n}[v_0^{\varphi^1,\beta} - v_0^{\pi,\beta}] \le \lim_{\beta \to 1-} (1-\beta)^{-n}[v_0^{\varphi^1,\beta}] = \begin{cases} 0, & \text{if } n < m; \\ 1, & \text{if } n = m; \\ \infty, & \text{if } n > m. \end{cases}$$

The next example shows that, if the model is not finite, then a strategy which is not Blackwell optimal can still be n-discount optimal for any $n \ge -1$.

Let $\mathbf{X} = \{\Delta, 0, 1, 2, \ldots\}$, $\mathbf{A} = \{1, 2\}$, $p(1|0,1) = 1$, $p(i+1|i,a) \equiv 1$ for all $i = 1, 2, \ldots$, $p(\Delta|0,2) = 1$, $p(\Delta|\Delta,a) \equiv 1$, with other transition probabilities zero. We put $c(0,2) = 0$, $c(\Delta, a) \equiv 0$, $c(0,1) = 1/e \stackrel{\Delta}{=} C_0$, $c(i,a) \equiv C_i$, where C_i is the ith coefficient in the Taylor expansion

$$g(\beta) = e^{-\frac{1}{(1-\beta)^2}} = \sum_{j=0}^{\infty} C_j \beta^j$$

(see Fig. 3.25). Since the function g is holomorphic everywhere except for the unique singular point $\beta = 1$, this series converges absolutely for all $\beta \in [0, 1)$ [Priestley(1990), Taylor's Theorem, p. 69].

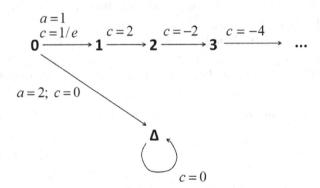

Fig. 3.25 Example 3.2.20: a strategy that is n-discount optimal for all $n \ge -1$ is not Blackwell optimal.

Here again we have only two essentially different strategies (stationary selectors) $\varphi^1(x) \equiv 1$ and $\varphi^2(x) \equiv 2$ (actions in states $\Delta, 1, 2, \ldots$ play no role).

$$v_\Delta^{\pi,\beta} \equiv v_\Delta^{*,\beta} = 0,$$

$$v_i^{\pi,\beta} \equiv v_i^{*,\beta} = \sum_{j=i}^{\infty} C_j \beta^{j-i} \quad \text{for } i = 1, 2, \ldots$$

$$v_0^{\varphi^1,\beta} = \sum_{j=0}^{\infty} C_j \beta^j = g(\beta) = e^{-\frac{1}{(1-\beta)^2}}, \quad v_0^{\varphi^2,\beta} = 0.$$

Therefore, φ^2 is Blackwell optimal, and φ^1 is not. At the same time, φ^1 is n-discount optimal for all $n = -1, 0, 1, 2, \ldots$:

$$\lim_{\beta \to 1-} (1-\beta)^{-n} [v_0^{\varphi^1,\beta} - v_0^{\pi,\beta}] \leq \lim_{\beta \to 1-} (1-\beta)^{-n} [v_0^{\varphi^1,\beta}] = 0.$$

3.2.21 No Blackwell (Maitra) optimal strategies

Maitra (1965) suggested the following definition, similar to, but weaker than, the Blackwell optimality.

Definition 3.9. [Maitra(1965)] A strategy π^* is (*Maitra*) optimal if, for any strategy π, for each $x \in \mathbf{X}$ there exists $\beta_0(x, \pi) \in (0, 1)$ such that $v_x^{\pi^*,\beta} \leq v_x^{\pi,\beta}$ for all $\beta \in [\beta_0(x, \pi), 1)$.

Counterexample 1 in [Hordijk and Yushkevich(2002)] confirms that, if the state and action spaces are countable, then the Blackwell optimality is stronger than the Maitra optimality.

The following example, based on [Maitra(1965), p. 246], shows that, if $\mathbf{X}$ is not finite, a Maitra optimal strategy may not exist.

Let $\mathbf{X} = \{1, 2, \ldots\}$, $\mathbf{A} = \{0, 1\}$, $p(i|i, 0) \equiv 1$, $p(i+1|i, 1) \equiv 1$, with all other transition probabilities zero; $c(i, 0) = C_i < 0$, $c(i, 1) \equiv 0$, and the sequence $\{C_i\}_{i=1}^{\infty}$ is strictly decreasing, $\lim_{i \to \infty} C_i = C$ (see Fig. 3.26).

The optimality equation (3.2) is given by

$$v^\beta(i) = \min\{C_i + \beta v^\beta(i); \quad \beta v^\beta(i+1)\}, \quad i = 1, 2, \ldots$$

Let $l(i) = \operatorname{argmin}_{k \geq i}\{\beta^{k-i} C_k\}$; if the minimum is provided by several values of k, take, say, the maximal one. To put it differently, $l(i) = \min\{j \geq i : C_j < \beta^k C_{j+k} \text{ for all } k \geq 1\}$. Note that $l(i) < \infty$ because $\lim_{k \to \infty} \beta^{k-i} C_k = 0$ and $C_k < 0$; obviously, the sequence $l(i)$ is not decreasing.

Now one can check that the only bounded solution to the optimality equation is given by

$$v^\beta(i) = \beta^{l(i)-i} \frac{C_{l(i)}}{1-\beta}.$$

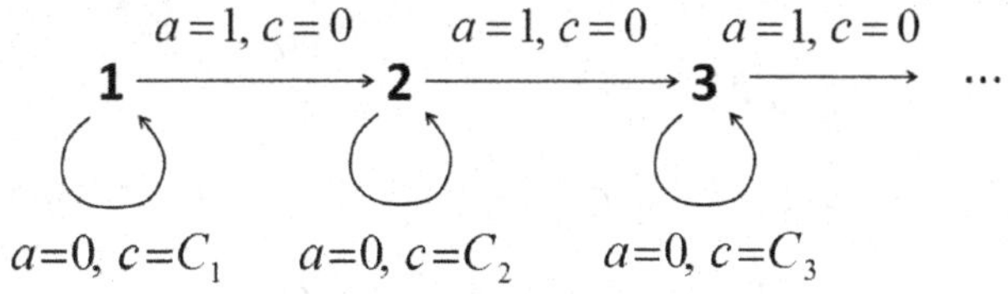

Fig. 3.26 Example 3.2.21: no Maitra optimal strategies.

Indeed, if $l(i) < l(i+1)$ then $l(i) = i$ and

$$v^\beta(i) = C_i + \beta v^\beta(i) = \frac{C_i}{1-\beta} < \frac{\beta^{k-i} C_k}{1-\beta}$$

for all $k > i$. In particular, for $k = l(i+1)$ we have

$$v^\beta(i) < \beta^{l(i+1)-i} \frac{C_{l(i+1)}}{1-\beta} = \beta v^\beta(i+1).$$

In the case where $l(i) = l(i+1)$ we have $l(i) > i$,

$$v^\beta(i) = \beta v^\beta(i+1) = \frac{\beta^{l(i)-i} C_{l(i)}}{1-\beta},$$

and

$$C_i + \beta v^\beta(i) \geq \beta^{l(i)-i} C_{l(i)} + \frac{\beta}{1-\beta} \beta^{l(i)-i} C_{l(i)} = v^\beta(i).$$

The Bellman function $v_i^{*,\beta}$ coincides with $v^\beta(i)$ [Bertsekas and Shreve(1978), Prop. 9.14].

Suppose now that π^* is a Maitra optimal strategy in this model, and fix an initial state $i \in \mathbf{X}$. Without loss of generality, we assume that π^* is Markov (see Lemma 3.1). Let $\varphi^0(x) \equiv 0$ and $\varphi^1(x) \equiv 1$ and consider the strategies

$$\pi^1 \triangleq \{\varphi^1, \ldots, \varphi^1, \pi_1^*, \pi_2^*, \ldots\} \quad (n \text{ copies of } \varphi^1); \quad \pi^0 \triangleq \{\varphi^1, \pi_1^*, \pi_2^*, \ldots\}:$$

the controls are initially deterministic and switch to π^* afterwards. For all β sufficiently close to 1 we have

$$v_i^{\pi^*,\beta} \leq v_i^{\pi^1,\beta} = \beta^n v_{i+n}^{\pi^*,\beta}$$

and

$$v_{i+n}^{\pi^*,\beta} \leq v_{i+n}^{\pi^0,\beta} = C_{i+n} + \beta v_{i+n}^{\pi^*,\beta},$$

so that
$$v_{i+n}^{\pi^*,\beta} \leq \frac{C_{i+n}}{1-\beta}; \qquad v_i^{\pi^*,\beta} \leq \frac{\beta^n C_{i+n}}{1-\beta}$$

and $\limsup_{\beta \to 1-}(1-\beta)v_i^{\pi^*,\beta} \leq C_{i+n}$ for any $n = 1, 2, \ldots$ Therefore,

$$\limsup_{\beta \to 1-} (1-\beta)v_i^{\pi^*,\beta} \leq C.$$

On the other hand, $v_i^{\pi^*,\beta} \geq \frac{C}{1-\beta}$ for all β, and therefore

$$\liminf_{\beta \to 1-}(1-\beta)v_i^{\pi^*,\beta} \geq C \implies \lim_{\beta \to 1-}(1-\beta)v_i^{\pi^*,\beta} = C \quad \text{for all } i.$$

Since $v_1^{\varphi^1,\beta} \equiv 0$ and $C_j < 0$, the selector φ^1 cannot be Maitra optimal. Therefore, at some decision epoch $T \geq 1$, $p = \pi_T^*(0|x) > 0$ at $x = T$. We assume that T is the minimal value: starting from initial state 1, the selector φ^1 is applied $(T-1)$ times, and in state T there is a positive probability p of using action $a = 0$.

Consider the Markov strategy π which differs from the strategy π^* at epoch T only: $\pi_T(0|x) \equiv 0$. Now

$$v_T^{\pi,\beta} = v_{T+1}^{\pi^*,\beta}; \qquad v_T^{\pi^*,\beta} = p\left[C_T + \beta v_T^{\pi^*,\beta}\right] + (1-p)\beta v_{T+1}^{\pi^*,\beta}$$

and

$$v_1^{\pi,\beta} = \beta^T v_{T+1}^{\pi^*,\beta}; \qquad v_1^{\pi^*,\beta} = \beta^{T-1}v_T^{\pi^*,\beta} = \beta^{T-1} \cdot \frac{pC_T + (1-p)\beta v_{T+1}^{\pi^*,\beta}}{1-p\beta}.$$

Consider the difference

$$\delta \triangleq v_1^{\pi^*,\beta} - v_1^{\pi,\beta} = \left[\beta^{T-1}C_T - \beta^T v_{T+1}^{\pi^*,\beta}(1-\beta)\right]\frac{p}{1-p\beta}.$$

If $p < 1$ then $\lim_{\beta \to 1-} \delta = \frac{p}{1-p}[C_T - C] > 0$, and the strategy π^* is not Maitra optimal. When $p = 1$,

$$(1-\beta)\delta = \beta^{T-1}C_T - \beta^T v_{T+1}^{\pi^*,\beta}(1-\beta)$$

and $\lim_{\beta \to 1-}(1-\beta)\delta = C_T - C > 0$, meaning that δ is again positive for all β close enough to 1.

Therefore, the Maitra optimal strategy does not exist. A Blackwell optimal strategy does not exist either. A sufficient condition for the existence of a Maitra optimal strategy is given in [Hordijk and Yushkevich(2002), Th. 8.7]. That theorem is not applicable here, because Assumption 8.5 in [Hordijk and Yushkevich(2002)] does not hold.

3.2.22 Optimal strategies as $\beta \to 1-$ and MDPs with the average loss – I

Very often, optimal strategies in MDPs with discounted loss also provide a solution to problems with the expected average loss, that is, with the performance functional

$$\limsup_{T \to \infty} \frac{1}{T} E^\pi \left[\sum_{t=1}^T c(X_{t-1}, A_t) \right] \to \inf_\pi . \qquad (3.17)$$

We need the following:

Condition 3.1. The state space $\mathbf{X}$ is discrete, the action space $\mathbf{A}$ is finite, and $\sup_{(x,a) \in \mathbf{X} \times \mathbf{A}} |c(x,a)| < \infty$.

Theorem 3.3. *[Ross(1983), Chapter V, Th. 2.2] Let Condition 3.1 be satisfied and suppose, for some $N < \infty$,*

$$|v_x^{*,\beta} - v_y^{*,\beta}| < N \quad \text{for all } x, y \in \mathbf{X}. \qquad (3.18)$$

Then there exist a bounded function $h(x)$ on $\mathbf{X}$ and a constant ρ satisfying the equation

$$\rho + h(x) = \min_{a \in \mathbf{A}} \left\{ c(x,a) + \sum_{y \in \mathbf{X}} p(y|x,a) h(y) \right\}. \qquad (3.19)$$

If $z \in \mathbf{X}$ is an arbitrary fixed state, then

$$h(x) = \lim_{n \to \infty} (v_x^{*,\beta_n} - v_z^{*,\beta_n})$$

for some sequence $\beta_n \to 1-$, and $\rho = \lim_{\beta \to 1-} (1-\beta) v_z^{*,\beta}$.

Moreover, ρ is the (initial state-independent) minimal value of the performance functional (3.17), and if the map $\varphi^* : \mathbf{X} \to \mathbf{A}$ provides the minimum in (3.19) then the stationary selector φ^* is the (uniformly) optimal strategy in problem (3.17) [Ross(1983), Chapter V, Th. 2.1].

Under additional conditions, this statement was generalized for arbitrary Borel spaces $\mathbf{X}$ and $\mathbf{A}$ in [Hernandez-Lerma and Lasserre(1996a), Th. 5.5.4]. If there is a Blackwell optimal stationary selector then, under the assumptions of Theorem 3.3, it is optimal in problem (3.17). Indeed that selector provides the minimum in the equation

$$(1-\beta) v_z^{*,\beta} + h^\beta(x) = \min_{a \in \mathbf{A}} \left\{ c(x,a) + \beta \sum_{y \in \mathbf{X}} p(y|x,a) h^\beta(y) \right\}$$

for all β sufficiently close to 1 and hence also in the limiting case $\beta_n \to 1-$. Here $h^\beta(x) \triangleq v_x^{*,\beta} - v_z^{*,\beta}$.

The following simple example shows that condition (3.18) is important even in finite models (see [Hernandez-Lerma and Lasserre(1996b), Ex. 6.1]).

Let $\mathbf{X} = \{1, 2, 3\}$, $\mathbf{A} = \{0\}$ (a dummy action), $p(1|1, a) = 1$, $p(2|2, a) = 1$, $p(1|3, a) = \alpha \in (0, 1)$, $p(2|3, a) = 1 - \alpha$, with all other transition probabilities zero; $c(x, a) = x$ (see Fig. 3.27).

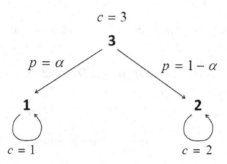

Fig. 3.27 Example 3.2.22: no solutions to equation (3.19).

Obviously,

$$v_1^{*,\beta} = \frac{1}{1-\beta}, \quad v_2^{*,\beta} = \frac{2}{1-\beta}, \quad v_3^{*,\beta} = 3 + \beta[\alpha v_1^{*,\beta} + (1-\alpha)v_2^{*,\beta}] = \frac{3-\beta-\alpha}{1-\beta}.$$

Condition (3.18) is violated and the (minimal) value of the expected average loss (3.17) depends on the initial state x:

$$\rho(1) = 1, \quad \rho(2) = 2, \quad \rho(3) = 2 - \alpha.$$

Equation (3.19) has no bounded solutions, because otherwise the value of ρ would have coincided with the state-independent (minimal) value of the performance functional (3.17) [Ross(1983), Chapter V, Th. 2.1].

3.2.23 Optimal strategies as $\beta \to 1-$ and MDPs with the average loss – II

Another theorem on the approach to MDPs with the expected average loss (3.17) via the vanishing discount factor is as follows [Ross(1968), Th. 3.3].

Theorem 3.4. *Let Condition 3.1 be satisfied. Suppose, for some sequence $\beta_k \to 1-$, there exists a constant $N < \infty$ such that*

$$|v_x^{*,\beta_k} - v_y^{*,\beta_k}| < N \qquad (3.20)$$

for all $k = 1, 2, \ldots$ and all $x, y \in \mathbf{X}$. Let $\varphi^{\beta_k}(x)$ be the uniformly optimal stationary selector in the corresponding discounted MDP. Then there exists a stationary selector solving problem (3.17) which is a limit point of φ^{β_k}. Moreover, for any $\varepsilon > 0$, for large k, the selectors φ^{β_k} are ε-optimal in the sense that

$$\limsup_{T \to \infty} \frac{1}{T} E^{\varphi^{\beta_k}} \left[\sum_{t=1}^T c(X_{t-1}, A_t) \right]$$

$$\leq \inf_\pi \left\{ \limsup_{T \to \infty} \frac{1}{T} E^\pi \left[\sum_{t=1}^T c(X_{t-1}, A_t) \right] \right\} + \varepsilon.$$

The existence of uniformly optimal selectors φ^{β_k} follows from Corollary 9.17.1 in [Bertsekas and Shreve(1978)]. Clearly, if there is a Blackwell optimal strategy then, under the assumptions of Theorem 3.4, it is optimal in problem (3.17). The following example, based on [Ross(1968), p. 417], shows that condition (3.20) is important.

Let $\mathbf{X} = \{(i,j) \in \mathbb{N}_0^2 : 0 \leq j \leq i, i \geq 1\} \cup \{\Delta\}$, $\mathbf{A} = \{1, 2\}$,

$$p((k,j)|(i,0),a) = \begin{cases} 1, & \text{if } a = 1, k = i+1, j = 0 \text{ or} \\ & \text{if } a = 2, k = i, j = 1; \\ 0 & \text{otherwise,} \end{cases}$$

$$p((i, j+1)|(i,j),a) \equiv 1 \text{ if } 0 < j < i,$$

$$p(\Delta|(i,i),a) = p(\Delta|\Delta,a) \equiv 1,$$

with all other transition probabilities zero; $c((i,0),a) \equiv 1$, $c(\Delta, a) \equiv 2$, with other one-step losses zero (see Fig. 3.28).
The optimality equation (3.2) is given by

$v^\beta(\Delta) = 2 + \beta v^\beta(\Delta),$
$v^\beta((i,i)) = \beta v^\beta(\Delta), \quad i = 1, 2, \ldots,$
$v^\beta((i,j)) = \beta v^\beta((i, j+1)), \quad i = 1, 2, \ldots, \quad j = 1, 2, \ldots, i-1,$
$v^\beta((i,0)) = \min\{1 + \beta v^\beta((i+1,0)); \quad 1 + \beta v^\beta((i,1))\}, \quad i = 1, 2, \ldots,$

so that

$$v^\beta(\Delta) = \frac{2}{1 - \beta}; \quad v^\beta((i,j)) = \frac{2\beta^{i+1-j}}{1 - \beta}, \quad i = 1, 2, \ldots, \quad j = 1, 2, \ldots, i,$$

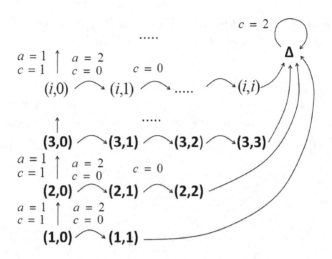

Fig. 3.28 Example 3.2.23: discount-optimal strategies are not ε-optimal in the problem with average loss.

and the last equation takes the form

$$v^\beta((i,0)) = \min\left\{1 + \beta v^\beta((i+1,0));\ 1 + \frac{2\beta^{i+1}}{1-\beta}\right\}. \quad (3.21)$$

Lemma 3.2. *Let* $n \stackrel{\triangle}{=} \min\{k:\ \beta^k(1+\beta) \leq \frac{1}{2}\}$. *Then*

$$v^\beta((i,0)) = \begin{cases} \dfrac{1 - \beta^{n-i+1} + 2\beta^{2n-i+1}}{1-\beta}, & \text{if } i < n; \\[2mm] 1 + \dfrac{2\beta^{i+1}}{1-\beta}, & \text{if } i \geq n, \end{cases}$$

and the stationary selector $\varphi^\beta((i,0)) = \begin{cases} 1, & \text{if } i < n; \\ 2, & \text{if } i \geq n \end{cases}$ *is uniformly optimal. (Actions in other states obviously play no role.)*

The proof is presented in Appendix B. The Bellman function $v_x^{*,\beta}$ coincides with $v^\beta(x)$ [Bertsekas and Shreve(1978), Prop. 9.14].

It is easy to show that condition (3.20) is violated. Let $i_1 < i_2 \in \mathbb{N}$ and put $\delta = i_2 - i_1$. Then, for large enough n (when β is close to 1),

$$v_{(i_1,0)}^{*,\beta} - v_{(i_2,0)}^{*,\beta} = \frac{\beta^{n+1-i_2}}{1-\beta}\{2\beta^{n+\delta} - \beta^\delta + 1 - 2\beta^n\} = \beta^{n+1-i_2}\sum_{k=0}^{\delta-1}\beta^k(1-2\beta^n).$$

We know that $\beta^n \leq \frac{1}{2(1+\beta)}$ and $\beta^{n-1} > \frac{1}{2(1+\beta)}$. Thus

$$v^{*,\beta}_{(i_1,0)} - v^{*,\beta}_{(i_2,0)} > \frac{\beta^{2-i_2}}{2(1+\beta)} \sum_{k=0}^{\delta-1} \beta^k \cdot \frac{\beta}{1+\beta},$$

so that

$$\lim_{\beta \to 1-} |v^{*,\beta}_{(i_1,0)} - v^{*,\beta}_{(i_2,0)}| \geq \frac{\delta}{8},$$

and there does not exist a constant N for which (3.20) holds for any δ.

When $\beta \to 1-$, the selector $\varphi(x) \equiv 1$ is the limit point of $\varphi^\beta(x)$ and is optimal in problem (3.17):

$$\limsup_{T \to \infty} \frac{1}{T} E^\varphi \left[\sum_{t=1}^{T} c(X_{t-1}, A_t) \right] = 1.$$

But, for any $\beta \in (0, 1)$,

$$\limsup_{T \to \infty} \frac{1}{T} E^{\varphi^\beta} \left[\sum_{t=1}^{T} c(X_{t-1}, A_t) \right] = 2,$$

because the chain gets absorbed at state Δ with $c(\Delta, a) \equiv 2$. Thus, discount-optimal selectors are not ε-optimal in problem (3.17) for $\varepsilon > 1$.

In this example, there are no Blackwell optimal strategies because, as β approaches 1, the uniformly optimal selector φ^β essentially changes. (The only flexibility in choosing the optimal actions appears in state n, if $\beta^n(1+\beta) = 1/2$; otherwise, the uniformly optimal strategy is unique: see the proof of Lemma 3.2).

Along with the Blackwell optimality, Maitra gives the following weaker definition (but stronger than the Maitra optimality).

Definition 3.10. [Maitra(1965)] A strategy is called *good* (we try to avoid the over-used term "optimal") if, for each $x \in \mathbf{X}$, there is $\beta_0(x) < 1$ such that $v^{\pi^*,\beta}_x = v^\beta_x$ for all $\beta \in [\beta_0(x), 1)$.

In this example, there are no good strategies, for the same reason as above: the optimal strategy for each initial state does not stop to change when β approaches 1.

Chapter 4

Homogeneous Infinite-Horizon Models: Average Loss and Other Criteria

4.1 Preliminaries

This chapter is about the following problem

$$v^\pi = \limsup_{T\to\infty} \frac{1}{T} E^\pi_{P_0}\left[\sum_{t=1}^T c(X_{t-1}, A_t)\right] \to \inf_\pi. \qquad (4.1)$$

As usual, v^π is called the *performance functional*. Under rather general conditions, problem (4.1) is well defined, e.g. if the loss function c is bounded below. As previously, we write P^π_x and v^π_x, if the initial distribution is concentrated at a single point $x \in \mathbf{X}$. In this connection,

$$v^*_x \stackrel{\triangle}{=} \inf_\pi v^\pi_x,$$

and v^π_x is defined similarly to (4.1). A strategy π^* is uniformly optimal if, for all $x \in \mathbf{X}$, $v^{\pi^*}_x = v^*_x$. In this context, such strategies will be called *AC-optimal*, i.e. average-cost-optimal [Hernandez-Lerma and Lasserre(1999), Section 10.1]. A strategy π is called *AC-ε-optimal* if $v^\pi_x \leq v^*_x + \varepsilon$ for all $x \in \mathbf{X}$, assuming $|v^*_x| < \infty$. If the model is finite then there exists a stationary AC-optimal selector [Puterman(1994), Th. 9.1.8]. The situation becomes more complicated if either space $\mathbf{X}$ or $\mathbf{A}$ is not finite. The dynamic programming approach leads to the following concepts.

Definition 4.1. Let ρ and h be real-valued measurable functions on $\mathbf{X}$, and φ^* a given stationary selector. Then $\langle \rho, h, \varphi^* \rangle$ is said to be a *canonical triplet* if $\forall x \in \mathbf{X}$, $\forall T = 0, 1, 2, \ldots$

$$E^{\varphi^*}_x\left[\sum_{t=1}^T c(X_{t-1}, A_t) + h(X_T)\right] = \inf_\pi E^\pi_x\left[\sum_{t=1}^T c(X_{t-1}, A_t) + h(X_T)\right]$$

$$= T\rho(x) + h(x).$$

Theorem 4.1. *[Arapostatis et al.(1993), Th. 6.2] Suppose the loss function c is bounded. Then the bounded measurable functions ρ and h and the stationary selector φ^* form a canonical triplet if and only if the following canonical equations are satisfied:*

$$\rho(x) = \inf_{a \in A} \int_X \rho(y) p(dy|x, a) = \int_X \rho(y) p(dy|x, \varphi^*(x))$$

$$\rho(x) + h(x) = \inf_{a \in A} \left\{ c(x, a) + \int_X h(y) p(dy|x, a) \right\} \qquad (4.2)$$

$$= c(x, \varphi^*(x)) + \int_X h(y) p(dy|x, \varphi^*(x)).$$

Remark 4.1. If the triplet $\langle \rho, h, \varphi^* \rangle$ solves equations (4.2) then so does the triplet $\langle \rho, h + const, \varphi^* \rangle$ for any value of *const*. Thus one can put $h(\hat{x}) = 0$ for an arbitrarily chosen state $\hat{x}$.

In the case where a stationary selector φ^* is an element of a canonical triplet, it is called *canonical*. Canonical triplets exist if the model is finite [Puterman(1994), Th. 9.1.4]; the corresponding canonical selector is AC-optimal.

Theorem 4.2. *[Hernandez-Lerma and Lasserre(1996a), Th. 5.2.4] Suppose the loss function c is bounded below, and $\langle \rho, h, \varphi^* \rangle$ is a canonical triplet.*

(a) *If, for any π and any $x \in X$,*

$$\lim_{T \to \infty} E_x^\pi [h(X_T)/T] = 0,$$

then φ^* is an AC-optimal strategy and

$$v_x^* = \rho(x) = v_x^{\varphi^*} = \lim_{T \to \infty} \frac{1}{T} E_x^{\varphi^*} \left[\sum_{t=1}^T c(X_{t-1}, A_t) \right]$$

(note the ordinary limit).

(b) *If $\forall x \in X$*

$$\lim_{T \to \infty} \sup_\pi E_x^\pi [h(X_T)/T] = 0$$

then, for all π, $x \in X$

$$v_x^{\varphi^*} \leq \liminf_{T \to \infty} \frac{1}{T} E_x^\pi \left[\sum_{t=1}^T c(X_{t-1}, A_t) \right]$$

and

$$\lim_{T \to \infty} \left\{ E_x^{\varphi^*} \left[\sum_{t=1}^T c(X_{t-1}, A_t) \right] - \inf_\pi E_x^\pi \left[\sum_{t=1}^T c(X_{t-1}, A_t) \right] \right\} / T = 0.$$

In case (b), if the loss function c is bounded, the stationary selector φ^* is optimal in problem (4.1) at any initial distribution P_0.

Sufficient conditions for the existence of canonical triplets (including the case $\rho(x) = const$) can be found in [Hernandez-Lerma and Lasserre(1996a), Section 5.5], [Hernandez-Lerma and Lasserre(1999), Section 10.3] and [Puterman(1994), Sections 8.4, 9.1].

Note also Remark 2.1 about Markov and semi-Markov strategies which concerns the (uniform) AC-optimality.

4.2 Examples

Two examples strongly connected with the discounted MDPs were presented in Sections 3.2.22 and 3.2.23.

4.2.1 Why lim sup?

As mentioned in [Puterman(1994), Section 8.1], one can also consider the following expected average loss criterion:

$$v^\pi = \liminf_{T \to \infty} \frac{1}{T} E_{P_0}^\pi \left[\sum_{t=1}^{T} c(X_{t-1}, A_t) \right] \to \inf_\pi. \tag{4.3}$$

Formula (4.1) corresponds to comparing strategies in terms of the worst-case limiting performance, while (4.3) corresponds to comparison in terms of the best-case performance. From the formal viewpoint, both of these define the maps $v : \mathcal{D} \to \mathbb{R}$ from the strategic measures space $\mathcal{D}$ to real numbers. But the theory of mathematical programming is better developed for the minimization of convex (rather than concave) functionals over convex sets (see, e.g., [Rockafellar(1987)]). Note that, when using the conventions about infinity described in Section 1.2, the mathematical expectation is a convex functional on the measures space; hence all problems discussed in chapters 1,2 and 3 were convex. In this connection, the performance functional (4.1) is also convex, while formula (4.3) defines the functional on $\mathcal{D}$ which is not necessarily convex. More about the convexity of performance functionals can be found in [Piunovskiy(1997), Section 2.1.2], see also Remark 4.5.

Here, we present an example illustrating that the lower limit leads to the degeneration of many concrete problems.

Let $\mathbf{X} = \{0\}$ (in fact, the controlled process is absent); $\mathbf{A} = \{0, 1\}$.

Suppose we are interested in minimizing two performance functionals, with one-step losses ${}^1c(x,a) = a$ and ${}^2c(x,a) = -a$. The objectives are contradictory, as when the first objective decreases, the second one is expected to increase; the decision maker is interested in the trade-off between these objectives. This trade-off appears consistent with intuition if we accept formula (4.1). If we use formula (4.3), it is possible to make both objectives minimal. Indeed, let

$$\varphi_t(x) = \begin{cases} 0, & \text{if } 2^{2^m} \le t < 2^{2^{m+1}}, \quad m = 0, 2, 4, \ldots, \text{ or if } t = 1; \\ 1 & \text{otherwise.} \end{cases}$$

Then, for any $N \ge 1$ and $\varepsilon > 0$, one can find $T = 2^{2^{m+1}} - 1 > N$ with an even value of m, such that

$$\frac{1}{T}\sum_{t=1}^{T} {}^1c(x,\varphi_t(x)) = \frac{1}{2^{2^{m+1}}-1} \sum_{t=1}^{2^{2^m}-1} {}^1c(x,\varphi_t(x)) \le \frac{2^{2^m}-1}{2^{2^{m+1}}-1} < \varepsilon,$$

because $\lim\limits_{m\to\infty} \dfrac{2^{2^m}-1}{2^{2^{m+1}}-1} = 0$. Therefore,

$$\inf_{T>N}\left\{\frac{1}{T}\sum_{t=1}^{T} {}^1c(x,\varphi_t(x))\right\} = 0.$$

Similarly, when taking $T = 2^{2^{m+1}} - 1 > N$ with a sufficiently large odd value of m, we obtain

$$\frac{1}{T}\sum_{t=1}^{T} {}^2c(x,\varphi_t(x)) \le -1 + \frac{2^{2^m}-1}{2^{2^{m+1}}-1} < -1 + \varepsilon,$$

so that

$$\inf_{T>N}\left\{\frac{1}{T}\sum_{t=1}^{T} {}^2c(x,\varphi_t(x))\right\} = -1.$$

Therefore, if we use formula (4.3) then the selector φ provides the minimal possible values for the both objectives:

$${}^1v^\varphi = 0, \qquad {}^2v^\varphi = -1.$$

The values of the objectives calculated using formula (4.1) give ${}^1v^\varphi = 1$ and ${}^2v^\varphi = 0$, but for any stationary strategy π^s (which are known to be sufficient for solving such minimization problems) we have ${}^1v^{\pi^s} + {}^2v^{\pi^s} = 0$ as expected.

A similar example can be found in [Altman and Shwartz(1991b), Counterexample 2.7].

4.2.2 AC-optimal non-canonical strategies

Let $\mathbf{X} = \{1,2\}$, $\mathbf{A} = \{1,2\}$, $p(1|1,1) = p(2|1,1) = 1/2$, $p(2|1,2) = 1$, $p(2|2,a) \equiv 1$, with all other transition probabilities zero. Let $c(1,1) = -5$, $c(1,2) = -10$, $c(2,a) \equiv 1$ (see Fig. 4.1). A similar example was presented in [Puterman(1994), Ex. 8.4.3]. This model is unichain in the following sense.

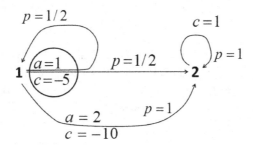

Fig. 4.1 Example 4.2.2: an AC-optimal non-canonical selector.

Definition 4.2. A model with a countable (or finite) state space is called (aperiodic) *unichain* if, for every stationary selector, the controlled process is a unichain Markov process with a single (aperiodic) positive-recurrent class plus a possibly empty set of transient states; absorption into the positive-recurrent class takes place in a finite expected time.

For such models, we can put $\rho(x) \equiv \rho$ in Definition 4.1 and in equations (4.2) [Puterman(1994), Th. 8.4.3]:

$$\rho + h(1) = \min\left\{-5 + \frac{1}{2}h(1) + \frac{1}{2}h(2), \quad -10 + h(2)\right\};$$

$$\rho + h(2) = 1 + h(2).$$

We see that $\rho = 1$, and we can put $h(1) = 0$. Now it is easy to see that $h(2) = 12$ and $\varphi^*(2) = 1$. The actions $\varphi^*(2)$ in state $x = 2$ play no role. The triplet $\langle \rho, h, \varphi^* \rangle$ is canonical according to Theorem 4.1. Thus, the stationary selector $\varphi^*(x) \equiv 1$ is canonical and hence AC-optimal, according to Theorem 4.2; the value of the infimum in (4.1) equals $\rho = 1$.

On the other hand, the stationary selector $\varphi(x) \equiv 2$ (as well as any other strategy) is also AC-optimal because, for any initial distribution, $v^\varphi = +1$

(the process will be ultimately absorbed at state 2). But this selector is not canonical.

Remark 4.2. In this example, all the conditions of Theorem 3.5 [Hernandez-Lerma and Vega-Amaya(1998)] are satisfied, but the AC-optimal stationary selector φ is not canonical. Hence, assertion (b) of that theorem, saying that a stationary selector is AC-optimal if and only if it is canonical, is wrong. For discrete models, the proof can be corrected by requiring that, for every stationary strategy, the controlled process X_t is positive recurrent.

If the model is not finite, then equations (4.2) may have no solutions. As an example, let $\mathbf{X} = \{1, 2\}$, $\mathbf{A} = [0, 1]$, $p(2|1, a) = 1 - p(1|1, a) = a^2$, $p(2|2, a) \equiv 1$, $c(1, a) = -a$, $c(2, a) \equiv 1$ (see Fig. 4.2). This model is semi-continuous in the following sense.

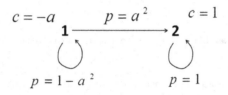

Fig. 4.2 Example 4.2.2: optimal non-canonical selector.

Definition 4.3. We say that the model is *semi-continuous* if

(a) the action space $\mathbf{A}$ is compact;

(b) the transition probability $p(dy|x, a)$ is strongly continuous, i.e. integral $\int_\mathbf{X} u(y) p(dy|x, a)$ is continuous for any measurable bounded function u;

(c) the loss function c is bounded below and lower semi-continuous.

Note that this definition is slightly different from those introduced in Chapters 1 and 2.

Equations (4.2) can be rewritten as follows

$$\rho(1) = \inf_{a \in \mathbf{A}} \{a^2 \rho(2) + (1-a^2)\rho(1)\};$$

$$\rho(1) + h(1) = \inf_{a \in \mathbf{A}} \{-a + a^2 h(2) + (1-a^2) h(1)\};$$

$$\rho(2) + h(2) = 1 + h(2).$$

We see that $\rho(2) = 1$ and, as usual, we put $h(1) = 0$ without loss of generality. From the first equation, which has the form $\inf_{a \in [0,1]}\{a^2[1-\rho(1)]\} = 0$, we conclude that either $\rho(1) = 1$ or $\rho(1) < 1$ and $\varphi^*(1) = 0$. Looking at the second equation, we see that if $\rho(1) = 1$ then $h(2) \neq 0$ and both assumptions $h(2) > 0$ and $h(2) < 0$ lead to a contradiction. Finally, if $\rho(1) < 1$ and $\varphi^*(1) = 0$, then again $h(2) \neq 0$, leading to a contradiction. The details are left to the reader.

Thus, in this example there are no canonical triplets, the stationary selector $\varphi^*(x) \equiv 0$ is AC-optimal, and $v_1^* = 0$, $v_2^* = 1$.

This model is very similar to the Blackmailer's Dilemma (Section 2.2.15). In this context, one can say that $c(2,a) \equiv 1$ is the cost of being in prison for one time interval (e.g. a day), after the victim refuses to yield to the blackmailer's demand and takes him to the police. In such a case, the optimal behaviour is not to blackmail at all.

Remark 4.3. In this example, Condition 4.1(b) is violated, so that Theorem 4.3 fails to hold; $v_1^{*,\beta} = \frac{1-\sqrt{2-\beta}}{2(1-\beta)}$; $v_2^{*,\beta} = \frac{1}{1-\beta}$. Note also that $v_1^* = \lim_{\beta \to 1^-}(1-\beta)v_1^{*,\beta} = 0$, $v_2^* = \lim_{\beta \to 1^-}(1-\beta)v_2^{*,\beta} = 1$. One can also check that the stationary AC-optimal selector $\varphi^*(x) \equiv 0$ is not Blackwell optimal.

4.2.3 Canonical triplets and canonical equations

It seems that the proof of sufficiency in Theorem 4.1 also holds for unbounded loss function c and unbounded functions ρ and h: if the finite-horizon Bellman function with the final loss h is well defined, and equations (4.2) are satisfied, then $\langle \rho, h, \varphi^* \rangle$ is a canonical triplet.

The first example shows that there can be many different canonical triplets and that not all canonical selectors are optimal.

Let $\mathbf{X} = \{\Delta, 0, 1, 2, \ldots\}$, $\mathbf{A} = \{1, 2\}$, $p(\Delta|\Delta, a) \equiv 1$, $p(\Delta|0,1) = 1$, $p(1|0,2) = 1$, $p(i+1|i,a) = 1 - p(i|i,a) = \frac{1}{2^i}$ for all $i \geq 1$; other transition probabilities are zero. Let $c(\Delta, a) = 1$, $c(0,a) = 0$, $c(i,a) = 1$ for all $i \geq 1$ (see Fig. 4.3).

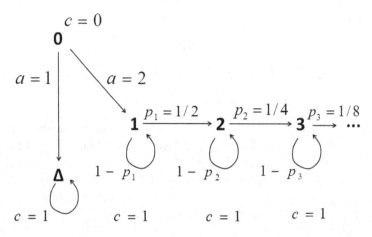

Fig. 4.3 Example 4.2.3: multiple canonical triplets.

The total loss in state $i \geq 1$ equals 2^i, and it is obvious that $\varphi^*(x) \equiv 1$ is the AC-optimal strategy, and that action $a = 2$ is not optimal if $X_0 = 0$. The corresponding canonical triplet can be defined in the following way:

$\rho(0) = \rho(\Delta) = 1, \quad \rho(i) \equiv \hat{\rho} > 1$ (any number for $i = 1, 2, \ldots$);

$h(0) = -1, \quad h(\Delta) = 0, \quad h(1) = \hat{h} > 0$ (any number),

$h(i+1) = h(i) + 2^i \hat{\rho} - 2^i$;

$\varphi^*(x) = 1.$

The canonical equations (4.2) are also satisfied.

On the other hand, if we put $\rho(i) \equiv \hat{\rho} < 1$ for $i = 1, 2, \ldots$ and $h(1) = \hat{h} < 0$, then $\langle \rho, h, \hat{\varphi} \rangle$, where $\hat{\varphi}(x) = 2$, is also a canonical triplet satisfying equations (4.2). Theorem 4.2 is not applicable because $\lim_{T \to \infty} E_0^{\hat{\varphi}}[h(X_T)/T] = -\infty$. Theorem 10.3.7 [Hernandez-Lerma and Lasserre(1999)], concerning the uniqueness of the solution to the canonical equations, is not applicable because the controlled process is not λ-irreducible under each control strategy.

Another example, which confirms that equations (4.2) can hold for ρ, h and φ^*, the stationary selector φ^* being not AC-optimal, can be found in [Robinson(1976), p. 161].

The second example shows that the boundedness of function h is important in Theorem 4.1: the canonical triplet can fail to satisfy equations (4.2).

Let $\mathbf{X} = \{s_0, \Delta, 0, 1, 2, \ldots\}$, $\mathbf{A} = \{1, 2,\}$, $p(\Delta|\Delta, a) = p(0|0, a) = p(0|i, a) \equiv 1$ for $i = 1, 2, \ldots$, $p(\Delta|s_0, 1) = 1$, $p(i|s_0, 2) = \left(\frac{1}{2}\right)^i$ for

$i = 1, 2, \ldots$, with the other transition probabilities zero. Let $c(s_0, a) = c(0, a) \equiv 0$, $c(\Delta, a) \equiv 1$, $c(i, a) \equiv 2^i$ for $i = 1, 2, \ldots$ (see Fig. 4.4). The loss function c can be made bounded by introducing loops, as in the previous example; see also Remark 2.6.

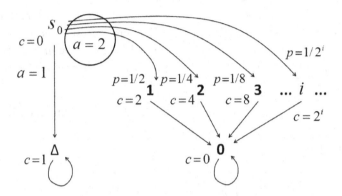

Fig. 4.4 Example 4.2.3: canonical equations have no solutions.

Let $\rho(x) = \begin{cases} 1, & \text{if } x = s_0 \text{ or } x = \Delta; \\ 0 & \text{otherwise,} \end{cases}$

$$h(x) = \begin{cases} 0, & \text{if } x = \Delta \text{ or } x = 0; \\ -1, & \text{if } x = s_0; \\ 2^i, & \text{if } x = i \in \{1, 2, \ldots\}, \end{cases} \quad \varphi^*(x) \equiv 1.$$

It is easy to check that $\langle \rho, h, \varphi^* \rangle$ is a canonical triplet. But equations (4.2) are not satisfied: for $x = s_0$,

$$\min\{\rho(\Delta), \sum_{i=1}^{\infty} p(i|s_0, 2)\rho(i)\} = \min\{1, 0\} = 0 \neq \rho(s_0) = 1.$$

Moreover, equations (4.2) have no finite solutions at all. Indeed, from the second equation under $x = 0, \Delta$ or $i \in \{1, 2, \ldots\}$, we deduce that $h(0)$ and $h(\Delta)$ are arbitrary numbers, $\rho(0) = 0$, $\rho(\Delta) = 1$ and $\rho(i) + h(i) = 2^i + h(0)$. The first equation at $x = i \in \{1, 2, \ldots\}$ shows that $\rho(i) = \rho(0) = 0$. Now $h(i) = h(0) + 2^i$, and from the second equation (4.2) under $x = s_0$ we see that $\varphi^*(s_0) = 1$ and $\rho(s_0) + h(s_0) = h(\Delta)$, because $\sum_{i=1}^{\infty} h(i) p(i|s_0, 2) = \infty$. But the first equation implies that $\rho(s_0) = 0$ and

$\varphi^*(s_0) = 2$. The resulting contradiction confirms that equations (4.2) have no finite solutions.

4.2.4 Multiple solutions to the canonical equations in finite models

Definition 4.4. Suppose R is the set of all states which are recurrent under some stationary selector, and any two states $i, j \in R$ communicate; that is, there is a stationary selector φ (depending on i and j) such that $P_i^\varphi(X_t = j) > 0$ for some t. Then the model is called *communicating*. (In [Puterman(1994), Section 9.5] such a model is called weakly communicating.)

If the model is unichain (see Section 4.2.2) or communicating, then in equations (4.2) (and in Definition 4.1) one should put $\rho(x) \equiv \rho$; the remainder equation

$$\rho + h(x) = \inf_{a \in \mathbf{A}} \left\{ c(x,a) + \sum_{y \in \mathbf{X}} h(y) p(y|x,a) \right\} \qquad (4.4)$$

$$= c(x, \varphi^*(x)) + \sum_{y \in \mathbf{X}} h(y) p(y|x, \varphi^*(x)).$$

is solvable according to [Puterman(1994), Section 8.4.2] and [Scweitzer(1987), Th. 1]. Moreover, the value of ρ is unique and, in the unichain case, if h^1 and h^2 are two solutions then $h^1(x) - h^2(x) = const$. The following example, first presented in [Scweitzer(1987), Ex. 3] shows that the unichain condition is important.

Let $\mathbf{X} = \{1,2\}$, $\mathbf{A} = \{1,2\}$, $p(1|1,1) = p(2|2,1) = 1$, $p(1|1,2) = p(2|1,2) = p(1|2,2) = p(2|2,2) = 1/2$, $c(1,1) = c(2,1) = 0$, $c(1,2) = 1$, $c(2,2) = 2$ (see Fig. 4.5).

Equation (4.4) takes the form

$$\rho + h(1) = \min\{h(1),\ 1 + 0.5(h(1) + h(2))\},$$
$$\rho + h(2) = \min\{h(2),\ 2 + 0.5(h(1) + h(2))\},$$

and, without loss of generality, we can put $h(1) = 0$.

Assume that $h(2) < -2$. Then $\rho = 1 + h(2)/2$ and, from the second equation, we see that $h(2) = -2$, which is a contradiction. Similarly, $h(2)$ cannot be greater than 4. But any value $h(2) \in [-2, 4]$ together with $\rho = 0$ solves the presented equations. All the corresponding triplets $\langle \rho = 0, h, \varphi^* \rangle$ with $\varphi^*(x) \equiv 1$ are canonical.

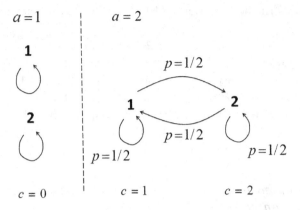

Fig. 4.5 Example 4.2.4: a communicating model which is not unichain.

If the state space is not finite then, even in communicating models, it can happen that equations (4.2) are solvable, but the corresponding stationary selector φ^* is not AC-optimal (see Section 4.2.10, Remark 4.4).

4.2.5 No AC-optimal strategies

Let $\mathbf{X} = \{1, 1', 2, 2', \ldots\}$, $\mathbf{A} = \{1, 2\}$; for all $i \geq 1$ we put $p(i'|i'a) \equiv 1$, $p(i+1|i,1) \equiv 1$, $p(i'|i,2) \equiv 1$, with all other transition probabilities zero; $c(i,a) \equiv 1$, $c(i',a) = \frac{1}{i}$ (see Fig. 4.6). A similar example was presented in [Sennott(2002), Ex. 5.1].

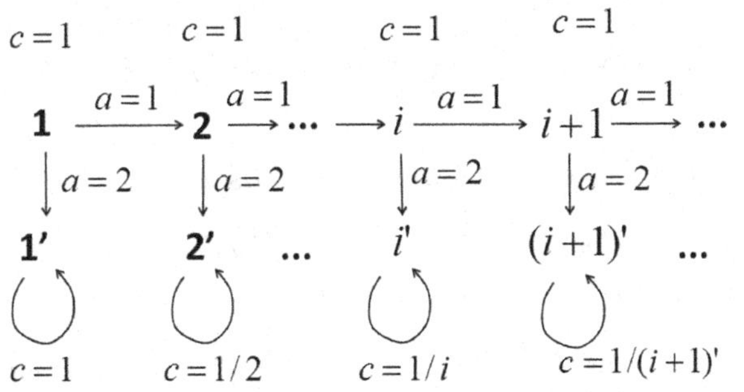

Fig. 4.6 Example 4.2.5: no AC-optimal strategies.

The performance functional (4.1) is non-negative, and, for any $\varepsilon > 0$, the stationary selector $\varphi^\varepsilon(x) = \begin{cases} 1, & \text{if } x = i < \frac{1}{\varepsilon}; \\ 2, & \text{otherwise} \end{cases}$ gives $v_i^{\varphi^\varepsilon} \leq \varepsilon$, so that $v_i^* \equiv 0$. (Obviously, $v_{i'}^* = \frac{1}{i}$.) On the other hand, if $x = i \in \mathbf{X}$ is fixed then, for any control strategy π, $v_i^\pi > 0$ because, if $P_i^\pi\{\forall t \geq 1\ A_t = 1\} = 1$, then $v_i^\pi = 1$; otherwise, if $P_i^\pi\{\exists t \geq 1 : A_t = 2\} = p > 0$, then $v_i^\pi > \frac{p}{i+t-1} > 0$.

The canonical equations (4.2) have no solution with a bounded function h: one can check that $\rho(i) \equiv 0$, $\rho(i') = \frac{1}{i}$ and $h(i) = 1 + h(i+1)$. Theorems 4.1 and 4.2 are not applicable.

4.2.6 Canonical equations have no solutions: the finite action space

Consider a positive model ($c(x,a) \geq 0$) assuming that the state space $\mathbf{X}$ is countable and the action space $\mathbf{A}$ is finite.

Condition 4.1.

(a) The Bellman function for the discounted problem (3.1) $v_x^{*,\beta} < \infty$ is finite and, for all β close to 1, the product $(1-\beta)v_z^{*,\beta} \leq M < \infty$ is uniformly bounded for a particular state $z \in \mathbf{X}$.

(b) There is a function $b(x)$ such that the inequality

$$-M \leq h^\beta(x) \stackrel{\triangle}{=} v_x^{*,\beta} - v_z^{*,\beta} \leq b(x) < \infty \quad (4.5)$$

holds for all $x \in \mathbf{X}$ and all β close to 1.

Condition 4.2. Condition 4.1 holds and, additionally,

(a) For each $x \in \mathbf{X}$ there is $a \in \mathbf{A}$ such that $\sum_{y \in \mathbf{X}} p(y|x,a)b(y) < \infty$.

(b) For all $x \in \mathbf{X}$ and $a \in \mathbf{A}$ $\sum_{y \in \mathbf{X}} p(y|x,a)b(y) < \infty$.

Theorem 4.3. *Suppose Condition 4.1 is satisfied.*

(a) *[Cavazos-Cadena(1991), Th. 2.1] Under Condition 4.2(a), there exists a triplet $\langle \rho, h, \varphi^* \rangle$ such that*

$$\rho + h(x) \geq c(x, \varphi^*(x)) + \sum_{y \in \mathbf{X}} h(y) p(y|x, \varphi^*(x)). \quad (4.6)$$

See also [Sennott(1989)].

(b) *[Hernandez-Lerma and Lasserre(1996a), Th. 5.5.4]* Under Condition 4.2(b), there exists a solution to the canonical equations (4.2) with $\rho(x) \equiv \rho$.

In both cases, the corresponding stationary selector φ^* is AC-optimal in problem (4.1): $v_x^* = v_x^{\varphi^*} = \rho$, and $\rho = \lim_{\beta \to 1-} (1-\beta)v_z^{*,\beta}$; for each $x \in \mathbf{X}$, $h(x)$ is a limiting point for $h^\beta(x)$ as $\beta \to 1-$.

The following unichain model, based on [Cavazos-Cadena(1991), Section 3] shows that, under Condition 4.2(a), it may happen that the canonical equations (4.2) have no solution.

Let $\mathbf{X} = \{0, 1, 2, \ldots\}$, $\mathbf{A} = \{1, 2\}$, $p(x-1|x, a) \equiv 1$ for all $x \geq 1$, $p(1|0, 2) = 1$, $p(y|0, 1) = q_y$, an arbitrary probability distribution on $\{1, 2, \ldots\}$. Other transition probabilities are zero. Finally, we put
$$c(x, a) = \begin{cases} 0, & \text{if } x = 0,\ a = 1; \\ 1 & \text{otherwise.} \end{cases}$$
See Fig. 4.7.

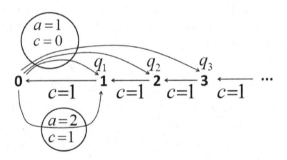

Fig. 4.7 Example 4.2.6: no canonical triplets.

First, we check the imposed conditions. The discounted optimality equation (3.2) can be explicitly solved: action $\varphi^*(0) = 1$ is optimal for any $\beta \in (0,1)$ (so that the stationary selector $\varphi^*(x) \equiv 1$ is Blackwell optimal),

$$v^\beta(x) = v_x^{*,\beta} = v_x^{\varphi^*,\beta} = \begin{cases} \dfrac{\beta}{1-\beta} \cdot \dfrac{1 - \sum_{y \geq 1} \beta^y q_y}{1 - \beta \sum_{y \geq 1} \beta^y q_y}, & \text{if } x = 0; \\[2mm] \dfrac{1-\beta^x}{1-\beta} + \dfrac{\beta^{x+1}}{1-\beta} \cdot \dfrac{1 - \sum_{y \geq 1} \beta^y q_y}{1 - \beta \sum_{y \geq 1} \beta^y q_y}, & \text{if } x \geq 1, \end{cases}$$

and, after we put $z = 0$,
$$h^\beta(x) = \frac{1 - \beta^x}{1 - \beta \sum_{y \geq 1} \beta^y q_y}.$$
After we apply L'Hôpital's rule,
$$\rho = \lim_{\beta \to 1-} (1 - \beta) v_0^{*,\beta} = \frac{\sum_{y \geq 1} y q_y}{1 + \sum_{y \geq 1} y q_y} < \infty,$$
and hence the product $(1 - \beta) v_0^{*,\beta}$ is uniformly bounded. Similarly, for any $x = 1, 2, \ldots,$
$$h(x) = \lim_{\beta \to 1-} h^\beta(x) = \lim_{\beta \to 1-} \frac{x \beta^{x-1}}{\sum_{y \geq 1} (y+1) \beta^y q_y} = \frac{x}{1 + \sum_{y \geq 1} y q_y} < \infty,$$
and inequality (4.5) holds for a finite function $b(x)$. Thus, Condition 4.1 is satisfied.

Condition 4.2(a) is also satisfied: one can take $a = 2$ in state $x = 0$. Now, the optimality inequality (4.6) holds for ρ, $h(\cdot)$ and $\varphi^*(x) \equiv 1$. Indeed, if $x \geq 1$ then inequality (4.6) takes the form
$$\frac{\sum_{y \geq 1} y q_y}{1 + \sum_{y \geq 1} y q_y} + \frac{x}{1 + \sum_{y \geq 1} y q_y} \geq 1 + \frac{x - 1}{1 + \sum_{y \geq 1} y q_y},$$
and in fact we have an equality. If $x = 0$ then $h(0) = 0$ and, in the case where $\sum_{y \geq 1} y q_y < \infty$,
$$\sum_{y \in \mathbf{X}} h(y) p(y|x, \varphi^*(x)) = \frac{\sum_{y \geq 1} y q_y}{1 + \sum_{y \geq 1} y q_y} = \rho;$$
when $\sum_{y \geq 1} y q_y = \infty$, we have $\rho = 1$ and $h(y) \equiv 0$. Thus, the strategy $\varphi^*(x) \equiv 1$ is AC-optimal in problem (4.1) for any distribution q_x.

If $\sum_{y \geq 1} y q_y < \infty$, we have an equality in formula (4.6), i.e. $\langle \rho, h, \varphi^* \rangle$ is a canonical triplet satisfying the canonical equations (4.2). Condition 4.2(b) is also satisfied in this case: for $x \geq 1$,
$$h^\beta(x) = \frac{1 - \beta}{1 - \beta \sum_{y \geq 1} \beta^y q_y} \cdot \frac{1 - \beta^x}{1 - \beta} \leq (1 + \beta + \beta^2 + \cdots + \beta^{x-1}) < x,$$
and one can take $b(x) = x$.

Consider now the case $\sum_{y \geq 1} y q_y = \infty$: here $h(x) \equiv 0$, $\rho = 1$, and inequality (4.6) holds strictly at $x = 0$ with $\varphi^*(0) = 1$. In fact, one can put $\tilde\varphi(0) = 2$: as a result we obtain another AC-optimal strategy $\tilde\varphi(x) \equiv 2$ for which (4.6) becomes an equality. But the stationary selector $\tilde\varphi$ is not canonical either, because on the right-hand side of (4.2) one has to take the minimum w.r.t. $a \in \mathbf{A}$, which is zero.

Proposition 4.1. *Suppose* $\sum_{y \geq 1} y q_y = \infty$. *Then*

(a) the canonical equations (4.2) have no solution,
(b) Condition 4.2(b) is not satisfied,
(c) under a control strategy $\varphi^*(x) \equiv 1$, there is no stationary distribution.

If $\sum_{y \geq 1} y q_y < \infty$ then there exists a stationary distribution $\eta(x)$ for the control strategy $\varphi^*(x) \equiv 1$, and

$$\sum_{x \in \mathbf{X}} \eta(x) c(x, \varphi^*(x)) = 1 - \eta(0) = \frac{\sum_{y \geq 1} y q_y}{1 + \sum_{y \geq 1} y q_y} = \rho.$$

The proof is given in Appendix B. Note that theorems about the existence of canonical triplets, for example [Puterman(1994), Cor. 8.10.8] and [Hernandez-Lerma and Lasserre(1996a), Th. 5.5.4] do not hold, since Condition 4.2(b) is violated.

4.2.7 No AC-ε-optimal stationary strategies in a finite state model

It is known that, in homogeneous models with total expected loss (Chapters 2 and 3), under very general conditions, if v_x^* is finite then there exists a stationary ε-optimal selector. The situation is different in the case of average loss: it can happen that, for any stationary strategy, the controlled process gets absorbed in a "bad" state; however one can make that absorption probability very small using a non-stationary strategy. The following example illustrating this point is similar to that published in [Dynkin and Yushkevich(1979), Chapter 7, Section 8].

Let $\mathbf{X} = \{0, 1\}$; $\mathbf{A} = \{1, 2, \ldots\}$; $p(0|0, a) \equiv 1$, $p(1|1, a) = 1 - p(0|1, a) = q_a$, where $\{q_a\}_{a=1}^\infty$ are given probabilities such that, for all $a \in \mathbf{A}$, $q_a \in (0, 1)$ and $\lim_{a \to \infty} q_a = 1$. We put $c(0, a) \equiv 1$, $c(1, a) \equiv 0$ (see Fig. 4.8).

For any stationary strategy π^{ms}, $\sum_{a=1}^\infty \pi^{\mathrm{ms}}(a|1) q_a < 1$, so that the controlled process will ultimately be absorbed at 0, and $v_1^{\pi^{\mathrm{ms}}} = 1$. On the other hand, for any number $Q \in (0, 1)$, there is a sequence $a_t \to \infty$ for which $\prod_{t=1}^\infty q_{a_t} \geq Q$: it is sufficient to take a_t such that $q_{a_t} \geq Q^{\frac{1}{2^t}}$. Now, if $\varphi_t(1) = a_t$, then the controlled process X_t starting from $X_0 = 1$ will never be absorbed at 0 with probability $\prod_{t=1}^\infty q_{a_t} \geq Q$, and $v_1^\varphi \leq 1 - Q$. Therefore, $v_1^* = \inf_\pi v_1^\pi = 0$ and no one stationary strategy is AC-ε-optimal if $\varepsilon < 1$.

It can easily be shown show that, for any $\beta \in (0, 1)$, $v_0^{*,\beta} = \frac{1}{1-\beta}$ and $v_1^{*,\beta} = 0$, so that inequality (4.5) is violated.

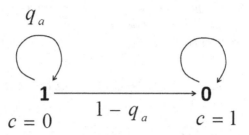

Fig. 4.8 Example 4.2.7: no AC-ε-optimal stationary strategies in a finite state model.

If we add point ∞ to the action space $\mathbf{A}$ (one-point compactification, i.e. $\lim_{a\to\infty} a \stackrel{\triangle}{=} \infty$) and put $p(0|0,\infty) = p(1|1,\infty) = 1$, $c(0,\infty) = 1$, $c(1,\infty) = 0$, then the stationary selector $\varphi^*(x) \equiv \infty$ is AC-optimal. In this connection, it is interesting to note that the sequence of stationary selectors $\varphi^i(x) \stackrel{\triangle}{=} i \in \{1, 2, \ldots\}$, as functions of $x \in \mathbf{X}$, converges to $\varphi^*(x)$. But $v_x^{\varphi^i} \equiv 1$ and $v_x^{\varphi^*} = 1 - x = v_x^*$. Convergence of strategies therefore does not imply the convergence of performance functionals. Note that the model under consideration is semi-continuous.

4.2.8 No AC-optimal strategies in a finite-state semi-continuous model

Theorem 4.4. *[Hernandez-Lerma and Lasserre(1996a), Th. 5.4.3]* *Suppose the model is semi-continuous (see Definition 4.3) and Condition 4.1 is satisfied. Assume that $v_x^\pi < \infty$ for some $x \in \mathbf{X}$ and some strategy π. Then there exists a triplet $\langle \rho, h, \varphi^* \rangle$ satisfying (4.6) and such that the stationary selector φ^* is AC-optimal and $v_x^* = v_x^{\varphi^*} = \rho$ for all $x \in \mathbf{X}$.*

The following example, published in [Dynkin and Yushkevich(1979), Chapter 7, Section 8] shows that Condition 4.1 cannot be omitted. Let $\mathbf{X} = \{0, 1, \Delta\}$; $\mathbf{A} = [0, \frac{1}{2}]$; $p(0|0,a) = p(\Delta|\Delta,a) \equiv 1$, $p(0|1,a) = a$, $p(\Delta|1,a) = a^2$, $p(1|1,a) = 1 - a - a^2$; $c(0,a) \equiv 0$, $c(1,a) = c(\Delta,a) \equiv 1$ (see Fig. 4.9).

Suppose $X_0 = 1$. For any control strategy π, either $A_t \equiv 0$ and the controlled process never leaves state 1, or at the first moment τ when $X_\tau \neq 1$, there is a strictly positive chance $A_\tau^2 > 0$ of absorption at state Δ. In all cases, $v_1^\pi > 0$. At the same time, for any stationary selector φ with

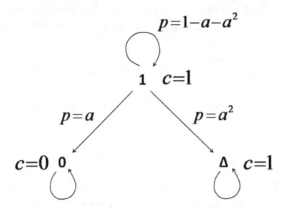

Fig. 4.9 Example 4.2.8: a semi-continuous model.

$\varphi(1) = a$, we have

$$P^{\varphi}(\tau < \infty, X_\tau = 0) = \sum_{i=1}^{\infty}(1 - a - a^2)^{i-1}a = \begin{cases} 0, & \text{if } a = 0; \\ \frac{1}{1+a}, & \text{if } a > 0 \end{cases}$$

and

$$\inf_{\pi} v_1^{\pi} = \inf_{\varphi}[1 - P^{\varphi}(\tau < \infty, X_\tau = 0)] = 0.$$

If we consider the discounted version then the problem is solvable. One can check that, for $\beta \in [\frac{4}{5}, 1)$, the stationary selector

$$\varphi^*(x) = \frac{5\beta - 4}{2[\beta - 2\sqrt{\beta(1-\beta)}]} - \frac{1}{2}$$

is optimal, and

$$(1 - \beta)v_x^{*,\beta} = \begin{cases} 0, & \text{if } c = 0; \\ 1, & \text{if } x = \Delta; \\ \frac{4(1-\beta) - 2\sqrt{\beta(1-\beta)}}{4 - 5\beta}, & \text{if } x = 1. \end{cases}$$

Theorem 4.4 is false because Condition 4.1(b) is not satisfied: the functions

$$\frac{1}{1-\beta}\left[\frac{4(1-\beta) - 2\sqrt{\beta(1-\beta)}}{4 - 5\beta}\right] \quad \text{and} \quad \frac{1}{1-\beta}\left[\frac{4(1-\beta) - 2\sqrt{\beta(1-\beta)}}{4 - 5\beta} - 1\right]$$

are not bounded when $\beta \to 1-$, and inequality (4.5) cannot hold for $z = 0, 1$, or Δ.

We now show that the canonical equations (4.2) have no solution. From the second equation at $x = 0, \Delta$ and 1, we deduce that $\rho(0) = 0$, $\rho(\Delta) = 1$ and

$$\rho(1) = \inf_{a \in \mathbf{A}} \{1 + a^2 h(\Delta) - (a + a^2) h(1)\} \tag{4.7}$$

correspondingly. (We have set $h(0) = 0$ following Remark 4.1.) Now the first equation in (4.2) at $x = 1$ gives

$$\inf_{a \in \mathbf{A}} \{a^2 - (a + a^2)\rho(1)\} = 0,$$

meaning that $\rho(1) \leq 0$, because otherwise the function in the parentheses decreases in the neighbourhood of $a = 0$. Hence either $\varphi^*(1) = 0$ and, according to (4.7), $\rho(1) = 1$; or $\varphi^*(1) = \frac{\rho(1)}{1-\rho(1)} < 0$. Both these cases lead to contradictions.

4.2.9 Semi-continuous models and the sufficiency of stationary selectors

Theorem 4.5. *[Fainberg(1977), Th. 3] Suppose the state space $\mathbf{X}$ is finite, the model is semi-continuous, and there is a strategy solving problem (4.1). Then there exists an AC-optimal stationary selector.*

The following example, based on [Fainberg(1977)], shows that this statement is false if the model (even a recurrent model) is not semi-continuous.

Let $\mathbf{X} = \{1, 2\}$, $\mathbf{A} = \{1, 2, \ldots\}$, $p(2|1, a) = 1 - p(1|1, a) = 1/a$, $p(1|2, a) \equiv 1$, $c(1, a) \equiv 0$, $c(2, a) \equiv 1$ (see Fig. 4.10).

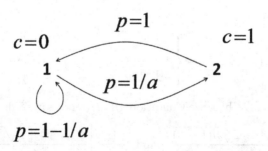

Fig. 4.10 Example 4.2.9: no AC-optimal stationary selectors.

For the stationary selector $\varphi^n(x) \equiv n$, the stationary probability of state 2 equals $\frac{1}{n+1}$, so that $v^{\varphi^n} = \frac{1}{n+1}$ and no one stationary selector is AC-optimal. On the other hand, the non-stationary Markov selector $\varphi_t^*(x) = t$ is AC-optimal because $v^{\varphi^*} = 0$. To show this, fix an arbitrary $\varepsilon > 0$ and ignore the first several decision epochs: without loss of generality, we accept that $\varphi_t^*(x) \geq 1/\varepsilon$ for all $t \geq 1$. Now

$$P_{P_0}^{\varphi^*}(X_t = 2) = P_{P_0}^{\varphi^*}(X_{t-1} = 1)/\varphi_t^*(1) \leq \varepsilon,$$

so that $v^{\varphi^*} \leq \varepsilon$ and, since ε is arbitrarily positive, $v^{\varphi^*} = 0$ (for any initial distribution P_0).

In the above example, **A** is not compact. The same example can illustrate that requirements (b) and (c) in Definition 4.3 are also important. For instance, add action "$+\infty$" to **A** (one-point compactification) and put $p(2|1, \infty) = 1 - p(1|1, \infty) = 0$; $c(1, \infty) = c(2, \infty) = 1$. The transition probability is strongly continuous, but the loss function c is not lower semi-continuous. The same non-stationary selector φ^* is AC-optimal, but $v^\varphi > 0$ for any stationary selector φ, as before.

4.2.10 No AC-optimal stationary strategies in a unichain model with a finite action space

This example, first published in [Fisher and Ross(1968)], illustrates that the requirement $-M \leq h^\beta(x)$ in Condition 4.1 is important. Moreover, this model is semi-continuous and unichain (and even recurrent and aperiodic).

Let $\mathbf{X} = \{0, 1, 1', 2, 2', \ldots\}$, $\mathbf{A} = \{1, 2\}$; for all $i > 0$, $p(i|0, a) = p(i'|0, a) \equiv \frac{3}{2} \cdot \left(\frac{1}{4}\right)^i$; $p(0|i, 1) = 1 - p(i'|i, 1) = \left(\frac{1}{2}\right)^i$; $p(0|i, 2) = p(i+1|i, 2) = \frac{1}{2}$; $p(0|i', a) = 1 - p(i'|i', a) \equiv \left(\frac{1}{2}\right)^i$. Other transition probabilities are zero. We put $c(0, a) \equiv 1$, and all the other costs are zero. See Fig. 4.11.

Proposition 4.2.

(a) *For any stationary strategy π^{ms}, for any initial distribution P_0, $v^{\pi^{\mathrm{ms}}} > \frac{1}{5}$ and, for any initial distribution P_0 such that $\sum_{i \geq 1}[P_0(i) + P_0(i')]2^i < \infty$, for an arbitrary control strategy π, the inequality $v^\pi \geq \frac{1}{5}$ holds. For any stationary strategy, the controlled process is positive recurrent.*

(b) *There exists a non-stationary selector $\varphi_t^*(x)$ such that $v^{\varphi^*} = \frac{1}{5}$ for an arbitrary initial distribution P_0. (Hence selector φ^* is AC-optimal.)*

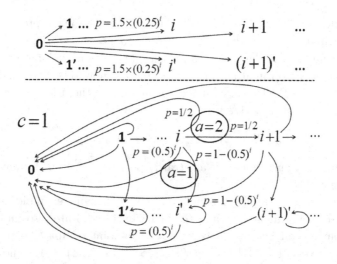

Fig. 4.11 Example 4.2.10: no AC-optimal stationary strategies in a unichain model. The transition probabilities are shown only to and from states i, i'.

The proof, presented in Appendix B, is based on the following statement: for any control strategy π, the mean recurrence time $M_{00}(\pi)$ from state 0 to 0 is strictly smaller than 5. The selector φ^* applies different stationary selectors of the form

$$\varphi^n(x) = \begin{cases} 2, & \text{if } x = i < n; \\ 1 & \text{otherwise} \end{cases} \qquad n = 1, 2, \ldots, \infty \qquad (4.8)$$

on longer and longer random time intervals $(0, T_1]$, $(T_1, T_2]$, ..., so that $\lim_{n \to \infty} M_{00}(\varphi^n) = 5$ and $v^{\varphi^*} = \frac{1}{5}$. Note that $M_{00}(\varphi^\infty) = 7/2 < 5$.

Consider the discounted version of this example with $\beta \in (0, 1)$. Obviously, $v_0^{*,\beta} \leq \frac{1}{1-\beta}$ because $|c(x, a)| \leq 1$. During the proof of Proposition 4.2, we established that $v_0^{*,\beta} \geq \frac{1}{1-\beta^5}$ and, for any $i \geq 1$,

$$v_i^{*,\beta}, \ v_{i'}^{*,\beta} \leq \frac{\beta \left(\frac{1}{2}\right)^i v_0^{*,\beta}}{1 - \beta \left[1 - \left(\frac{1}{2}\right)^i\right]}.$$

Now it is obvious that $\forall x \in \mathbf{X}$ $(1 - \beta) v_x^{*,\beta} \leq 1$, so that Condition 4.1(a) is satisfied.

We now show that Condition 4.1(b) is violated. Clearly, if we take $z = 0$ then $h^\beta(x) < 0$ (see the proof of Proposition 4.2: $v_x^{*,\beta} \leq v_0^{*,\beta}$), but, for $x = i$

or i' with $i \geq 1$,

$$h^\beta(x) \leq \frac{\beta\left(\frac{1}{2}\right)^i v_0^{*,\beta}}{1-\beta\left[1-\left(\frac{1}{2}\right)^i\right]} - \frac{1}{1-\beta^5} \quad (4.9)$$

$$\leq \frac{1}{1-\beta}\left\{\frac{\beta\left(\frac{1}{2}\right)^i}{1-\beta\left[1-\left(\frac{1}{2}\right)^i\right]} - \frac{1}{1+\beta+\cdots+\beta^4}\right\}.$$

For arbitrary M, we fix $\beta > 1 - \frac{1}{12M}$ such that $\frac{1}{1+\beta+\cdots+\beta^4} > \frac{1}{6}$ and i such that $\frac{\beta\left(\frac{1}{2}\right)^i}{1-\beta\left[1-\left(\frac{1}{2}\right)^i\right]} < \frac{1}{12}$. Now we have

$$h^\beta(i) < 12M\left\{\frac{1}{12} - \frac{1}{6}\right\} = -M.$$

The left-hand inequality (4.5) is also violated for any other value of z. In such cases (if $z = j$ or j' with $j \geq 1$), $v_z^{*,\beta} \geq \beta^{2+2j}\frac{1}{1-\beta^5}$: see the proof of Proposition 4.2. Hence, in a similar manner to (4.9),

$$h^\beta(x) \leq \frac{1}{1-\beta}\left\{\frac{\beta\left(\frac{1}{2}\right)^i}{1-\beta\left[1-\left(\frac{1}{2}\right)^i\right]} - \frac{\beta^{2+2j}}{1+\beta+\cdots+\beta^4}\right\},$$

and the above reasoning shows that a finite value of M, for which $-M \leq h^\beta(x)$, does not exist.

Note that Theorems 3.3 and 3.4 are not applicable here, because inequalities (3.18) and (3.20) fail to hold.

A simpler example, showing that stationary selectors are not sufficient if the state space is not finite, is given in [Ross(1970), Section 6.6] and in [Puterman(1994), Ex. 8.10.2]; see also [Sennott(2002), Ex. 5.2]. But this model is not unichain: $\mathbf{X} = \{1, 2, \ldots\}$, $\mathbf{A} = \{1, 2\}$, $p(i+1|i, 1) = p(i|i, 2) \equiv 1$, with all other transition probabilities zero; $c(i, 1) = 1$, $c(i, 2) = 1/i$ (see Fig. 4.12).

For the stationary selector $\varphi^1(x) \equiv 1$, $v^{\varphi^1} = 1$. If a stationary selector φ chooses an action $a = 2$ in state $i > x$, then $v_x^\varphi = 1/i$. In all cases $v_x^\varphi > 0$. But $v_x^{\varphi^*} = 0$ for the following non-stationary AC-optimal selector: when the process X_t enters state i, φ^* chooses action 2 i consecutive times, and then chooses action 1.

Remark 4.4. One can slightly modify the model and make it communicating: introduce the third action $a = 3$ and put $p(1|1, 3) = 1$, $p(i-1|i, 3) = 1$ for $i \geq 2$; $c(i, 3) = 0$. Now equation (4.4) has a solution $\rho = 0$, $h(x) = 1-x$, $\varphi^*(x) \equiv 1$, but the stationary selector φ^* is not AC-optimal; the conditions of Theorem 4.2 are not satisfied.

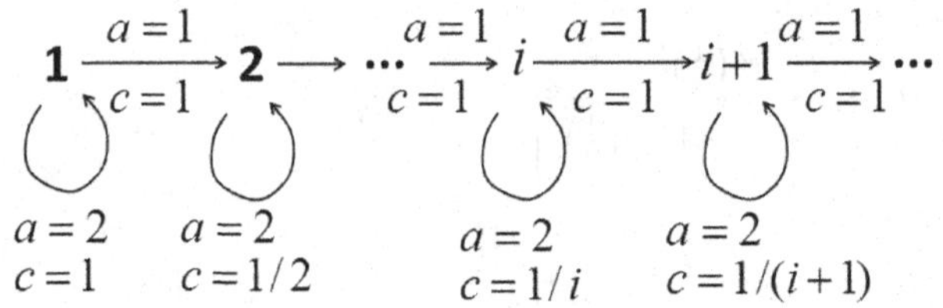

Fig. 4.12 Example 4.2.10: only a non-stationary selector is AC-optimal.

4.2.11 No AC-ε-optimal stationary strategies in a finite action model

This example is based on [Ross(1971)]; the model is semi-continuous.

Let $\mathbf{X} = \{0, 1, 1', 2, 2', \ldots\}$, $\mathbf{A} = \{1, 2\}$; for $x = i \geq 1$, $p(i+1|i, 1) = 1$, $p(i'|i, 2) = 1 - p(0|i, 2) = q_i$, where $\{q_i\}_{i=1}^{\infty}$ are given probabilities such that $q_i \in (0, 1)$ and $\prod_{j=1}^{\infty} q_j = Q > 0$. We put $p((i-1)'|i', a) \equiv 1$ for all $i > 1$, $p(1|1', a) \equiv 1$, and $p(0|0, a) \equiv 1$. Other transition probabilities are zero. Finally, $c(0, a) \equiv 2$, $c(i', a) \equiv 0$ and $c(i, a) \equiv 2$ for all $i \geq 1$. See Fig. 4.13.

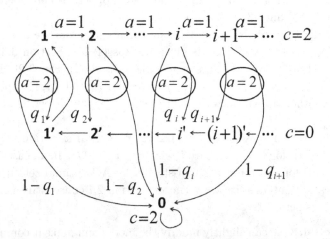

Fig. 4.13 Example 4.2.11: no AC-ε-optimal stationary strategies in a finite action model.

Starting from any $X_0 = x$, under an arbitrary fixed stationary strategy π^{ms}, the controlled process will be either absorbed at 0, or it will go to infinity along trajectory $1 \to 2 \to 3 \to \cdots$ (if $A_t \equiv 1$). Thus, $v_x^{\pi^{\mathrm{ms}}} \equiv 2$.

On the other hand, consider a control strategy π which initially chooses action 2, and then on the nth return to state 1 (if any), chooses action 1 n times and then chooses action 2. The process starting from $X_0 = i'$ (from $X_0 = i$) will be never absorbed at 0 with probability $\prod_{j=2}^{\infty} q_j$ ($q_i \prod_{j=2}^{\infty} q_j$). In this case, after the initial return to state 1, the trajectory and associated losses are as follows:

X_t	1	2	2'	1'	1	2	3	3'	2'	1'	1	2	3	...
$c(X_t, A_{t+1})$	2	2	0	0	2	2	2	0	0	0	2	2	2	...

This shows that the average loss equals 1. The complementary probability corresponds to absorption at 0, with the average loss being 2. Therefore, for example, if $X_0 = 1$ then $v_1^\pi = Q + 2(1-Q) = 2 - Q$ and no one stationary strategy is AC-ε-optimal if $\varepsilon < Q$.

In this model, the left-hand inequality in (4.5) is violated. For instance, take $z = 0$. Clearly, $v_0^{*,\beta} = \frac{2}{1-\beta}$ and, from the discounted optimality equation (3.2), we obtain:
$v_{i'}^{*,\beta} = \beta^i \cdot v_1^{*,\beta}$ for all $i \geq 1$;
$v_i^{*,\beta} \leq 2 + \beta(1-q_i)\frac{2}{1-\beta} + \beta q_i \beta^i v_1^{*,\beta}$ for all $i \geq 1$.
Hence
$$v_1^{*,\beta} \leq \frac{2 - 2\beta q_1}{(1-\beta)(1-q_1\beta^2)}$$
and
$$h^\beta(x) = v_i^{*,\beta} - v_0^{*,\beta} < 2\left[\frac{q_i\beta^{i+1}(1-\beta q_1)}{(1-\beta)(1-q_1\beta^2)} - \frac{\beta q_i}{1-\beta}\right].$$

Now, for any fixed M, we can take β such that $\frac{\beta q_i}{1-\beta} > M$ and afterwards take i such that $\frac{q_i\beta^{i+1}(1-\beta q_1)}{(1-\beta)(1-q_1\beta^2)} < \frac{M}{2}$. Then $h^\beta(x) < -M$.

Theorems 3.3 and 3.4 are not applicable here, because inequalities (3.18) and (3.20) fail to hold.

4.2.12 No AC-ε-optimal Markov strategies

According to Remark 2.1, Markov strategies are sufficient for solving almost all optimization problems if the initial distribution is fixed, but an AC-ε-optimal strategy π must satisfy the inequality $v_x^\pi \leq v_x^* + \varepsilon$ for all $x \in \mathbf{X}$ simultaneously. If the state space $\mathbf{X}$ is finite and the loss function c is bounded below then, for any $\varepsilon > 0$, there is an AC-ε-optimal

Markov selector [Fainberg(1980), Th. 1]. The following example, based on [Fainberg(1980), Section 5], shows that, if $\mathbf{X}$ is not finite, it can happen that only a semi-Markov strategy is AC-optimal, and no one Markov strategy is AC-ε-optimal if $\varepsilon < 1/2$.

Let $\mathbf{X} = \{0, 1, 1', 2, 2', \ldots\}$, $\mathbf{A} = \{1, 2\}$, $p(3|0, 1) = p(3'|0, 2) = 1$, $p(0|1, a) = p(2|1, a) = p(0|1', a) = p(2'|1', a) \equiv 1/2$; for $j \geq 2$ $p(j+1|j, a) = p((j+1)'|j', a) \equiv 1$, with all other transition probabilities zero. We put $c(0, a) = c(1, a) = c(1', a) = c(2, a) = c(2', a) = 0$ and

$$c(j, a) = q_j = \begin{cases} +1, & \text{if } 2^{2^m} < j \leq 2^{2^{m+1}}, \quad m = 0, 2, 4, \ldots; \\ -1, & \text{for other } j > 2, \end{cases}$$

$c(j', a) = q_{j'} = -q_j = -c(j, a)$ for all $j > 2$ (see Fig. 4.14).

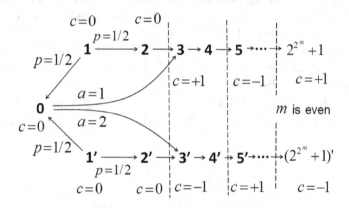

Fig. 4.14 Example 4.2.12: only a semi-Markov strategy is AC-optimal.

For arbitrary $N \geq 1$ and $\varepsilon > 0$, one can find a $T = 2^{2^{m+1}} > N$ with an even value of $m \geq 0$, such that $\frac{2 \cdot 2^{2^m}}{2^{2^{m+1}}} < \varepsilon$. Now

$$\frac{1}{T} \sum_{j=3}^{T} q_j \geq \frac{1}{2^{2^{m+1}}} \left[2^{2^{m+1}} - 2 \cdot 2^{2^m} \right] \geq 1 - \varepsilon.$$

(We estimated the first 2^{2^m} terms from below: $q_j \geq -1$ for all $j \leq 2^{2^m}$.) Therefore, $\limsup_{T \to \infty} \frac{1}{T} \sum_{j=3}^{T} q_j = 1$. Similarly, one can show that $\limsup_{T \to \infty} \frac{1}{T} \sum_{j=3}^{T} q_{j'} = 1$.

Now it is clear that, for any strategy π, $v_3^\pi = \limsup_{T \to \infty} \frac{1}{T} \sum_{j=3}^{T+2} q_j = 1 = v_3^*$ and $v_{3'}^\pi = \limsup_{T \to \infty} \frac{1}{T} \sum_{j=3}^{T+2} q_{j'} = 1 = v_{3'}^*$; the same equalities hold for all initial states i, i' with $i \geq 2$.

When starting from $X_0 = 0$, the next states X_t in the sequence $t = 1, 2, \ldots$ can be $(2+t)$ or $(2+t)'$. If they appear equiprobably (i.e. $\pi_1(1|0) = \pi_1(2|0) = 1/2$) then $E_0^\pi[c(X_{t-1}, A_t)] = \frac{1}{2}q_{t+1} + \frac{1}{2}q_{(t+1)'} = 0$ for all $t \geq 0$, and $v_0^\pi = 0$. Otherwise, $v_0^\pi > 0$: for instance, if $\pi_1(1|0) = \alpha > 1/2$ then

$$E_0^\pi[c(X_{t-1}, A_t)] = \alpha q_{t+1} + (1-\alpha)q_{(t+1)'} = (2\alpha - 1)q_{t+1} \quad (\text{for } t \geq 2),$$

and we know that $\limsup_{T \to \infty} \frac{1}{T} \sum_{t=2}^{T} q_{t+1} = 1$.

Remark 4.5. The performance functional (4.1) is convex, but not linear on the space of strategic measures: for the $(1/2, 1/2)$ mixture $\hat{\pi}$ of strategies $\pi_t^1(1|x) \equiv 1$ and $\pi_t^2(2|x) \equiv 1$, we have $v_0^{\hat{\pi}} = 0$ while $v_0^{\pi^1} = v_0^{\pi^2} = 1$.

Suppose that $X_0 = 1$ or $1'$. The next states X_t in the sequence $t = 2, 3, \ldots$ can be $(1+t)$ or $(1+t)'$. The above reasoning implies that, for any strategy π, $v_1^\pi, v_{1'}^\pi \geq 0$, and, for the semi-Markov strategy $\tilde{\pi}$ satisfying

$$\tilde{\pi}_2(2|x_0 = 1) = \tilde{\pi}(1|x_0 = 1') = 1, \quad \tilde{\pi}_1(1|x_0 = 0) = \tilde{\pi}_1(2|x_0 = 0) = 1/2,$$

we have

$$v_1^{\tilde{\pi}} = v_{1'}^{\tilde{\pi}} = 0 = v_1^* = v_{1'}^*; \quad v_0^{\tilde{\pi}} = 0 = v_0^*,$$

meaning that $\tilde{\pi}$ is (uniformly) AC-optimal.

On the other hand, consider an arbitrary Markov strategy π^m with $\pi_2^m(1|0) = \alpha \in [0, 1]$:

$$\begin{aligned}
v_1^{\pi^m} &= \limsup_{T \to \infty} \frac{1}{T} E_1^{\pi^m} \left[\sum_{t=1}^{T} c(X_{t-1}, A_t) \right] \\
&= \limsup_{T \to \infty} \frac{1}{T} \sum_{i=3}^{T} \left[\left(\frac{1}{2} + \frac{1}{2}\alpha \right) q_i + \frac{1}{2}(1-\alpha) q_{i'} \right] \\
&= \limsup_{T \to \infty} \frac{1}{T} \sum_{i=3}^{T} \alpha q_i = \alpha,
\end{aligned}$$

and, similarly, $v_{1'}^{\pi^m} = (1 - \alpha)$. It is clear that one cannot have $v_1^{\pi^m} = v_{1'}^{\pi^m} = 0$; the strategy π^m is not (uniformly) AC-ε-optimal if $\varepsilon < 1/2$. But there certainly exists an optimal Markov strategy for a fixed initial distribution.

4.2.13 Singular perturbation of an MDP

Suppose the state space is finite (or countable). We say that an MDP is *perturbed* if the transition probabilities (and possibly the loss function) change slightly according to $p(y|x,a) + \varepsilon d(y|x,a)$, where ε is a small parameter. A perturbation is *singular* if it changes the ergodic structure of the underlying Markov chain. In such cases it can happen that one stationary selector φ^ε is AC-optimal for all small enough $\varepsilon > 0$, but an absolutely different selector φ^* is AC-optimal for $\varepsilon = 0$. The following example is based on [Avrachenkov et al.(2002), Ex. 2.1].

Let $\mathbf{X} = \{1,2\}$, $\mathbf{A} = \{1,2\}$, $p(1|1,a) \equiv p(2|2,a) \equiv 1$, with all other transition probabilities zero. Let $d(1|1,2) = -1$, $d(2|1,2) = 1$, $d(1|2,a) \equiv 1$, $d(2|2,a) \equiv -1$, with other values of function d being zero. We put $c(1,1) = 1$, $c(1,2) = 1.5$, $c(2,a) \equiv 0$ (see Fig. 4.15).

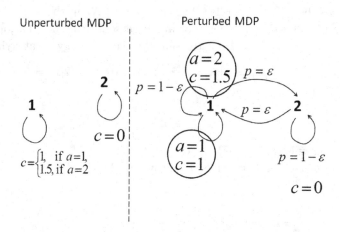

Fig. 4.15 Example 4.2.13: singularly perturbed MDP.

The solution of the unperturbed MDP (when $\varepsilon = 0$) is obvious: the stationary selector $\varphi^*(x) \equiv 1$ is optimal; $\rho(1) = 1$, $\rho(2) = 0$.

When $\varepsilon > 0$, after we put $h^\varepsilon(1) = 0$, the canonical equations (4.2) take the form

$$\rho^\varepsilon = \min\{1,\ 1.5 + \varepsilon h^\varepsilon(2)\};$$
$$\rho^\varepsilon + h^\varepsilon(2) = (1-\varepsilon)h^\varepsilon(2)$$

and have a single solution $\rho^\varepsilon = 0.75$, $h^\varepsilon(2) = -0.75/\varepsilon$ leading to the canonical triplet $\langle \rho^\varepsilon, h^\varepsilon, \varphi^\varepsilon \rangle$, where $\varphi^\varepsilon(x) \equiv 2$ is the only AC-optimal stationary selector for all $\varepsilon \in (0,1)$ (one can ignore any actions at the uncontrolled state 2). We also see that the limit $\lim_{\varepsilon\to 0}\rho^\varepsilon=0.75$ is different from $\rho(1)$ and $\rho(2)$.

4.2.14 Blackwell optimal strategies and AC-optimality

It is known that, in a finite model, a Blackwell optimal strategy exists and is AC-optimal [Bertsekas(2001), Prop. 4.2.2].

The first example shows that, even in finite models, an AC-optimal strategy may be not Blackwell optimal.

Let $\mathbf{X}=\{0,1\}$, $\mathbf{A}=\{0,1\}$, $p(0|0,a)\equiv 1$, $p(0|1,0)=1$, $p(1|1,1)=1$, with all other transition probabilities zero; $c(0,a)\equiv 0$, $c(1,1)=0$, $c(1,0)=1$ (see Fig. 4.16).

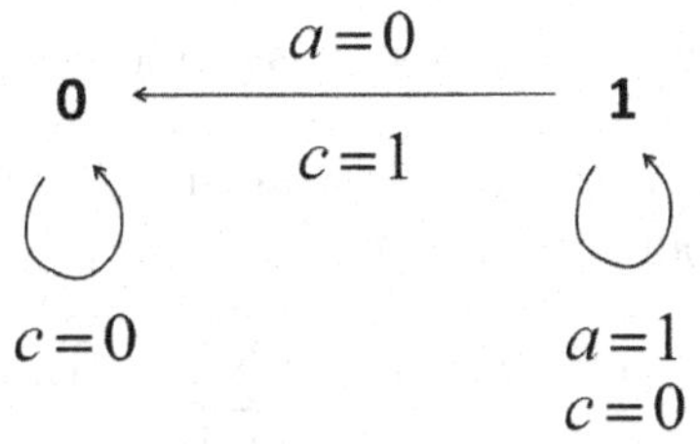

Fig. 4.16 Example 4.2.14: an AC-optimal stationary selector $\varphi(x)\equiv 0$ is not Blackwell optimal.

Clearly, $v_x^{*,\beta}\equiv 0$ and the stationary selector $\varphi(x)\equiv x$ is Blackwell optimal (and also AC-optimal). At the same time, for any control strategy π we have $v_x^\pi = v_x^* = 0$, because only trajectories $(0,0,\ldots)$, $(1,1,\ldots)$, $(1,1,\ldots,0,0,\ldots)$ can be realized. Thus, all strategies are AC-optimal, but the stationary selector $\varphi(x)\equiv 0$ is not Blackwell optimal: it is not optimal for any value $\beta\in(0,1)$, if $X_0=1$.

The second example, based on [Flynn(1974)], shows that, if the state space $\mathbf{X}$ is not finite, then a Blackwell optimal strategy may not be AC-optimal.

Let $C_1,C_2,\ldots$ be a bounded sequence such that

$$C^* \stackrel{\triangle}{=} \limsup_{n\to\infty} \frac{1}{n}\sum_{i=1}^n C_i > \limsup_{\beta\to 1-}(1-\beta)\sum_{i=1}^\infty \beta^{i-1}C_i \stackrel{\triangle}{=} C_*$$

(see Appendix A.4). Suppose $\mathbf{X}=\{1,2,\ldots\}$; $\mathbf{A}=\{0,1\}$; $p(1|1,0)=1$, $p(2|1,1)=1$, $p(i+1|i,a)\equiv 1$ for all $i>1$; all other transition probabilities are zero. We put $c(1,0) \triangleq (C_* + C^*)/2$, $c(1,1) = C_1$, $c(i,a) \equiv C_i$ for all $i > 1$ (see Fig. 4.17).

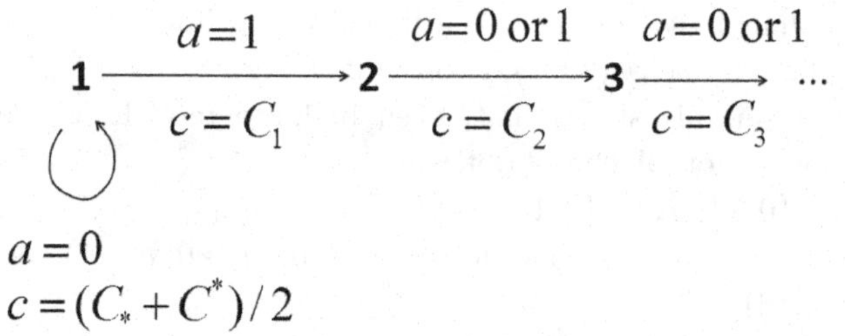

```
       a=1         a=0 or 1     a=0 or 1
  1 ─────────► 2 ─────────► 3 ─────────► ...
  ↻   c = C₁        c = C₂        c = C₃

  a = 0
  c = (C_* + C^*)/2
```

Fig. 4.17 Example 4.2.14: a Blackwell optimal strategy is not AC-optimal.

It is sufficient to consider only two strategies $\varphi^0(x) \equiv 0$ and $\varphi^1(x) \equiv 1$ and initial state 1. For all β close to 1,

$$(1-\beta)v_1^{\varphi^0,\beta} = (C_* + C^*)/2 > (1-\beta)\sum_{i=1}^{\infty} \beta^{i-1}C_i = (1-\beta)v_1^{\varphi^1,\beta},$$

meaning that the stationary selector φ^1 is Blackwell optimal. But

$$v_1^{\varphi^0} = (C_* + C^*)/2 < C^* = v_1^{\varphi^1},$$

so that the stationary selector φ^0 is AC-optimal. The Blackwell optimal strategy φ^1 is not AC-optimal.

4.2.15 Strategy iteration in a unichain model

The basic *strategy iteration algorithm* can be written as follows [Puterman(1994), Section 8.6.1].

1. Set $n = 0$ and select a stationary selector φ^0 arbitrarily enough.
2. Obtain a scalar ρ_n and a bounded function h_n on $\mathbf{X}$ by solving the equation

$$\rho_n + h_n(x) = c(x, \varphi^n(x)) + \int_{\mathbf{X}} h_n(y) p(dy|x, \varphi^n(x)).$$

(Clearly, $h_n + const$ is also a solution for any value of $const$; we leave aside the question of the measurability of v^{n+1}.)

3. Choose $\varphi^{n+1} : \mathbf{X} \to \mathbf{A}$ such that

$$c(x, \varphi^{n+1}(x)) + \int_\mathbf{X} h_n(y) p(dy|x, \varphi^{n+1}(x))$$

$$= \inf_{a \in \mathbf{A}} \left\{ c(x, a) + \int_\mathbf{X} h_n(y) p(dy|x, a) \right\},$$

setting $\varphi^{n+1}(x) = \varphi^n(x)$ whenever possible.
4. If $\varphi^{n+1} = \varphi^n$, then stop and set $\varphi^* = \varphi^n$; $\rho = \rho_n$; $h = h_n$. Otherwise increment n by 1 and return to step 2.

It is known that, in a finite unichain model, this algorithm converges in a finite number of iterations to a solution of the canonical equations (4.2); furthermore, $\langle \rho, h, \varphi^* \rangle$ is a canonical triplet and the stationary selector φ^* is AC-optimal [Puterman(1994), Th. 8.6.6].

The following example shows that the stationary selector returned by the strategy iteration need not be bias optimal. Let $\mathbf{X} = \{1, 2\}$, $\mathbf{A} = \{1, 2\}$, $p(1|x, 1) \equiv 1$, $p(2|1, 2) = p(1|2, 2) = 1$, with all other transition probabilities zero; $c(1, 1) = -4$, $c(1, 2) = 0$, $c(2, a) \equiv -8$ (see Fig. 4.18).

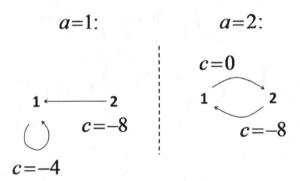

Fig. 4.18 Example 4.2.15: the strategy iteration does not return a bias optimal stationary selector.

The stationary selectors $\hat{\varphi}(x) \equiv 1$ and $\hat{\hat{\varphi}}(x) \equiv 2$ are equally AC-optimal, but only the selector $\hat{\varphi}$ is bias optimal (0-discount optimal – see Definition 3.8) because

$$v_x^{\hat{\varphi},\beta} - v_x^{\varphi,\beta} = \begin{cases} \dfrac{-8\beta}{1-\beta^2} - \dfrac{-4}{1-\beta} = \dfrac{4}{1+\beta} > 0, & \text{if } x = 1; \\ \\ \dfrac{-8}{1-\beta^2} - \dfrac{4\beta-8}{1-\beta} = \dfrac{4\beta}{1+\beta} > 0, & \text{if } x = 2. \end{cases}$$

At the same time, the strategy iteration starting from $\varphi^0 = \hat{\varphi}$ gives the following:

$$\rho_0 = -4, \quad h_0(x) = \begin{cases} 0, & \text{if } x = 1; \\ -4, & \text{if } x = 2, \end{cases} \quad \varphi^1 = \hat{\varphi},$$

and we terminate the algorithm concluding that $\hat{\varphi}$ is AC-optimal and $\langle \rho_0, h_0, \varphi^0 \rangle$ is the associated canonical triplet. Note that $\langle \rho_0, h_0, \check{\varphi} \rangle$ is another canonical triplet. A similar example was considered in [Puterman(1994), Ex. 8.6.2].

In discrete unichain models, if $\mathbf{X} = \mathbf{X}^r \cup \mathbf{X}^t$, where, under any stationary selector, $\mathbf{X}^r$ is the (same) set of recurrent states, $\mathbf{X}^t$ being the transient subset, then one can apply the strategy iteration algorithm to the recurrent subset $\mathbf{X}^r$. The transient states can be ignored.

There was a conjecture [Hordijk and Puterman(1987), Th. 4.2] that if, in a finite unichain model, for some number ρ and function h,

$$c(x, \varphi(x)) + \sum_{y \in \mathbf{X}} h(y) p(y|x, \varphi(x)) \qquad (4.10)$$

$$= \inf_{a \in A} \left\{ c(x, a) + \sum_{y \in \mathbf{X}} h(y) p(y|x, a) \right\} \text{ for all } x \in \mathbf{X}$$

and

$$\rho + h(x) = \inf_{a \in A} \left\{ c(x, a) + \sum_{y \in \mathbf{X}} h(y) p(y|x, a) \right\} \text{ for all } x \in \mathbf{X}^{r,\varphi}, \quad (4.11)$$

then the stationary selector φ is AC-optimal. Here $\mathbf{X}^{r,\varphi}$ is the set of recurrent states under strategy φ.

The following example, based on [Golubin(2003), Ex. 1], shows that this statement is incorrect. We consider the same model as before (Fig. 4.18), but with a different loss function: $c(1,1) = 1$, $c(2,1) = 3$, $c(x,2) \equiv 0$ (see Fig. 4.19).

Take $\varphi(x) = x$, $\rho = 1$ and $h(x) = \begin{cases} 0, & \text{if } x = 1; \\ 2, & \text{if } x = 2. \end{cases}$ Then $\mathbf{X}^{r,\varphi} = \{1\}$ and equations (4.10) and (4.11) are satisfied, but the stationary selector φ is obviously not AC-optimal: $v_x^* = v_x^{\varphi^*} \equiv 0$ for $\varphi^*(x) \equiv 2$.

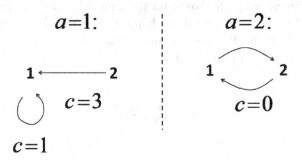

Fig. 4.19 Example 4.2.15: different recurrent subsets.

The same example shows that it is insufficient to apply step 3 of the algorithm only to the subset $\mathbf{X}^{r,\varphi^n}$. Indeed, if we take $\varphi^0(x) \equiv 1$, then direct calculations show that $\rho_0 = 1$ and $h_0(x) = \begin{cases} 0, & \text{if } x = 1; \\ 2, & \text{if } x = 2. \end{cases}$ The equation

$$c(x,\varphi^0(x)) + \sum_{y \in \mathbf{X}} h_0(y) p(y|x, \varphi^0(x)) = \inf_{a \in \mathbf{A}} \left\{ c(x,a) + \sum_{y \in \mathbf{X}} h_0(y) p(y|x, a) \right\}$$

holds for all $x \in \mathbf{X}^{r,\varphi^0}$, i.e. for $x = 1$. But the iterations are unfinished because, after further steps, we will obtain $\varphi^1(2) = 2$ and $\varphi^2(x) = \varphi^*(x) \equiv 2$.

Of course, one can ignore the transient states if the recurrent subset $\mathbf{X}^{r,\varphi^n}$ does not increase with n.

4.2.16 Unichain strategy iteration in a finite communicating model

According to [Puterman(1994), Section 9.5.1], if a finite model is communicating, then there exists an AC-optimal stationary selector such that the controlled process has a single communicating class, i.e. it is a unichain Markov process. Therefore, one might conjecture that we can solve such a problem using the unichain strategy iteration algorithm described in Section 4.2.15.

The following example, based on the same idea as [Puterman(1994), Ex. 9.5.1] shows that this need not happen. Let $\mathbf{X} = \{1,2\}$, $\mathbf{A} = \{1,2\}$, $p(a|x,a) \equiv 1$, with all other transition probabilities zero; $c(1,a) = a$, $c(2,1) = 2$, $c(2,2) = 0$ (see Fig. 4.20).

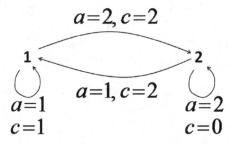

Fig. 4.20 Example 4.2.16: the unichain strategy iteration algorithm is not applicable for communicating models.

We try to apply the unichain strategy iteration algorithm starting from $\varphi^0(x) \equiv 1$. This stationary selector gives a unichain Markov process, and one can find $\rho_0 = 1$, $h_0(x) = x - 1$. At step 3, we obtain the improved stationary selector $\varphi^1(x) = x$. On the next iteration, step 2, we have the equations:

$$\rho_1 + h_1(1) = 1 + h_1(1); \qquad \rho_1 + h_1(2) = 0 + h_1(2),$$

which have no solutions.

There exist special strategy iteration algorithms applicable to communicating and general finite models [Puterman(1994), Sections 9.5 and 9.2]; see also Section 4.2.17.

4.2.17 Strategy iteration in semi-continuous models

If the model is multichain (i.e. Definition 4.2 is violated), then this algorithm, in the case of finite or countable *ordered* state space $\mathbf{X}$, can be written as follows [Puterman(1994), Section 9.2.1].

1. Set $n = 0$ and select a stationary selector φ^0 arbitrarily enough.
2. Obtain bounded functions ρ_n and h_n on $\mathbf{X}$ by solving the equations

$$\rho_n(x) = \sum_{y \in \mathbf{X}} \rho_n(y) p(y|x, \varphi^n(x));$$

$$\rho_n(x) + h_n(x) = c(x, \varphi^n(x)) + \sum_{y \in \mathbf{X}} h_n(y) p(y|x, \varphi^n(x)).$$

At this step, one must determine the structure of the controlled process under the selector φ^n, denote its recurrent classes by $R_1, R_2, \ldots$, and put $h_n(x_i) = 0$, where x_i is the minimal state in R_i.

3. Choose $\varphi^{n+1} : \mathbf{X} \to \mathbf{A}$ such that

$$\sum_{y \in \mathbf{X}} \rho_n(y) p(y|x, \varphi^{n+1}(x)) = \inf_{a \in \mathbf{A}} \left\{ \sum_{y \in \mathbf{X}} \rho_n(y) p(y|x, a) \right\},$$

setting $\varphi^{n+1}(x) = \varphi^n(x)$ whenever possible. If $\varphi^{n+1} = \varphi^n$, go to step 4; otherwise, increment n by 1 and return to step 2.

4. Choose $\varphi^{n+1} : \mathbf{X} \to \mathbf{A}$ such that

$$c(x, \varphi^{n+1}(x)) + \sum_{y \in \mathbf{X}} h_n(y) p(y|x, \varphi^{n+1}(x))$$

$$= \inf_{a \in \mathbf{A}} \left\{ c(x, a) + \sum_{y \in \mathbf{X}} h_n(y) p(y|x, a) \right\},$$

setting $\varphi^{n+1}(x) = \varphi^n(x)$ whenever possible. If $\varphi^{n+1} = \varphi^n$, stop and set $\varphi^* = \varphi^n$, $\rho = \rho_n$, $h = h_n$. Otherwise increment n by 1 and return to step 2.

It is known that, in a finite model, this algorithm converges in a finite number of iterations to a solution of the canonical equations (4.2); $\langle \rho, h, \varphi^* \rangle$ is a canonical triplet and the stationary selector φ^* is AC-optimal [Puterman(1994), Cor. 9.2.7].

The following example, based on [Dekker(1987), Th. 1], shows that, if the action space is not finite, then this algorithm may fail to converge, even in a semi-continuous model (Definition 4.3).

Let $\mathbf{X} = \{1, 2, 3, 4\}$, $\mathbf{A} = \{\hat{0}\} \cup \{a : 0 \leq \alpha \leq \frac{\sqrt{3}-1}{2}\}$, $p(1|1, a) =$
$p(2|2, a) = p(3|4, a) \equiv 1$, $p(4|3, \hat{0}) = 1$, $p(y|3, \alpha) = \begin{cases} \frac{1}{4} + \alpha^2, & \text{if } y = 1; \\ \frac{1}{4} + \alpha, & \text{if } y = 2; \\ \frac{1}{2} - \alpha - \alpha^2, & \text{if } y = 3, \end{cases}$

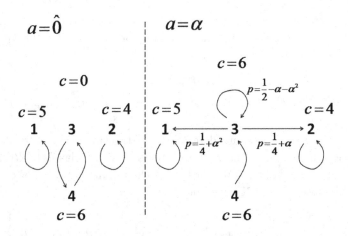

Fig. 4.21   Example 4.2.17: the strategy iteration algorithm does not converge.

with all other transition probabilities zero; $c(1,a) \equiv 5$, $c(2,a) \equiv 4$, $c(4,a) \equiv 6$, $c(3,\hat{0}) = 0$, $c(3,\alpha) = 6$ (see Fig. 4.21).

The canonical equations (4.2) have the following solution:

$$\rho(x) = \begin{cases} 5, & \text{if } x = 1; \\ 4, & \text{if } x = 2; \\ 3, & \text{if } x = 3 \text{ or } 4 \end{cases} \qquad h(x) = \begin{cases} 3, & \text{if } x = 4; \\ 0 & \text{otherwise} \end{cases} \qquad \varphi^*(x) \equiv \hat{0},$$

and the canonical stationary selector φ^* is the only AC-optimal strategy. We ignore the actions in the uncontrolled states $x = 1, 2, 4$. If $P_3^\pi\{A_t = \alpha\} = p > 0$ for some $t \geq 1$, then $v_3^\pi \geq 4p + 3(1-p)$, but $v_3^{\varphi^*} = v_3^* = 3$.

The multichain strategy iteration starting from $\varphi^0(x) = \tilde{\alpha} \in \mathbf{A} \setminus \{\hat{0}\}$ results in the following. For all $n = 0, 1, 2, \ldots$,

$$\rho_n(1) \equiv 5, \quad \rho_n(2) \equiv 4, \quad \rho_n(4) = \rho_n(3), \quad h_n(1) \equiv 0, \quad h_n(2) \equiv 0.$$

The value of $\rho_0(3)$ comes from the equation $\rho_0(3) = 5(\frac{1}{4} + \tilde{\alpha}^2) + 4(\frac{1}{4} + \tilde{\alpha}) + \rho_0(3)(\frac{1}{2} - \tilde{\alpha} - \tilde{\alpha}^2)$:

$$\rho_0(3) = \frac{9 + 20\tilde{\alpha}^2 + 16\tilde{\alpha}}{2 + 4\tilde{\alpha}^2 + 4\tilde{\alpha}} \in (4,5) \text{ for all } \tilde{\alpha} \in \mathbf{A} \setminus \{\hat{0}\} \qquad (4.12)$$

Finally, the values of $h_0(3)$ and $h_0(4)$ come from the equations

$$\rho_0(3) + h_0(3) = 6 + h_0(3)\left(\frac{1}{2} - \tilde{\alpha} - \tilde{\alpha}^2\right);$$

$$\rho_0(4) + h_0(4) = 6 + h_0(3).$$

At step 3, we need to minimize the expression

$$F(\alpha) = 5\left(\frac{1}{4} + \alpha^2\right) + 4\left(\frac{1}{4} + \alpha\right) + \rho_0(3)\left(\frac{1}{2} - \alpha - \alpha^2\right)$$

$$= \frac{9}{4} + \frac{1}{2}\rho_0(3) + (5 - \rho_0(3))\alpha^2 + (4 - \rho_0(3))\alpha.$$

Therefore, the minimum $\min_{\alpha \in \left[0, \frac{\sqrt{3}-1}{2}\right]} F(\alpha)$ is provided by

$$\alpha^* = \min\left\{\frac{\rho_0(3) - 4}{2(5 - \rho_0(3))}, \frac{\sqrt{3}-1}{2}\right\}.$$

In any case,

$$\min_{\alpha \in \mathbf{A}\setminus\{\hat{0}\}} F(\alpha) \leq 5\left(\frac{1}{4} + \tilde{\alpha}^2\right) + 4\left(\frac{1}{4} + \tilde{\alpha}\right) + \rho_0(3)\left(\frac{1}{2} - \tilde{\alpha} - \tilde{\alpha}^2\right) = \rho_0(3),$$

and the equality is attained iff $\alpha^* = \tilde{\alpha} = \frac{1}{4}(\sqrt{5} - 1)$. Note that $\sum_{y \in \mathbf{X}} \rho_0(y) p(y|3, \hat{0}) = \rho_0(4) = \rho_0(3)$. Therefore, if $\tilde{\alpha} \neq \frac{1}{4}(\sqrt{5}-1)$, then

$$\varphi^1(3) = \alpha^* \neq \tilde{\alpha}, \qquad \sum_{y \in \mathbf{X}} \rho_0(y) p(y|3, \alpha^*) < \sum_{y \in \mathbf{X}} \rho_0(y) p(y|3, \tilde{\alpha}),$$

and we return to step 2. Similar reasoning applies to the further loops of the algorithm. If $\tilde{\alpha}$ is rational then α^* is a rational function of $\tilde{\alpha}$ or $\sqrt{3}$. Therefore, $\varphi^n(3)$ will never reach the value of $\frac{1}{4}(\sqrt{5}-1)$ and the algorithm never terminates. The value of $\rho_n(3)$ decreases at each step, but remains greater than 4.

If a semi-continuous model is unichain then, again, it can happen that the (unichain) strategy iteration never terminates: see Theorem 3 in [Dekker(1987)].

4.2.18 When value iteration is not successful

The basic value iteration algorithm can be written as follows [Puterman(1994), Section 8.5.1].

1. Set $n = 0$, specify a small enough $\varepsilon > 0$, and select a bounded measurable function $v^0(x) \in \mathbf{B}(\mathbf{X})$.
2. Compute

$$v^{n+1}(x) = \inf_{a \in \mathbf{A}} \left\{ c(x,a) + \int_{\mathbf{X}} v^n(y) p(dy|x,a) \right\} \qquad (4.13)$$

(we leave aside the question of the measurability of v^{n+1}).

3. If
$$\sup_{x \in \mathbf{X}}[v^{n+1}(x) - v^n(x)] - \inf_{x \in \mathbf{X}}[v^{n+1}(x) - v^n(x)] < \varepsilon,$$
stop and choose $\varphi^*(x)$ providing the infimum in (4.13). Otherwise increment n by 1 and return to step 2.

In what follows, for $v \in \mathbf{B}(\mathbf{X})$,
$$sp(v) \triangleq \sup_{x \in \mathbf{X}}[v(x)] - \inf_{x \in \mathbf{X}}[v(x)]$$
is the so-called *span* of the (bounded) function v; it exhibits all the properties of a seminorm and is convenient for the comparison of classes of equivalence when we do not distinguish between two bounded functions v_1 and v_2 if $v_2(x) \equiv v_1(x) + const$. See Remark 4.1 in this connection. It can easily happen that $\sup_{x \in \mathbf{X}} |v^{n+1}(x) - v^n(x)|$ does not approach zero, but $\lim_{n \to \infty} sp(v^{n+1} - v^n) = 0$, and the value iteration returns an AC-optimal selector in a finite number of steps, if ε is small enough.

The example presented in Section 4.2.2 (Fig. 4.1) confirms this statement. Value iteration starting from $v^0(x) \equiv 0$ results in the following values for $v^n(x)$:

n	0	1	2	3	4	5	...
$x=1$	0	-10	-9.5	-8.75	-4.875	-6.9375	
$x=2$	0	1	2	3	4	5	

One can prove by induction that, starting from $n = 2$,
$$v^n(1) = -12 + n + (0.5)^{n-1}, \qquad v^n(2) = n,$$
and the minimum in (4.13) is provided by $\varphi^*(x) \equiv 1$. We see that $\sup_{x \in \mathbf{X}} |v^{n+1}(x) - v^n(x)| = 1$ and $sp(v^{n+1} - v^n) = 0.5^n$. The stationary selector φ^* is AC-optimal, just like any other control strategy, because the process will ultimately be absorbed at state 2; $\langle \rho = 1, h(x) = \begin{cases} 0, & \text{if } x = 1; \\ 12, & \text{if } x = 2, \end{cases} \varphi^* \rangle$ is the canonical triplet.

Condition 4.3. The model is finite and either

(a) $\max\limits_{x \in \mathbf{X}, a \in \mathbf{A}, y \in \mathbf{X}, b \in \mathbf{A}} [1 - \sum\limits_{z \in \mathbf{X}} \min\{p(z|x,a), p(z|y,b)\}] < 1$;

(b) there exists a state $\hat{x} \in \mathbf{X}$ and an integer K such that, for any stationary selector φ, for all $x \in \mathbf{X}$, the K-step transition probability satisfies $p^K(\hat{x}|x, \varphi(x)) > 0$, or

(c) the model is unichain and $p(x|x,a) > 0$ for all $x \in \mathbf{X}$, $a \in \mathbf{A}$.

It is known that, under Condition 4.3, value iteration achieves the stopping criterion for any $\varepsilon > 0$ [Puterman(1994), Th. 8.5.3]. In the above example, all the Conditions 4.3(a,b,c) are satisfied.

In the next example [Puterman(1994), Ex. 8.5.1], Condition 4.3 is violated and the value iteration never terminates. Let $\mathbf{X} = \{1,2\}$, $\mathbf{A} = \{0\}$ (dummy action); $p(2|1,0) = p(1|2,0) = 1$, with all other transition probabilities zero. We put $c(1,0) = c(2,0) = 0$. If $v^0(1) = r^1$ and $v^0(2) = r^2$ then $v^1(1) = r^2$, $v^1(2) = r^1$, $v^2(1) = r^1$, $v^2(2) = r^2$, and so on: v^n oscillates with period 2, $\sup_{x \in \mathbf{X}} |v^{n+1}(x) - v^n(x)| = |r^1 - r^2|$, and $sp(v^{n+1} - v^n) = 2|r^1 - r^2|$. Value iteration is unsuccessful unless $r^1 = r^2 = 0$.

4.2.19 The finite-horizon approximation does not work

One might think that, by solving the finite-horizon problem (1.1) with the final loss $C(x) \equiv 0$ for a large enough value of T, then an AC-optimal control strategy will be approximated in some sense and $V_x^T/T \stackrel{\triangle}{=} \inf_\pi E_x^\pi \left[\sum_{t=1}^T c(X_{t-1}, A_t) \right]/T$ will converge to v_x^* as $T \to \infty$. The following example, based on [Flynn(1980), Ex. 1], shows that this conjecture is false in general.

Let $\mathbf{X} = \{0, 0', 1, 1', \ldots\}$, $\mathbf{A} = \{0, 1, 2\}$, $p(0|0,a) = p(0'|0',a) \equiv 1$, $p((i-1)'|i',a) \equiv 1$ for all $i \geq 1$, $p(0|i,0) = p(i+1|i,1) = p(i'|i,2) = 1$ for all $i \geq 1$, $c(0,a) \equiv 0$, $c(0',a) \equiv 1$, $c(i,a) \equiv 1$, $c(i',a) \equiv -3$ (see Fig. 4.22).

If the initial state is 0, or $0'$, or i' ($i \geq 1$), then in fact the process is uncontrolled and, for any $T < \infty$, the values

$$\frac{V_0^T}{T} = 0, \quad \frac{V_{0'}^T}{T} = 1, \quad \frac{V_{i'}^T}{T} = \begin{cases} -3, & \text{if } T \leq i; \\ 1 - 4i/T, & \text{if } T > i \end{cases}$$

indeed approach the corresponding long-run average losses.

Suppose the initial state is $i \geq 1$ and the time horizon T is finite. Then the optimal control strategy π^* prescribes moving right (applying action 1) s times and applying action 2 afterwards, where $s = \lfloor \frac{T-i-1}{2} \rfloor$ is the integer part. As a result,

$$V_i^T = \begin{cases} 1 - 3(T-1), & \text{if } T \leq i+1; \\ -3i - 2s + 1, & \text{if } T - i - 1 = 2s \quad \text{for } s = 1, 2, \ldots; \\ -3i - 2s + 2, & \text{if } T - i - 1 = 2s+1 \quad \text{for } s = 1, 2, \ldots. \end{cases}$$

Therefore,

$$\lim_{T \to \infty} \frac{1}{T} V_i^T = -1.$$

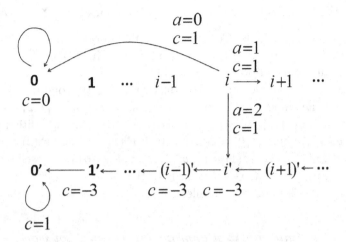

Fig. 4.22 Example 4.2.19: an optimal strategy in a finite-horizon model is not AC-optimal.

On the other hand, the expected average loss v_i^π equals 0 if action $a = 0$ appears before action 2, and equals $+1$ in all other cases. In other words, the stationary selector $\varphi^*(i) \equiv 0$ is AC-optimal, leading to $v_i^{\varphi^*} = v_i^* = 0$. The finite-horizon optimal strategy has nothing in common with φ^* and $\lim_{T \to \infty} \frac{1}{T} V_i^T \neq v_i^*$. When T goes to infinity, the difference between the performance functionals under φ^* and the control strategy π^* described above also goes to infinity, meaning that the AC-optimal selector φ^* becomes progressively less desirable as T increases. In this example, the canonical triplet $\langle \rho, h, \varphi^* \rangle$ exists:

$$\rho(x) = \begin{cases} 0, & \text{if } x = i \geq 0; \\ 1, & \text{if } x = i' \text{ with } i \geq 0, \end{cases}$$

$$h(x) = \begin{cases} 0, & \text{if } x = 0 \text{ or } 0'; \\ 1, & \text{if } x = i \geq 1; \\ i, & \text{if } x = i' \text{ with } i \geq 1, \end{cases} \qquad \varphi^*(x) \equiv 0.$$

Note that the stationary selector φ^* satisfies the following very strong condition of optimality: for any π, for all $x \in \mathbf{X}$,

$$\limsup_{T \to \infty} \left\{ E_x^{\varphi^*} \left[\sum_{t=1}^T c(X_{t-1}, A_t) \right] - E_x^\pi \left[\sum_{t=1}^T c(X_{t-1}, A_t) \right] \right\} \leq 0. \quad (4.14)$$

If (4.14) holds then the strategy φ^* is AC-optimal.

4.2.20 The linear programming approach to finite models

If the model is finite then problem (4.1) can be solved using the linear programming approach. In this approach, one needs to solve the problem

$$\sum_{x \in \mathbf{X}} \sum_{a \in \mathbf{A}} c(x,a)\eta(x,a) \to \inf_{\eta, \tilde{\eta}} \qquad (4.15)$$

$$\sum_{a \in \mathbf{A}} \eta(x,a) = \sum_{y \in \mathbf{X}} \sum_{a \in \mathbf{A}} p(x|y,a)\eta(y,a), \qquad x \in \mathbf{X}$$

$$\eta(x,a) \geq 0 \qquad (4.16)$$

$$\sum_{a \in \mathbf{A}} \eta(x,a) + \sum_{a \in \mathbf{A}} \tilde{\eta}(x,a) - \sum_{y \in \mathbf{X}} \sum_{a \in \mathbf{A}} p(x|y,a)\tilde{\eta}(y,a) = \alpha(x), \quad x \in \mathbf{X}$$

$$\tilde{\eta}(x,a) \geq 0, \qquad (4.17)$$

where $\alpha(x) > 0$ are arbitrarily fixed numbers such that $\sum_{x \in \mathbf{X}} \alpha(x) = 1$. The following stationary strategy is then AC-optimal [Puterman(1994), Th. 9.3.8]:

$$\pi^s(a|x) = \begin{cases} \eta(x,a)/\sum_{a \in \mathbf{A}} \eta(x,a), & \text{if } \sum_{a \in \mathbf{A}} \eta(x,a) > 0; \\ \tilde{\eta}(x,a)/\sum_{a \in \mathbf{A}} \tilde{\eta}(x,a), & \text{if } \sum_{a \in \mathbf{A}} \eta(x,a) = 0. \end{cases} \qquad (4.18)$$

We say that the strategy π^s in (4.18) is *induced* by a feasible solution $(\eta, \tilde{\eta})$. Note that at least one of equations in (4.16) is redundant.

It is known that, for any stationary strategy π^s, one can construct a feasible solution $(\eta, \tilde{\eta})$ to problem (4.15), (4.16), and (4.17), such that π^s is induced by $(\eta, \tilde{\eta})$. Moreover, if the policy $\pi^s(a|x) = I\{a = \varphi(x)\}$ is actually a selector, then $(\eta, \tilde{\eta})$ is a *basic* feasible solution (see [Puterman(1994), Section 9.3 and Th. 9.3.5]). In other words, some basic feasible solutions induce all the stationary selectors. The following example, based on [Puterman(1994), Ex. 9.3.2], shows that a basic feasible solution can induce a randomized strategy π^s.

Let $\mathbf{X} = \{1,2,3,4\}$, $\mathbf{A} = \{1,2,3\}$, $p(2|1,2) \equiv 1$, $p(1|2,1) = 1$, $p(3|2,2) = 1$, $p(4|2,3) = 1$, $p(4|4,a) \equiv 1$, $c(1,a) \equiv 1$, $c(2,1) = 4$, $c(2,2) = 3$, $c(2,3) = 0$, $c(4,1) = 3$, $c(4,2) = c(4,3) = 4$. See Fig. 4.23 (all transitions are deterministic). The linear program (4.15), (4.16), and (4.17) at $\alpha(1) = 1/6$, $\alpha(2) = 1/3$, $\alpha(3) = 1/6$, $\alpha(4) = 1/3$ can be rewritten in the following basic form:

Objective to be minimized: $\dfrac{8}{3} + \eta(4,2) + \eta(4,3) + \dfrac{1}{2}\tilde{\eta}(2,3);$

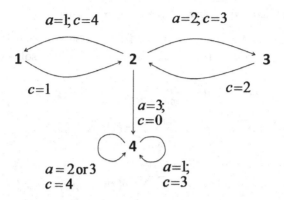

Fig. 4.23 Example 4.2.20: the basic solution gives a randomized strategy.

$$\eta(1,1) + \eta(1,2) + \eta(1,3) - \sum_{a=1}^{3} \tilde{\eta}(3,a) + \tilde{\eta}(2,2) + \frac{1}{2}\tilde{\eta}(2,3) = \frac{1}{6};$$

$$\eta(2,1) - \sum_{a=1}^{3} \tilde{\eta}(3,a) + \tilde{\eta}(2,2) + \frac{1}{2}\tilde{\eta}(2,3) = \frac{1}{6};$$

$$\eta(2,2) + \sum_{a=1}^{3} \tilde{\eta}(3,a) - \tilde{\eta}(2,2) = \frac{1}{6};$$

$$\eta(2,3) = 0;$$

$$\eta(3,1) + \eta(3,2) + \eta(3,3) + \sum_{a=1}^{3} \tilde{\eta}(3,a) - \tilde{\eta}(2,2) = \frac{1}{6};$$

$$\eta(4,1) + \eta(4,2) + \eta(4,3) - \tilde{\eta}(2,3) = \frac{1}{3};$$

$$\tilde{\eta}(2,1) + \tilde{\eta}(2,2) + \frac{1}{2}\tilde{\eta}(2,3) - \sum_{a=1}^{3}[\tilde{\eta}(1,a) + \tilde{\eta}(3,a)] = 0.$$

The variables $\eta(1,1), \eta(2,1), \eta(2,2), \eta(2,3), \eta(3,1), \eta(4,1)$, and $\tilde{\eta}(2,1)$ are basic, and this solution is in fact optimal. Now, according to (4.18), $\pi^s(1|1) = \pi^s(1|3) = \pi^s(1|4) = 1$, but $\pi^s(1|2) = \pi^s(2|2) = 1/2$.

One can make $\tilde{\eta}(2,2)$ basic, at level 0, instead of $\tilde{\eta}(2,1)$. That new basic solution will still be optimal, leading to the same strategy π^s. Of course, there exist many other optimal basic solutions leading to AC-optimal stationary selectors $\varphi(1) = 1$ or 2 or 3, $\varphi(2) = 1$ or 2, $\varphi(3) = 1$ or 2 or 3, $\varphi(4) = 1$. If one takes another distribution $\alpha(x)$, the linear program will produce AC-optimal strategies, but the value of the infimum in (4.15) will change. In fact, that infimum coincides with the long-run expected average loss (4.1) under an AC-optimal strategy if $P_0(x) = \alpha(x)$.

If the finite model is *recurrent*, i.e. under any stationary strategy, the controlled process has a single recurrent class and no transient states, then the linear program has the form (4.15), (4.16) complemented with the equation

$$\sum_{x \in \mathbf{X}} \sum_{a \in \mathbf{A}} \eta(x, a) = 1, \qquad (4.19)$$

where the variables $\tilde\eta$ are absent; for any feasible solution, $\sum_{a \in \mathbf{A}} \eta(x, a) > 0$, and formula (4.18) provides an AC-optimal strategy induced by the optimal solution η [Puterman(1994), Section 8.8.1]. In this case, the map (4.18) is a 1–1 correspondence between stationary strategies and feasible solutions to the linear program; moreover, that is a 1–1 correspondence between stationary *selectors* and *basic* feasible solutions. The inverse mapping to (4.18) looks like the following: $\eta(x, a) = \hat\eta^{\pi^s}(x) \pi^s(a|x)$, where $\hat\eta^{\pi^s}$ is the stationary distribution of the controlled process under strategy π^s [Puterman(1994), Th. 8.8.2 and Cor. 8.8.3].

In the example presented above, the model is not recurrent. One can still consider the linear program (4.15), (4.16), and (4.19). It can be rewritten in the following basic form:

Objective to be minimized: $\dfrac{5}{2} + \dfrac{1}{2}\eta(4,1) + \dfrac{3}{2}\eta(4,2) + \dfrac{3}{2}\eta(4,3);$

$\eta(1,1) + \eta(1,2) + \eta(1,3) + \eta(2,2) + \dfrac{1}{2}\sum_{a=1}^{3}\eta(4,a) = \dfrac{1}{2};$

$\eta(2,1) + \eta(2,2) + \dfrac{1}{2}\sum_{a=1}^{3}\eta(4,a) = \dfrac{1}{2};$

$\eta(2,3) = 0;$

$\eta(3,1) + \eta(3,2) + \eta(3,3) - \eta(2,2) = 0.$

The variables $\eta(1,1), \eta(2,1), \eta(2,3),$ and $\eta(3,1)$ are basic; this solution is in fact optimal. Now, the stationary selector satisfying $\varphi^*(1) = \varphi^*(2) = 1$ and the corresponding stationary distribution

$$\hat\eta(1) = \sum_{a=1}^{3} \eta(1, a) = \dfrac{1}{2}, \quad \hat\eta(2) = \sum_{a=1}^{3} \eta(2, a) = \dfrac{1}{2},$$

$$\hat{\eta}(3) = \sum_{a=1}^{3} \eta(3,a) = 0, \quad \hat{\eta}(4) = \sum_{a=1}^{3} \eta(4,a) = 0$$

solve the problem

$$\limsup_{T\to\infty} \frac{1}{T} E_{P_0}^{\pi}\left[\sum_{t=1}^{T} c(X_{t-1}, A_t)\right] \to \inf_{P_0, \pi}.$$

See [Hernandez-Lerma and Lasserre(1999), Th. 12.3.3] and Theorem 4.6 below. Note that the actions in states 3 and 4 can be arbitrary, as those states will never be visited under strategy φ^* because $P_0(x) = \hat{\eta}(x)$. Conversely, the original linear program (4.15), (4.16), (4.17) provides all the optimal actions: it solves problem (4.1) for all initial states.

The dual linear program to (4.15), (4.16), (4.17) can be written as follows:

$$\sum_{x \in \mathbf{X}} \alpha(x) \rho(x) \to \sup_{\rho, h}$$

$$\rho(x) \leq \sum_{y \in \mathbf{X}} \rho(y) p(y|x, a), \qquad x \in \mathbf{X}, a \in \mathbf{A} \qquad (4.20)$$

$$\rho(x) + h(x) \leq c(x,a) + \sum_{y \in \mathbf{X}} h(y) p(y|x,a), \qquad x \in \mathbf{X}, a \in \mathbf{A}$$

(compare this with the canonical equations (4.2)).

In the above example, one of the optimal solutions is written as follows:

$$\rho^*(1) = \rho^*(2) = \rho(3) = \frac{5}{2}, \quad \rho^*(4) = 3,$$

$$h^*(1) = -5, \quad h^*(2) = -\frac{7}{2}, \quad h^*(3) = -4, \quad h^*(4) = 0,$$

and all the constraints in (4.20) are satisfied as equalities. In fact, $\langle \rho^*, h^*, \varphi^*(x) \equiv 1 \rangle$ is a canonical triplet.

We can also consider the dual problem to (4.15), (4.16), and (4.19):

$$\rho \to \sup_{\rho, h}$$

$$\rho + h(x) \leq c(x, a) + \sum_{y \in \mathbf{X}} h(y) p(y|x,a), \qquad x \in \mathbf{X}, a \in \mathbf{A}.$$

In the above example, one of the optimal solutions is written as follows:

$$\rho^* = \frac{5}{2}, \quad h^*(1) = -5, \quad h^*(2) = -\frac{7}{2}, \quad h^*(3) = -4, \quad h^*(4) = 0.$$

Some of the constraints-inequalities in the last program remain strict; there are no canonical triplets with $\rho \equiv const$.

4.2.21 Linear programming for infinite models

The linear programming proved to be effective in finite models [Puterman(1994), Section 9.3]; [Kallenberg(2010), Section 5.8]. In the general case, this approach was developed in [Hernandez-Lerma and Lasserre(1996a), Chapter 6]; [Hernandez-Lerma and Lasserre(1999), Section 12.3], but under special conditions such as the following.

Condition 4.4.

(a) $v_{\hat{x}}^{\hat{\pi}} < \infty$ for some strategy $\hat{\pi}$ and some initial state $\hat{x}$.
(b) The loss function c is non-negative and lower semi-continuous; moreover, it is *inf-compact*; that is, the set $\{(x,a) \in \mathbf{X} \times \mathbf{A} : c(x,a) \leq r\}$ is compact for every number $r \in \mathbb{R}$.
(c) The transition probability p is a weakly continuous stochastic kernel.
(d) $\int_{\mathbf{X}} \min_{a \in \mathbf{A}} c(y,a) p(dy|x,a) \leq k \cdot [1 + c(x,a)]$ for some constant k, for all $(x,a) \in \mathbf{X} \times \mathbf{A}$.

Theorem 4.6. *[Hernandez-Lerma and Lasserre(1999), Th. 12.3.3] Under Condition 4.4, there exists a solution η^* to the following linear program on the space of measures on $\mathbf{X} \times \mathbf{A}$:*

$$\int_{\mathbf{X} \times \mathbf{A}} c(x,a) d\eta(y,a) \to \inf,$$

$$\eta(\Gamma \times \mathbf{A}) = \int_{\mathbf{X} \times \mathbf{A}} p(\Gamma|x,a) d\eta(y,a) \text{ for all } \Gamma \in \mathcal{B}(\mathbf{X}), \quad (4.21)$$

$$\eta(\mathbf{X} \times \mathbf{A}) = 1,$$

and its minimal value $\int_{\mathbf{X} \times \mathbf{A}} c(x,a) d\eta^*(y,a)$ *coincides with* $\inf_{P_0} \inf_\pi v^\pi$.

Moreover, there is a stationary strategy π^s and a corresponding invariant probability measure $\hat{\eta}$ on $\mathbf{X}$ such that

$$\hat{\eta}(\Gamma) = \int_{\mathbf{X}} \int_{\mathbf{A}} p(\Gamma|x,a) \pi^s(da|x) \hat{\eta}(dx) \text{ for all } \Gamma \in \mathcal{B}(\mathbf{X});$$

the pair $(\hat{\eta}, \pi^s)$ solves the problem

$$\limsup_{T \to \infty} \frac{1}{T} E_{P_0}^\pi \left[\sum_{t=1}^T c(X_{t-1}, A_t) \right] \to \inf_{P_0, \pi}, \quad (4.22)$$

and the measure

$$\eta^*(\Gamma^{\mathbf{X}} \times \Gamma^{\mathbf{A}}) \stackrel{\triangle}{=} \int_{\Gamma^{\mathbf{X}}} \pi^s(\Gamma^{\mathbf{A}}|x) \hat{\eta}(dx)$$

on $\mathbf{X} \times \mathbf{A}$ solves the linear program (4.21).

Consider Example 4.2.7, Fig. 4.8. Condition 4.4 is satisfied apart from item (b). The loss function is not inf-compact: the set $\{(x,a) \in \mathbf{X} \times \mathbf{A} : c(x,a) \leq 1\} = \mathbf{X} \times \mathbf{A}$ is not compact (we have a discrete topology in the space $\mathbf{A}$). If we introduce another topology in $\mathbf{A}$, say, $\lim_{i \to \infty} i \stackrel{\triangle}{=} 1$, then the transition probability p becomes not (weakly) continuous. The only admissible solution to (4.21) is $\eta^*(1,a) \equiv 0$, $\eta^*(0,\mathbf{A}) = 1$, so that $\sum_{x \in \mathbf{X}} \sum_{a \in \mathbf{A}} c(x,a)\eta^*(x,a) = 1$. At the same time, we know that $\inf_\pi v_1^\pi = 0$ and $\inf_{P_0} \inf_\pi v^\pi = 0$. Example 4.2.9 can be investigated in a similar way.

In Example 4.2.8, Condition 4.4 is satisfied and Theorem 4.6 holds: $\eta^*(0,\mathbf{A}) = 1$; $\sum_{x \in \mathbf{X}} c(x,a)\eta^*(x,a) = 0 = \inf_{P_0} \inf_\pi v^\pi$. Note that Theorem 4.6 deals with problem (4.22) which is different from problem (4.1). (The latter concerns a specified initial distribution P_0.)

In Examples 4.2.10 and 4.2.11, the loss function c is not inf-compact, and Theorem 4.6 does not hold.

Statements similar to Theorem 4.6 were proved in [Altman(1999)] and [Altman and Shwartz(1991b)] for discrete models, but under the following condition.

Condition 4.5. For any control strategy π, the set of *expected frequencies* $\{\bar{f}_{\pi,x}^T\}_{T \geq 1}$, defined by the formula

$$\bar{f}_{\pi,x}^T(y,a) = \frac{1}{T} \sum_{t=1}^T P_x^\pi(X_{t-1} = y, A_t = a),$$

is tight.

Theorem 4.7. *[Altman and Shwartz(1991b), Cor. 5.4, Th. 7.1]; [Altman(1999), Th. 11.10]. Suppose the spaces $\mathbf{X}$ and $\mathbf{A}$ are countable (or finite), the loss function c is bounded below, and, under any stationary strategy π, the controlled process X_t has a single positive recurrent class coincident with $\mathbf{X}$. Then, under Condition 4.5, there is an AC-optimal stationary strategy. Moreover, if a stationary strategy π^s is AC-optimal and $\hat{\eta}^{\pi^s}$ is the corresponding invariant probability measure on $\mathbf{X}$, then the matrix $\eta^*(x,a) \stackrel{\triangle}{=} \pi^s(a|x)\hat{\eta}^{\pi^s}(x)$ solves the linear program (4.21).*

Conversely, if η^* solves that linear program, then the stationary strategy

$$\pi^s(x|a) \stackrel{\triangle}{=} \frac{\eta^*(x,a)}{\sum_{a \in \mathbf{A}} \eta^*(x,a)} \quad (4.23)$$

is AC-optimal. (If the denominator is zero, the distribution $\pi^s(\cdot|x)$ is chosen arbitrarily.)

We emphasize that Condition 4.5 only holds if the action space $\mathbf{A}$ is compact (see [Altman and Shwartz(1991b), p. 799 and Counterexample 3.5]). Indeed, let $\{a_t\}_{t=1}^{\infty}$ be a sequence in $\mathbf{A}$ having no convergent subsequences, and consider the selector $\varphi_t(y) \equiv a_t$. Then, for any compact set $K \subset \mathbf{X} \times \mathbf{A}$, only a finite number of values a_t can appear as the second components of elements $k = (y, a) \in K$. Thus, starting from some τ, $P_x^{\varphi}(X_{t-1} = y, A_t = a) = 0$ for $(y,a) \in K$ and $\bar{f}_{\varphi,x}^{n\tau}(K) \leq 1/n$.

In Example 4.2.10, Condition 4.5 is violated and no one stationary strategy is AC-optimal, unless all the other conditions of Theorem 4.7 are satisfied.

Consider now the following simple example. $\mathbf{X} = \{0, 1, 2, \ldots\}$, $\mathbf{A} = \{0,1\}$, $p(0|0,a) \equiv 1$, $p(0|x,0) = p(x+1|x,1) = 1$ for all $x > 0$, with all other transition probabilities zero; $c(0,a) \equiv 1$, $c(x,a) \equiv 0$ for $x > 0$ (see Fig. 4.24).

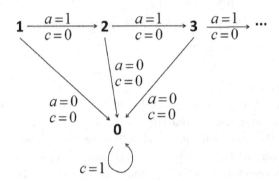

Fig. 4.24   Example 4.2.21: the linear programming approach is not applicable.

The linear program (4.21) can be written as follows
$$\eta(0, \mathbf{A}) \to \inf$$
$$\eta(0, \mathbf{A}) = \eta(0, \mathbf{A}) + \sum_{x>0} \eta(x, 0);$$
$$\eta(1, \mathbf{A}) = 0;$$
$$\eta(x, \mathbf{A}) = \eta(x-1, 1) \text{ for all } x > 1;$$
$$\eta(\mathbf{X} \times \mathbf{A}) = 1.$$

The only admissible solution is $\eta^*(0, \mathbf{A}) = 1$, $\eta^*(x, \mathbf{A}) = 0$ for all $x > 0$, so that $\sum_{x \in \mathbf{X}} \sum_{a \in \mathbf{A}} c(x,a) \eta^*(x,a) = \eta^*(0, \mathbf{A}) = 1$. But $\inf_{P_0} \inf_{\pi} v^{\pi} = 0$

and is provided by $P_0(x) = I\{x = 1\}$, $\varphi^*(x) \equiv 1$. Theorem 4.6 fails to hold because the loss function c is not inf-compact.

It is interesting to look at the dual linear program to (4.21) assuming that $c(x, a) \geq 0$:

$$\rho \to \sup$$
$$\rho + h(x) \leq c(x, a) + \int_{\mathbf{X}} h(y) p(dy|x, a) \qquad (4.24)$$
$$\rho \in \mathbb{R}, \qquad \sup_{x \in \mathbf{X}} \frac{|h(x)|}{1 + \inf_{a \in \mathbf{A}} c(x, a)} < \infty$$

(see [Hernandez-Lerma and Lasserre(1999), Section 12.3]; compare with (4.2)). In the current example, the dual program takes the form:

$$\rho \to \sup$$
$$\rho + h(0) \leq 1 + h(0);$$
$$\rho + h(x) \leq \min\{h(x+1), h(0)\} \text{ for all } x > 0;$$
$$\sup_{x \in \mathbf{X}} |h(x)| < \infty.$$

Since $h(x+1) \geq h(x) + \rho \geq \cdots \geq h(1) + x \cdot \rho$ and $\sup_{x \in \mathbf{X}} |h(x)| < \infty$, we conclude that $\rho \leq 0$. Actually, $\rho^* = 0$ and $h(x) \equiv 0$ provides a solution to the dual linear program, so that the duality gap equals 1.

Note that the canonical equations (4.2) have solution $\rho(0) = 1$, $\rho(x) \equiv 0$ for all $x > 0$, and $h(x) \equiv 0$. Thus, $\langle \rho, h, \varphi^* \equiv 1 \rangle$ is a canonical triplet, and the stationary selector φ^* is AC-optimal (see Theorem 4.2).

If Condition 4.4 is satisfied then the duality gap is absent [Hernandez-Lerma and Lasserre(1999), Th. 12.3.4].

In this example (Fig. 4.24), Theorem 4.7 also fails to hold.

We can modify the model in such a way that Theorem 4.6 becomes true. Cancel state 0 and action 0 and add a state "∞" making the state space $\mathbf{X}$ compact: $\lim_{x \to \infty} x = \infty$. To make the transition probability p weakly continuous, we have to put $p(\infty|\infty, 1) = 1$; we also put $c(\infty, 1) = 0$. Now the measure $\eta^*(\infty, 1) = 1$, $\eta^*(x, 1) = 0$ for all $x < \infty$ solves the linear program (4.21).

Incidentally, if we consider $\mathbf{X} = \{1, 2, \ldots\}$, $\mathbf{A} = \{1\}$, then the linear program (4.21) has no admissible solutions and the dual program (4.24) still gives the canonical triplet $\langle \rho = 0, h \equiv 0, \varphi^* \equiv 1 \rangle$. This modification appeared in [Altman and Shwartz(1991b), Counterexample 2.1].

4.2.22 Linear programs and expected frequencies in finite models

The definition of an expected frequency $\bar{f}^T_{\pi,x}(y,a)$ was given in Condition 4.5. We also consider the case of an arbitrary initial distribution P_0 replacing the initial state $x \in \mathbf{X}$. Let F_{π,P_0} be the set of all accumulation (or limit) points of the vectors $\bar{f}^1_{\pi,P_0}, \bar{f}^2_{\pi,P_0}, \ldots$. As usual, we write $F_{\pi,x}$ if $P_0(x) = 1$.

Theorem 4.8.

(a) *[Altman(1999), Th. 4.2]. If the model is unichain, then the sets* $\bigcup_{\pi \in \Delta^{\text{All}}} F_{\pi,P_0} = \bigcup_{\pi \in \Delta^S} F_{\pi,P_0}$ *do not depend on the initial distribution P_0, and coincide with the collection of all feasible solutions to the linear program (4.21), which we explicitly rewrite below for a finite (countable) model.*

$$\sum_{x \in \mathbf{X}} \sum_{a \in \mathbf{A}} c(x,a) \eta(x,a) \to \inf_\eta,$$

$$\sum_{a \in \mathbf{A}} \eta(x,a) = \sum_{y \in \mathbf{X}} \sum_{a \in \mathbf{A}} p(x|y,a) \eta(y,a), \quad (4.25)$$

$$\sum_{x \in \mathbf{X}} \sum_{a \in \mathbf{A}} \eta(x,a) = 1, \qquad \eta(x,a) \geq 0.$$

(b) *[Derman(1964), Th. 1(a)]. For each $x \in \mathbf{X}$ the closed convex hull of the set $\bigcup_{\pi \in \Delta^S} F_{\pi,x}$ contains the set $\bigcup_{\pi \in \Delta^{\text{All}}} F_{\pi,x}$.*

(c) *[Kallenberg(2010), Th. 9.4]. The set $\bigcup_{\pi \in \Delta^{\text{All}}} F_{\pi,P_0}$ coincides with the collection*

$$\{\eta : (\eta, \tilde{\eta}) \text{ is feasible in the linear program } (4.15), (4.16), (4.17)$$

$$\text{with } \alpha(x) = P_0(x)\}.$$

The following simple example illustrates that, in the multichain case, $\bigcup_{\pi \in \Delta^S} F_{\pi,P_0} \subset \bigcup_{\pi \in \Delta^{\text{All}}} F_{\pi,P_0}$, the inclusion being strict.

Let $\mathbf{X} = \{1, 2\}$, $\mathbf{A} = \{1, 2\}$, $P_0(1) = 1$, $p(1|1, 1) = 1$, $p(2|1, 2) = 1$, $p(2|2, a) \equiv 1$, with all other transition probabilities zero (see Fig. 4.26). For any stationary strategy π^s, if $f_{\pi^s,a} \in F_{\pi^s,1}$ then the sum $f_{\pi^s,1}(1,1) + f_{\pi^s,1}(1,2)$ equals either 1 (if $\pi^s(1|1) = 1$) or 0 (otherwise). At the same time, there exists a non-stationary strategy π, in the form of a mixture of two stationary selectors, such that $f_{\pi,1} \in F_{\pi,1}$ and $f_{\pi^s,1}(1,1) + f_{\pi^s,1}(1,2) = 1/2$. More details are given in Section 4.2.23.

The next example, based on [Derman(1964), Section 4], shows that statement (a) of Theorem 4.8 may fail to hold even if the model is communicating.

Let $\mathbf{X} = \{1, 2, 3\}$, $\mathbf{A} = \{1, 2\}$, $p(2|1, a) \equiv 1$, $p(2|2, 1) = 1$, $p(3|2, 2) = 1$, $p(3|3, 1) = 1$, $p(1|3, 2) = 1$ with all other transition probabilities zero (see Fig. 4.25).

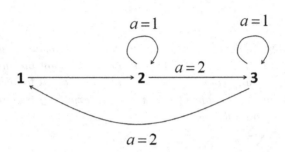

Fig. 4.25 Example 4.2.22: expected frequencies which are not generated by a stationary strategy.

Suppose that $P_0(1) = 1$, and π^s is a stationary strategy. Below we use the notation

$$\hat{f}_{\pi,1} \triangleq \left(\sum_{a \in A} f_{\pi,1}(1, a), \sum_{a \in A} f_{\pi,1}(2, a), \sum_{a \in A} f_{\pi,1}(3, a) \right)$$

for the vectors $f_{\pi,1} \in F_{\pi,1}$. The vector $\hat{f}_{\pi^s,1}$ coincides with the stationary distribution of the controlled process X_t governed by the control strategy π^s and has the form

$$\hat{f}_{\pi^s,1} = \begin{cases} (0, 1, 0), & \text{if } \pi^s(1|2) = 1; \\ (0, 0, 1), & \text{if } \pi^s(1|2) < 1 \text{ and } \pi^s(1|3) = 1; \\ \left[1 + \frac{1}{p_2} + \frac{1}{p_3}\right]^{-1} \left(1, \frac{1}{p_2}, \frac{1}{p_3}\right), & \text{if } \pi^s(2|2) = p_2 > 0 \\ & \text{and } \pi^s(2|3) = p_3 > 0. \end{cases}$$

In reality, $f_{\pi^s,1}(x, a) = \hat{f}_{\pi^s,1}(x) \pi^s(a|x)$. No one vector $\hat{f}_{\pi^s,1}$ has components $\hat{f}_{\pi^s,1}(1) = 0$ and $\hat{f}_{\pi^s,1}(2) \in (0, 1)$ simultaneously.

On the other hand, consider the following Markov strategy:

$$\pi_t^m(1|2) = 1 - \pi_t^m(2|2) = \begin{cases} p \in (0, 1), & \text{if } t = 2; \\ 1 & \text{otherwise,} \end{cases} \quad \pi_t^m(1|3) \equiv 1,$$

where $\pi_t^m(a|1)$ can be arbitrary as state 1 is uncontrolled. The (marginal) expected frequencies are

$$\hat{f}^1_{\pi^m,1}(2) \triangleq \sum_{a\in A} \bar{f}^1_{\pi^m,1}(2,a) = 0, \qquad \hat{f}^2_{\pi^m,1}(2) \triangleq \sum_{a\in A} \bar{f}^2_{\pi^m,1}(2,a) = 1/2,$$

$$\hat{f}^3_{\pi^m,1}(2) \triangleq \sum_{a\in A} \bar{f}^3_{\pi^m,1}(2,a) = \frac{1}{3}(1+p),$$

$$\hat{f}^4_{\pi^m,1}(2) \triangleq \sum_{a\in A} \bar{f}^4_{\pi^m,1}(2,a) = \frac{1}{4}(1+p+p), \ldots,$$

so that the only limit point equals

$$\hat{f}_{\pi^m,1}(2) = \lim_{T\to\infty} \hat{f}^T_{\pi^m,1}(2) = p.$$

Obviously, $\hat{f}_{\pi^m,1}(1) = 0$ $\hat{f}_{\pi^m,1}(3) = 1 - p$ and $f_{\pi^m,1} \notin \bigcup_{\pi\in\Delta^s} F_{\pi,1}$. Incidentally, $F_{\pi^m,1}$ contains a single point $f_{\pi^m,1}$: $f_{\pi^m,1}(1,a) \equiv 0$, $f_{\pi^m,1}(2,1) = p$, $f_{\pi^m,1}(2,2) = 0$, $f_{\pi^m,1}(3,1) = 1-p$, $f_{\pi^m,1}(3,2) = 0$.

Note that

$$\eta(1,a) \equiv 0, \quad \eta(2,1) = p, \quad \eta(2,2) = 0, \quad \eta(3,1) = 1-p, \quad \eta(3,2) = 0$$

is a feasible solution to the linear program (4.25), but the induced strategy

$$\pi^s(\cdot|1) \text{ is arbitrary}, \ \pi^s(1|2) = \pi^s(1|3) = 1$$

results in a stationary distribution on $\mathbf{X}$ dependent on the initial distribution. That stationary distribution coincides with

$$\hat{f}_{\pi^s,P_0} = \left(\sum_{a\in A} \eta(1,a) = 0, \ \sum_{a\in A} \eta(2,a) = p, \ \sum_{a\in A} \eta(3,a) = 1-p\right)$$

if and only if $P_0(1) + P_0(2) = p$ and $P_0(3) = 1 - p$. The controlled process is not ergodic under strategy π^s.

4.2.23 Constrained optimization

Suppose we have two loss functions $^1c(x,a)$ and $^2c(x,a)$. Then every control strategy π results in two performance functionals $^1v^\pi$ and $^2v^\pi$ defined according to (4.1). The *constrained* problem can be expressed as

$$^1v^\pi \to \inf_\pi; \quad ^2v^\pi \leq d, \tag{4.26}$$

where d is a chosen number. Strategies satisfying the above inequality are called *admissible*.

If a finite model is unichain and there is at least one admissible strategy, then there exists a stationary strategy solving problem (4.26); see [Altman(1999), Th. 4.3].

Remark 4.6. One should complement the linear program (4.15), (4.16), (4.19) with the obvious inequality, and build the stationary strategy using formula (4.18).

The following example from [Piunovskiy(1997), p. 149] shows that the unichain condition is important.

Let $\mathbf{X} = \{1,2\}$; $\mathbf{A} = \{1,2\}$; $p(1|1,1) = 1$, $p(2|1,2) = 1$, $p(2|2,a) \equiv 1$, with all other transition probabilities zero; $^1c(x,a) = I\{x = 2\}$, $^2c(x,a) = I\{x = 1\}$ (see Fig. 4.26).

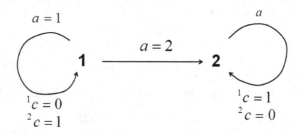

Fig. 4.26   Example 4.2.23: constrained MDP.

Suppose $d = \frac{1}{2}$ and $P_0(1) = 1$. For a stationary strategy with $\pi^s(1|1) = 1$ we have $^1v^{\pi^s} = 0$, $^2v^{\pi^s} = 1$, so that such strategies are not admissible for (4.26). If $\pi^s(1|1) < 1$ then the process will be absorbed at state 2, leading to $^1v^{\pi^s} = 1$, $^2v^{\pi^s} = 0$. At the same time, the solution to (4.26) is given by a strategy π^* with $^1v^{\pi^*} = 1/2$, $^2v^{\pi^*} = 1/2$.

Definition 4.5. The $(\alpha, 1-\alpha)$-*mixture* of two strategies π^1 and π^2 is a strategy π such that the corresponding strategic measures satisfy $P_{P_0}^{\pi} = \alpha P_{P_0}^{\pi^1} + (1-\alpha) P_{P_0}^{\pi^2}$ with $\alpha \in [0,1]$.

The existence of a mixture follows from the convexity of $\mathcal{D}$, the space of all strategic measures [Piunovskiy(1997), Th. 8]. One can say that the decision maker tosses a coin at the very beginning and applies strategies

π^1 or π^2 with probabilities α and $(1-\alpha)$. In finite models, one can replace lim sup in formula (4.1) with lim, if π is a stationary strategy, so that the functional v^π is linear w.r.t. $P_{P_0}^\pi$ if these strategic measures correspond to finite mixtures of stationary strategies.

If π^1 and π^2 are stationary strategies then the $(\alpha, 1-\alpha)$-mixture π is usually non-stationary, but there always exists an equivalent Markov strategy π^m with $v^{\pi^m} = v^\pi$ (the one-step loss c can be arbitrary). See [Piunovskiy(1997), Lemma 2].

In the example considered, π^* can be taken as the $(1/2, 1/2)$-mixture of the stationary selectors $\varphi^1(x) \equiv 1$ and $\varphi^2(x) \equiv 2$.

Proposition 4.3. *[Piunovskiy(1997), Th. 13] Let the functionals $^1v^\pi$ and $^2v^\pi$ be finite for any control strategy π, and let the Slater condition be satisfied, i.e. the inequality in (4.26) is strict for at least one strategy π. A strategy π^* solves the constrained problem (4.26) if and only if there is a Lagrange multiplier $\lambda^* \geq 0$ such that*

$$^1v^{\pi^*} + \lambda^*(\,^2v^{\pi^*} - d) = \min_\pi \{\,^1v^\pi + \lambda^*(\,^2v^\pi - d)\}$$

and $\lambda^*(\,^2v^{\pi^*} - d) = 0$, $^2v^{\pi^*} \leq d$.

Note that this proposition holds for any performance functionals $^1v^\pi$, $^2v^\pi$ which can be expressed as *convex* functionals on the space of strategic measures $\mathcal{D}$; the functional $^2v^\pi$ can be multi-dimensional.

In the example under consideration, $\lambda^* = 1$,

$$^1v^{\pi^*} = \limsup_{T\to\infty} \frac{1}{T} \left\{ \frac{1}{2} E_{P_0}^{\varphi^1}\left[\sum_{t=1}^T {}^1c(X_{t-1}, A_t)\right] + \frac{1}{2} E_{P_0}^{\varphi^2}\left[\sum_{t=1}^T {}^1c(X_{t-1}, A_t)\right] \right\}$$

$$= \frac{1}{2}\lim_{T\to\infty}\frac{1}{T} E_{P_0}^{\varphi^1}\left[\sum_{t=1}^T {}^1c(X_{t-1}, A_t)\right] + \frac{1}{2}\lim_{T\to\infty}\frac{1}{T} E_{P_0}^{\varphi^2}\left[\sum_{t=1}^T {}^1c(X_{t-1}, A_t)\right]$$

$$= 0 + 1/2 = 1/2$$

(note the usual limits). Similarly, $^2v^{\pi^*} = 1/2$ and, for any strategy π,

$$^1v^\pi + \lambda^*\,^2v^\pi = {}^1v^\pi + {}^2v^\pi$$

$$\geq \limsup_{T\to\infty}\frac{1}{T} E_{P_0}^\pi\left[\sum_{t=1}^T (\,^1c(X_{t-1}, A_t) + {}^2c(X_{t-1}, A_t))\right]$$

$$\equiv 1 = {}^1v^{\pi^*} + \lambda^*\,^2v^{\pi^*}.$$

Therefore, the $(1/2, 1/2)$-mixture of stationary selectors φ^1 and φ^2 really does solve problem (4.26).

A solution to a finite unichain model can be found in the form of a *time-sharing* strategy [Altman and Shwartz(1993)]. Such a strategy switches between several stationary selectors in such a way that the expected frequencies (see Condition 4.5) converge, as $T \to \infty$, to the values of $\eta^*(y, a)$ solving the corresponding linear program. If, for example, $\eta^*(x, 1) = \eta^*(x, 2) = 1/2$ for a particular recurrent state x then one applies actions $\varphi^1(x) = 1$ and $\varphi^2(x) = 2$ in turn, every time the controlled process visits state x.

The above example (Fig. 4.26) shows that a time-sharing strategy π cannot solve the constrained problem if the model is not unichain. If action $a = 2$ is applied (in state $X_0 = 1$) at least once, then $^1v^\pi = 1$. Otherwise $^2v^\pi = 1$, meaning that π is not admissible.

An algorithm for solving constrained problems for general finite models, based on the linear programming approach, is presented in [Kallenberg(2010), Alg. 9.1]. If the model is not unichain, it results in a complicated Markov (non-stationary) strategy. At the first step, one complements the linear constraints (4.16) and (4.17), where $\alpha(x) = P_0(x)$ is the initial distribution, with the main objective $\sum_{x \in \mathbf{X}} \sum_{a \in \mathbf{A}} {}^1c(x,a)\eta(x,a) \to \inf$ and with the additional constraint $\sum_{x \in \mathbf{X}} \sum_{a \in \mathbf{A}} {}^2c(x,a)\eta(x,a) \leq d$. The optimal control strategy is then built using the solution to the linear program obtained. It is a mistake to think that formula (4.18) provides the answer. Indeed, the induced strategy is stationary, and we know that in the above example (Fig. 4.26) only a non-stationary strategy can solve the constrained problem. If the finite model is recurrent, then variables $\tilde{\eta}$ and constraint (4.17) are absent, equation (4.19) is added, and the induced strategy solves the constrained problem. In this case, $\sum_{a \in \mathbf{A}} \eta(x, a) > 0$ for all states $x \in \mathbf{X}$.

It is interesting to consider the constrained MDP with discounted loss and to see what happens when the discount factor β goes to 1. Consider the same example illustrated by Fig. 4.26. In line with the Abelian Theorem, we normalize the discounted loss, multiplying it by $(1 - \beta)$:

$$(1 - \beta)\ {}^1v^{\pi,\beta} \to \inf_{\pi}; \qquad (1 - \beta)\ {}^2v^{\pi,\beta} \leq d. \qquad (4.27)$$

It is known that (under a fixed initial distribution, namely $P_0(1) = 1$) stationary strategies are sufficient for solving a constrained discounted MDP of the form (4.27); see [Piunovskiy(1997), Section 3.2.3.2]. In the current example, any such strategy π^s is characterized by the single number

$\pi^s(1|1) = p$, and one can compute
$$(1-\beta)\,^1 v^{\pi^s,\beta} = \frac{\beta(1-p)}{1-\beta p}; \qquad (1-\beta)\,^2 v^{\pi^s,\beta} = \frac{1-\beta}{1-\beta p},$$
so that $(1-\beta)[\,^1 v^{\pi^s,\beta} + \,^2 v^{\pi^s,\beta}] = 1$, and the strategy π^{s*} with $p^* = 2 - \frac{1}{\beta}$ solves problem (4.27). If β approaches 1, the optimal strategy π^{s*} does not stop to change: we have nothing similar to the Blackwell optimality. Moreover, $\lim_{\beta \to 1^-} p^* = 1$, but we already know that the stationary selector φ^1 is not admissible in problem (4.26). Note that
$$\lim_{\beta \to 1^-}(1-\beta)\,^1 v^{\pi^{s*},\beta} = \frac{1}{2} \neq \,^1 v^{\varphi^1} = 0,$$
$$\lim_{\beta \to 1^-}(1-\beta)\,^2 v^{\pi^{s*},\beta} = \frac{1}{2} \neq \,^2 v^{\varphi^1} = 1.$$

Similar analysis was performed in [Altman et al.(2002), Ex. 1].

4.2.24 AC-optimal, bias optimal, overtaking optimal and opportunity-cost optimal strategies: periodic model

Definition 4.6. [Puterman(1994), Section 5.4.2] A strategy π^* is *overtaking optimal* if, for each strategy π,
$$\limsup_{T \to \infty}\left\{E_x^{\pi^*}\left[\sum_{t=1}^T c(X_{t-1}, A_t)\right] - E_x^\pi\left[\sum_{t=1}^T c(X_{t-1}, A_t)\right]\right\} \leq 0, \quad x \in \mathbf{X}.$$

Any overtaking optimal strategy π^* is also AC-optimal, because
$$\limsup_{T \to \infty} \frac{1}{T} E_x^{\pi^*}\left[\sum_{t=1}^T c(X_{t-1}, A_t)\right] \leq \limsup_{T \to \infty} \frac{1}{T}\left\{E_x^{\pi^*}\left[\sum_{t=1}^T c(X_{t-1}, A_t)\right]\right.$$
$$\left. - E_x^\pi\left[\sum_{t=1}^T c(X_{t-1}, A_t)\right]\right\} + \limsup_{T \to \infty}\frac{1}{T} E_x^\pi\left[\sum_{t=1}^T c(X_{t-1}, A_t)\right]$$
$$\leq \limsup_{T \to \infty}\frac{1}{T} E_x^\pi\left[\sum_{t=1}^T c(X_{t-1}, A_t)\right].$$

Similar reasoning confirms that any overtaking optimal strategy minimizes the opportunity loss; that is, it solves problem (3.16) (such strategies are called *opportunity-cost optimal*):
$$\forall \pi \; \forall x \in \mathbf{X} \; \limsup_{T \to \infty}\left\{E_x^{\pi^*}\left[\sum_{t=1}^T c(X_{t-1}, A_t)\right] - V_x^T\right\}$$

$$\leq \limsup_{T\to\infty}\left\{E_x^{\pi^*}\left[\sum_{t=1}^T c(X_{t-1},A_t)\right] - E_x^\pi\left[\sum_{t=1}^T c(X_{t-1},A_t)\right]\right\}$$

$$+ \limsup_{T\to\infty}\left\{E_x^\pi\left[\sum_{t=1}^T c(X_{t-1},A_t)\right] - V_x^T\right\}$$

$$\leq \limsup_{T\to\infty}\left\{E_x^\pi\left[\sum_{t=1}^T c(X_{t-1},A_t)\right] - V_x^T\right\}.$$

See also [Flynn(1980), Corollary 1], where this assertion is formulated for discrete models.

Definition 4.7. A 0-discount optimal strategy (see Definition 3.8) is called *bias optimal* [Puterman(1994), Section 5.4.3].

Below, we consider finite models. It is known that a bias optimal strategy does exist; incidentally, any Blackwell optimal strategy is also bias optimal. A stationary selector φ^* is -1-discount optimal if and only if

$$\rho(x) = v_x^{\varphi^*} \leq \liminf_{T\to\infty}\frac{1}{T}E_x^\pi\left[\sum_{t=1}^T c(X_{t-1},A_t)\right]$$

for any strategy π, for all states $x \in \mathbf{X}$. All such selectors are AC-optimal. A stationary selector φ^* is 0-discount optimal (bias optimal) if and only if, for any -1-discount optimal stationary selector φ,

$$\lim_{T\to\infty}\frac{1}{T}\sum_{t=1}^T E_x^{\varphi^*}\left[\sum_{\tau=1}^t \{c(X_{\tau-1},A_\tau) - \rho(X_{\tau-1})\}\right]$$

$$\leq \lim_{T\to\infty}\frac{1}{T}\sum_{t=1}^T E_x^\varphi\left[\sum_{\tau=1}^t \{c(X_{\tau-1},A_\tau) - \rho(X_{\tau-1})\}\right]$$

[Puterman(1994), Th. 10.1.6].

Suppose the model is finite and aperiodic unichain; assume that, if the stationary selectors φ^1 and φ^2 are bias optimal, then $\varphi^1(x) = \varphi^2(x)$ for each state $x \in \mathbf{X}$ which is (positive) recurrent under strategy φ^1. Then any bias optimal stationary selector is also overtaking optimal [Denardo and Rothblum(1979), Corollary 2].

The following example, based on [Denardo and Miller(1968), p. 1221], shows that an AC-optimal strategy may be not overtaking optimal, moreover an overtaking optimal strategy may not exist even in the simplest finite models. Finally, this example illustrates that the aperiodicity assumption in the previous paragraph is important.

Let $\mathbf{X} = \{1,2,3\}$, $\mathbf{A} = \{1,2\}$, $p(2|1,1) = p(3|1,2) = 1$, $p(3|2,a) \equiv p(2|3,a) \equiv 1$, with all other transition probabilities zero. We put $c(1,1) = 1$, $c(1,2) = 0$, $c(2,a) \equiv 0$, $c(3,a) \equiv 2$ (see Fig. 4.27).

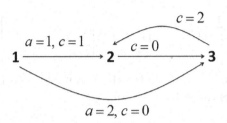

Fig. 4.27 Example 4.2.24: no overtaking optimal strategies.

In fact, there are only two essentially different strategies (stationary selectors) $\varphi^1(x) \equiv 1$ and $\varphi^2(x) \equiv 2$ (actions in states 2 and 3 play no role). Suppose $X_0 = 1$. Then the values of $E_1^{\varphi^{1,2}}\left[\sum_{t=1}^T c(X_{t-1}, A_t)\right]$ for different values of T are given in the following table:

	$T=1$	2	3	4	5	6	7	8	...
φ^1	1	1	3	3	5	5	7	7	...
φ^2	0	2	2	4	4	6	6	8	...

Therefore, for any strategy π^* there exists a strategy π such that

$$\limsup_{T\to\infty}\left\{E_1^{\pi^*}\left[\sum_{t=1}^T c(X_{t-1}, A_t)\right] - E_1^{\pi}\left[\sum_{t=1}^T c(X_{t-1}, A_t)\right]\right\} > 0,$$

and no one strategy is overtaking optimal. At the same time, all strategies are equally AC-optimal. The $(1/2, 1/2)$ mixture of selectors φ^1 and φ^2 is an opportunity-cost optimal strategy: it solves problem (3.16). This mixture is also *D-optimal* [Hernandez-Lerma and Lasserre(1999), p. 119], i.e. it provides the minimum to

$$D(\pi, x) \stackrel{\triangle}{=} \limsup_{T\to\infty}\left\{E_x^{\pi}\left[\sum_{t=1}^T c(X_{t-1}, A_t)\right] - Tv_x^*\right\}$$

for all $x \in \mathbf{X}$; v_x^* comes from Section 4.1. One can also verify that the stationary selector φ^1 is Blackwell optimal and hence bias optimal, but it is not overtaking optimal.

Theorem 10.3.11 in [Hernandez-Lerma and Lasserre(1999)] says that, under appropriate conditions, many types of optimality are equivalent. For instance, any stationary selector φ^*, D-optimal *among stationary selectors*, is also *weakly overtaking optimal* among stationary selectors; that is, for each $\varepsilon > 0$ and any stationary selector φ,

$$E_x^{\varphi^*}\left[\sum_{t=1}^{T} c(X_{t-1}, A_t)\right] \leq E_x^{\varphi}\left[\sum_{t=1}^{T} c(X_{t-1}, A_t)\right] + \varepsilon$$

as soon as $T \geq N(\varphi^*, \varphi, x, \varepsilon)$. In the above example, φ^1 is D-optimal among stationary selectors (as well as φ^2), but

$$\limsup_{T\to\infty}\left\{E_1^{\varphi^1}\left[\sum_{t=1}^{T} c(X_{t-1}, A_t)\right] - E_1^{\varphi^2}\left[\sum_{t=1}^{T} c(X_{t-1}, A_t)\right]\right\} = 1 > 0.$$

Theorem 10.3.11 from [Hernandez-Lerma and Lasserre(1999)] is not applicable here, because the controlled process X_t is neither geometric ergodic nor λ-irreducible under strategies φ^1 and φ^2.

4.2.25 AC-optimal and average-overtaking optimal strategies

The standard average loss (4.1) is under-selective because it does not take into account the finite-horizon accumulated loss $E_x^\pi\left[\sum_{t=1}^{T} c(X_{t-1}, A_t)\right]$. Consider the following example: $\mathbf{X} = \{\Delta, 1\}$, $\mathbf{A} = \{1, 2\}$, $p(\Delta|\Delta, a) \equiv 1$, $p(\Delta|1, a) \equiv 1$, $c(1, 1) = 0$, $c(1, 2) = 1$, $c(\Delta, a) \equiv 0$ (see Fig. 4.28).

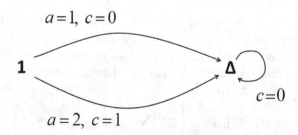

Fig. 4.28 Example 4.2.25: the AC-optimal selector φ^2 is not average-overtaking optimal.

In fact, there are only two essentially different strategies (stationary selectors) $\varphi^1(x) \equiv 1$ and $\varphi^2(x) \equiv 2$ (the actions in state Δ play no role). Suppose $X_0 = 1$. Then, for any finite $T \geq 1$,

$$E_1^{\varphi^1}\left[\sum_{t=1}^T c(X_{t-1}, A_t)\right] = 0 \text{ and } E_1^{\varphi^2}\left[\sum_{t=1}^T c(X_{t-1}, A_t)\right] = 1,$$

so that it is natural to say that selector φ^1 is better than φ^2. But formula (4.1) gives $v_x^{\varphi^1} \equiv v_x^{\varphi^2} \equiv 0$, meaning that all strategies are equally AC-optimal. On the other hand, as Section 4.2.24 shows, Definition 4.6 gives an over-selective notion of optimality. That is why the following definition is sometimes used:

Definition 4.8. [Puterman(1994), Section 5.4.2] A strategy π^* is *average-overtaking optimal* if, for each strategy π, for all $x \in \mathbf{X}$,

$$\limsup_{T\to\infty} \frac{1}{T} \sum_{t=1}^T \left\{ E_x^{\pi^*}\left[\sum_{\tau=1}^t c(X_{\tau-1}, A_\tau)\right] - E_x^\pi\left[\sum_{\tau=1}^t c(X_{\tau-1}, A_\tau)\right] \right\} \leq 0.$$

In the above example (Fig. 4.28) the selector φ^1 is average-overtaking optimal, while φ^2 is not:

$$\limsup_{T\to\infty} \frac{1}{T} \sum_{t=1}^T \left\{ E_1^{\varphi^2}\left[\sum_{\tau=1}^t c(X_{\tau-1}, A_\tau)\right] - E_1^{\varphi^1}\left[\sum_{\tau=1}^t c(X_{\tau-1}, A_\tau)\right] \right\} = 1.$$

The following example shows that an average-overtaking optimal strategy may be not AC-optimal (and hence not overtaking optimal).

Let $\mathbf{X} = \{\Delta, 0, 1, 2, \ldots\}$, $\mathbf{A} = \{1, 2\}$, $p(1|0, 1) = p(\Delta|0, 2) = 1$, $p(i+1|i, a) \equiv 1$ for all $i \geq 1$, with all other transition probabilities zero; $c(\Delta, a) \equiv 0$, $c(0, 1) = -1$, $c(0, 2) = 0$. To describe the loss function $c(i, a)$ for $i \geq 1$, we introduce the following increasing sequence $\{m_j\}_{j=0}^\infty$: $m_0 \stackrel{\Delta}{=} 0$; for each $j \geq 0$, let n_j be the first integer satisfying the inequality $0.9 n_j > 1.1 + 0.1 m_j$, and put $m_{j+1} \stackrel{\Delta}{=} m_j + 2n_j$. Now, for $i \geq 1$,

$$c(i, a) \equiv \begin{cases} 2n_j - 1, & \text{if } i = m_j + n_j \text{ for some } j \geq 0; \\ -1 & \text{otherwise} \end{cases}$$

(see Fig. 4.29).

In fact, there are only two essentially different strategies (stationary selectors) $\varphi^1(x) \equiv 1$ and $\varphi^2(x) \equiv 2$ (the actions in states $\Delta, 1, 2, \ldots$ play no role). Suppose $X_0 = 1$. Then $E_0^{\varphi^2}\left[\sum_{t=1}^T c(X_{t-1}, A_t)\right] = 0$.

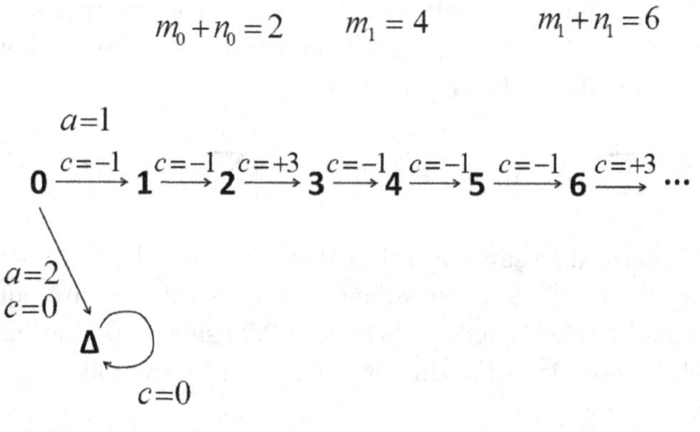

$$m_0 + n_0 = 2 \qquad m_1 = 4 \qquad m_1 + n_1 = 6$$

$$\begin{array}{c} a=1 \\ 0 \xrightarrow{c=-1} 1 \xrightarrow{c=-1} 2 \xrightarrow{c=+3} 3 \xrightarrow{c=-1} 4 \xrightarrow{c=-1} 5 \xrightarrow{c=-1} 6 \xrightarrow{c=+3} \cdots \\ \begin{array}{c} a=2 \\ c=0 \end{array} \searrow \\ \Delta \circlearrowright \\ c=0 \end{array}$$

Fig. 4.29 Example 4.2.25: the average-overtaking optimal strategy is not AC-optimal.

For the strategy φ^1, $E_0^{\varphi^1}\left[\sum_{t=1}^{m_j} c(X_{t-1}, A_t)\right] = 0$ for all $j \geq 0$. To see this, we notice that this assertion holds at $j = 0$. If it holds for some $j \geq 0$, then

$$E_0^{\varphi^1}\left[\sum_{t=1}^{m_{j+1}} c(X_{t-1}, A_t)\right] = -n_j + (2n_j - 1) - [(m_{j+1} - 1) - (m_j + n_j)] = 0.$$

As a consequence,

$$E_0^{\varphi^1}\left[\sum_{t=1}^{m_j+n_j+1} c(X_{t-1}, A_t)\right] = -n_j + (2n_j - 1) = n_j - 1$$

and

$$\frac{1}{m_j + n_j + 1} E_0^{\varphi^1}\left[\sum_{t=1}^{m_j+n_j+1} c(X_{t-1}, A_t)\right] = \frac{0.9(n_j - 1)}{0.9(m_j + n_j + 1)}$$

$$> \frac{(1.1 + 0.1 m_j) - 0.9}{0.9 m_j + (1.1 + 0.1 m_j) + 0.9} = 0.1,$$

so that the strategy φ^1 is not AC-optimal: remember, $v_0^{\varphi^2} = 0$.

On the other hand, $\sum_{t=1}^{m_j} E_0^{\varphi^1}\left[\sum_{\tau=1}^{t} c(X_{\tau-1}, A_\tau)\right] \leq 0$. Indeed, this assertion holds at $j = 0$. If it holds for some $j \geq 0$, then

$$\sum_{t=1}^{m_{j+1}} E_0^{\varphi^1}\left[\sum_{\tau=1}^{t} c(X_{\tau-1}, A_\tau)\right] \leq \sum_{t=m_j+1}^{m_{j+1}} E_0^{\varphi^1}\left[\sum_{\tau=1}^{t} c(X_{\tau-1}, A_\tau)\right]$$

$$= \frac{-n_j(n_j+1)}{2} + (-n_j + 2n_j - 1)$$

$$+ \frac{[(m_{j+1}-1) - (m_j+n_j)] \cdot [-n_j + 2n_j - 2]}{2}$$

$$= -\frac{n_j(n_j+1)}{2} + \frac{(n_j-1)n_j}{2} < 0.$$

Since $E_0^{\varphi^1}\left[\sum_{\tau=1}^{t} c(X_{\tau-1}, A_\tau)\right]$ is negative for $t = m_j + 1, \ldots, m_j + n_j$ and positive afterwards, up to $t = m_{j+1}$, we conclude that

$$\sum_{t=1}^{T} E_0^{\varphi^1}\left[\sum_{\tau=1}^{t} c(X_{\tau-1}, A_\tau)\right] \leq 0 \text{ for all } T \geq 0,$$

and hence the stationary selector φ^1 is average-overtaking optimal.

4.2.26 Blackwell optimal, bias optimal, average-overtaking optimal and AC-optimal strategies

The following example, based on [Flynn(1976), Ex. 1], shows that a Blackwell optimal strategy may be not average-overtaking optimal.

Let $\mathbf{X} = \{\Delta, 0, 1, 2, \ldots\}$, $\mathbf{A} = \{1, 2\}$, $p(\Delta|\Delta, a) \equiv 1$, $p(1|0, 1) = 1$, $p(\Delta|0, 2) = 1$, $p(i+1|i, a) \equiv 1$ for all $i \geq 1$. Let $\{C_j\}_{j=1}^{\infty}$ be a bounded sequence such that

$$\limsup_{n \to \infty} \frac{1}{n} \sum_{i=1}^{n} \sum_{j=1}^{i} C_j = 1 \text{ and } \lim_{\beta \to 1-} \sum_{j=1}^{\infty} \beta^{j-1} C_j = \liminf_{n \to \infty} \frac{1}{n} \sum_{i=1}^{n} \sum_{j=1}^{i} C_j = 0$$

(see Appendix A.4). We put $c(0,1) = C_1$, $c(i, a) \equiv C_{i+1}$ for all $i \geq 1$, $c(0, 2) = 1/4$, $c(\Delta, a) \equiv 0$ (see Fig. 4.30).

In fact, there are only two essentially different strategies (stationary selectors) $\varphi^1(x) \equiv 1$ and $\varphi^2(x) \equiv 2$ (actions in states $\Delta, 1, 2, \ldots$ play no role). The selector φ^1 is Blackwell optimal because

$$\lim_{\beta \to 1-} v_0^{\varphi^1, \beta} = \lim_{\beta \to 1-} \sum_{j=1}^{\infty} \beta^{j-1} C_j = 0 \text{ and } v_0^{\varphi^2, \beta} = 1/4.$$

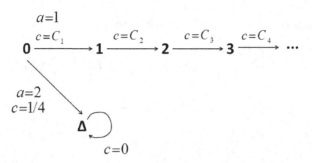

Fig. 4.30   Example 4.2.26: a Blackwell optimal strategy is not average-overtaking optimal.

At the same time,

$$\limsup_{T\to\infty} \frac{1}{T}\sum_{t=1}^{T}\left\{ E_0^{\varphi^1}\left[\sum_{\tau=1}^{t} c(X_{\tau-1},A_\tau)\right] - E_0^{\varphi^2}\left[\sum_{\tau=1}^{t} c(X_{\tau-1},A_\tau)\right]\right\}$$

$$= \limsup_{T\to\infty} \frac{1}{T}\sum_{t=1}^{T}\sum_{\tau=1}^{t} C_\tau - 1/4 = 3/4,$$

and hence φ^1 is not average-overtaking optimal. Note that selector φ^2 is not average-overtaking either, because

$$\limsup_{T\to\infty} \frac{1}{T}\sum_{t=1}^{T}\left\{ E_0^{\varphi^2}\left[\sum_{\tau=1}^{t} c(X_{\tau-1},A_\tau)\right] - E_0^{\varphi^1}\left[\sum_{\tau=1}^{t} c(X_{\tau-1},A_\tau)\right]\right\}$$

$$= 1/4 - \liminf_{T\to\infty} \frac{1}{T}\sum_{t=1}^{T}\sum_{\tau=1}^{t} C_\tau = 1/4.$$

Now, in the same example (Fig. 4.30), we can put $c(0,2) = 0$, so that $v_0^{\varphi^2,\beta} = 0$. The stationary selector φ^1 is 0-discount optimal (i.e. bias optimal), but still not average-overtaking optimal, while φ^2 is both 0-discount and average-overtaking optimal. A similar example was presented in [Flynn(1976), Ex. 2].

Remark 4.7. Theorem 1 in [Lippman(1969)] states that, in finite models, any strategy is 0-discount optimal if and only if it is average-overtaking optimal. The first part of the proof (sufficiency) holds also for any model with bounded one-step loss: if a strategy is average-overtaking optimal then it is also 0-discount optimal. This example shows that the second part (necessity) can fail if the model is not finite.

Now consider an MDP with the same state and action spaces and the same transition probabilities (see Fig. 4.30). Let $\{C_j\}_{j=1}^{\infty}$ be a bounded sequence such that

$$\lim_{n\to\infty} \frac{1}{n} \sum_{i=1}^{n} \sum_{j=1}^{i} C_j = -\infty \text{ and } \limsup_{n\to\infty} \frac{1}{n} \sum_{j=1}^{n} C_j > 0$$

(see Appendix A.4). We put $c(0,1) = C_1$, $c(0,2) = 0$, $c(i,a) \equiv C_{i+1}$ for all $i \geq 1$, and $c(\Delta, a) \equiv 0$. As previously, it is sufficient to consider only two strategies $\varphi^1(x) \equiv 1$ and $\varphi^2(x) \equiv 2$ and the initial state 0. The selector φ^1 is average-overtaking optimal, because

$$\limsup_{T\to\infty} \frac{1}{T} \sum_{t=1}^{T} \left\{ E_0^{\varphi^1} \left[\sum_{\tau=1}^{t} c(X_{\tau-1}, A_\tau) \right] - E_0^{\varphi^2} \left[\sum_{\tau=1}^{t} c(X_{\tau-1}, A_\tau) \right] \right\}$$

$$= \limsup_{T\to\infty} \frac{1}{T} \sum_{t=1}^{T} \sum_{\tau=1}^{t} C_\tau = -\infty;$$

we will show that φ^1 is also Blackwell optimal. Indeed, $\sum_{j=1}^{i} C_j < 0$ for all sufficiently large i; now, from the Abelian Theorem,

$$\lim_{\beta\to 1-} (1-\beta) \sum_{i=1}^{\infty} \beta^{i-1} \left[\sum_{j=1}^{i} C_j \right] = -\infty,$$

but

$$\sum_{i=1}^{\infty} \beta^{i-1} \left[\sum_{j=1}^{i} C_j \right] = \sum_{j=1}^{\infty} \sum_{i=j}^{\infty} \beta^{i-1} C_j = \sum_{j=1}^{\infty} C_j \frac{\beta^{j-1}}{1-\beta},$$

so that

$$\lim_{\beta\to 1-} \sum_{j=1}^{\infty} \beta^{j-1} C_j = \lim_{\beta\to 1-} v_0^{\varphi^1,\beta} = -\infty,$$

while $v_0^{\varphi^2,\beta} \equiv 0$.

On the other hand, the stationary selector φ^1 is not AC-optimal, because

$$v_0^{\varphi^1} = \limsup_{n\to\infty} \frac{1}{n} \sum_{j=1}^{n} C_j > 0 \text{ and } v_0^{\varphi^2} = 0.$$

For finite models, the following statement holds [Kallenberg(2010), Cor. 5.3]: if a stationary selector is Blackwell optimal then it is AC-optimal. In the current example, the state space $\mathbf{X}$ is not finite.

The selector φ^2 is AC-optimal, but not Blackwell optimal and not average-overtaking optimal.

A similar example was presented in [Flynn(1976), Ex. 3].

4.2.27 Nearly optimal and average-overtaking optimal strategies

The following example, based on [Flynn(1976), Ex. 7], illustrates that an average-overtaking optimal strategy (hence 0-discount optimal in accordance with Remark 4.7) may be not nearly optimal.

Let $\mathbf{X} = \{0, 1, (1,1), (1,2), 2, (2,1), \ldots, (2,4), 3, \ldots, k, (k,1), \ldots, (k, 2k), k+1, \ldots\}$, $\mathbf{A} = \{0, 1\}$, $p(0|0, a) \equiv 1$, $p(k+1|k, 0) = p((k,1)|k, 1) \equiv 1$, $p((k, i+1)|(k, i), a) \equiv 1$ for all $i < 2k$; $p(0|(k, 2k), a) \equiv 1$, with all other transition probabilities zero. For all $k \geq 1$, we put

$$c(k, a) \equiv 0, \quad c((k, i), a) = \begin{cases} -1, & \text{if } 1 \leq i \leq k; \\ 2, & \text{if } k+1 \leq i \leq 2k. \end{cases}$$

Finally, $c(0, a) \equiv 0$. See Fig. 4.31; note that Condition 2.1 is satisfied in this model.

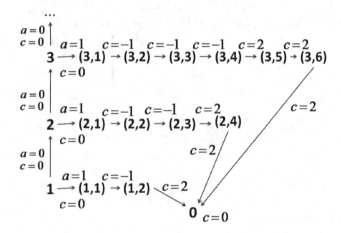

Fig. 4.31 Example 4.2.27: the average-overtaking optimal strategy is not nearly optimal.

Proposition 4.4.

(a) For any $k \geq 1$, for any control strategy π,

$$\liminf_{T \to \infty} \frac{1}{T} \sum_{t=1}^{T} E_k^\pi \left[\sum_{\tau=1}^{t} c(X_{\tau-1}, A_\tau) \right] \geq 0.$$

(b) There is an $\varepsilon > 0$ such that, for all $\beta \in (0,1)$ sufficiently close to 1, $v_1^{*,\beta} < -\varepsilon$.

The proof is given in Appendix B.

Consider a stationary selector $\varphi^0(x) \equiv 0$. Proposition 4.4(a) implies that φ^0 is average-overtaking optimal. Note that it is sufficient to consider only the initial states $k \geq 1$. At the same time, according to Proposition 4.4(b),

$$\lim_{\beta \to 1-} [v_1^{\varphi^0,\beta} - v_1^{*,\beta}] = -\lim_{\beta \to 1-} v_1^{*,\beta} > \varepsilon > 0,$$

so that φ^0 is not nearly optimal.

4.2.28 Strong-overtaking/average optimal, overtaking optimal, AC-optimal strategies and minimal opportunity loss

Definition 4.9. [Flynn(1980), Equation (7)] A strategy π^* is strong-overtaking optimal if, for all $x \in \mathbf{X}$,

$$\lim_{T \to \infty} \left\{ E_x^{\pi^*} \left[\sum_{\tau=1}^{T} c(X_{\tau-1}, A_\tau) \right] - V_x^T \right\} = 0$$

(recall that $V_x^T = \inf_\pi E_x^\pi \left[\sum_{\tau=1}^{T} c(X_{\tau-1}, A_\tau) \right]$). Such a strategy provides the minimal possible value to the (limiting) opportunity loss (3.16).

Definition 4.10. [Flynn(1980), Equation (5)] A strategy π^* is strong-average optimal if, for all $x \in \mathbf{X}$,

$$\lim_{T \to \infty} \frac{1}{T} \left\{ E_x^{\pi^*} \left[\sum_{\tau=1}^{T} c(X_{\tau-1}, A_\tau) \right] - V_x^T \right\} = 0.$$

Any strong-overtaking optimal strategy is overtaking optimal (and hence AC-optimal and opportunity-cost optimal); any strong-average optimal strategy is AC-optimal [Hernandez-Lerma and Vega-Amaya(1998), Remark 3.3]. The proofs of these statements are similar to those given after Definition 4.6. If the model is finite then the canonical stationary selector φ^*, the element of the canonical triplet $\langle \rho, h, \varphi^* \rangle$, is also strong-average optimal, because

$$V_x^T \geq E_x^\pi \left[\sum_{t=1}^{T} c(X_{t-1}, A_t) + h(X_T) \right] + \inf_\pi E_x^\pi [-h(X_T)]$$

$$\geq T\rho(x) + h(x) - \sup_{x \in \mathbf{X}} |h(x)|$$

and

$$E_x^{\varphi^*}\left[\sum_{\tau=1}^{T} c(X_{\tau-1}, A_\tau)\right] - V_x^T = T\rho(x) + h(x) - E_x^{\varphi^*}[h(X_T)] - V_x^T$$

$$\leq 2 \sup_{x \in \mathbf{X}} |h(x)|.$$

The following example, based on [Flynn(1980), Ex. 2], shows that a strong-overtaking optimal strategy may not exist, even in finite models, and even when an overtaking optimal strategy does exist.

Let $\mathbf{X} = \mathbf{A} = \{0, 1\}$, $p(0|x, 0) \equiv 1$, $p(1|x, 1) \equiv 1$, with all other transition probabilities zero; $c(0,0) = 5$, $c(0,1) = 10$, $c(1,0) = -5$, $c(1,1) = 0$ (see Fig. 4.32).

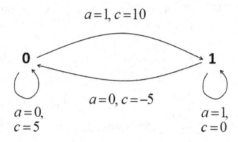

Fig. 4.32 Example 4.2.28: the overtaking optimal strategy is not strong-overtaking optimal.

Equations (1.4) give the following:

$$V_0^1 = 5, \quad V_1^1 = -5,$$
$$V_0^2 = 5, \quad V_1^2 = -5, \ldots,$$
$$V_0^T = 5, \quad V_1^T = -5, \ldots.$$

Starting from $X_0 = 0$ and from $X_0 = 1$, the trajectories ($x_0 = 0, a_1 = 1, x_1 = 1, a_2 = 1, \ldots$) and ($x_0 = 1, a_1 = 1, x_1 = 1, a_2 = 1, \ldots$) result in total losses 10 and 0 respectively, over any time interval $T \geq 1$. One can check that any other trajectory gives greater losses for all large enough values of T. Therefore, the stationary selector $\varphi^*(x) \equiv 1$ is the unique overtaking optimal strategy (and hence is opportunity-cost optimal and AC-optimal). But the selector φ^* is not strong-overtaking optimal, because

$$E_x^{\varphi^*}\left[\sum_{\tau=1}^T c(X_{\tau-1}, A_\tau)\right] - V_x^T \equiv 5 > 0 \text{ for all } T \geq 1.$$

At the same time, it is strong-average optimal. One can also show that $\rho = 0$, $h(x) = \begin{cases} 0, & \text{if } x = 0; \\ -10, & \text{if } x = 1, \end{cases}$ and $\varphi^*(x) \equiv 1$ form the single canonical triplet.

Now consider exactly the same example as in Section 2.2.4 (Fig. 2.3). One can check that $\rho(x) \equiv 0$, $h(x) = \begin{cases} 0, & \text{if } x = 0; \\ -1, & \text{if } x > 0, \end{cases}$ and $\varphi^*(x) \equiv 2$ form a canonical triplet $\langle \rho, h, \varphi^* \rangle$ (see Theorem 4.1). According to Theorem 4.2(a), the selector φ^* is AC-optimal. In fact, all strategies in this MDP are equally AC-optimal and strong-average optimal. (Note that the total expected loss in any time interval T is non-positive and uniformly bounded below by -1.)

Straightforward calculations lead to the expressions:

$$V_0^T = 0, \quad V_x^T = (x + T - 1)^{-1} - 1 \text{ for all } x > 0, T \geq 1.$$

No one strategy is overtaking optimal. Indeed, for each strategy π,

$$\lim_{T \to \infty} E_1^\pi \left[\sum_{t=1}^T c(X_{t-1}, A_t)\right] \triangleq F(\pi) > -1$$

(see Section 2.2.4), so that for the stationary selector

$$\varphi(x) \triangleq \begin{cases} 2, & \text{if } x \leq \frac{1}{F(\pi)+1}; \\ 1, & \text{if } x > \frac{1}{F(\pi)+1} \end{cases}$$

we have

$$F(\varphi) = \frac{1}{\lfloor \frac{1}{F(\pi)+1} \rfloor + 1} - 1 < F(\pi),$$

and π is not overtaking optimal (see Definition 4.6; $\lfloor \cdot \rfloor$ is the integer part). Exactly the same reasoning shows that no one strategy is opportunity-cost optimal or D-optimal. Finally, no one strategy is strong-overtaking optimal, because $\lim_{T \to \infty} V_1^T = -1$ and $F(\pi) > -1$.

This model was discussed in [Hernandez-Lerma and Vega-Amaya(1998), Ex. 4.14] and in [Hernandez-Lerma and Lasserre(1999), Section 10.9].

4.2.29 Strong-overtaking optimal and strong*-overtaking optimal strategies

In [Fernandez-Gaucherand et al.(1994)] and in [Nowak and Vega-Amaya(1999)] a strategy π^* was called overtaking optimal if, for any strategy π, there is $N(\pi^*, \pi, x)$ such that

$$E_x^{\pi^*}\left[\sum_{t=1}^T c(X_{t-1}, A_t)\right] \leq E_x^{\pi}\left[\sum_{t=1}^T c(X_{t-1}, A_t)\right] \quad (4.28)$$

as soon as $T \geq N(\pi^*, \pi, x)$. (Compare this with weak-overtaking optimality, introduced at the end of Section 4.2.24.) This definition is stronger than Definition 4.6. Thus, we shall call such a strategy π^* *strong*-overtaking optimal*, to distinguish it from Definitions 4.6 and 4.9.

Remark 4.8. If inequality (4.28) holds for all strategies π from a specified class Δ, then π^* is said to be strong*-overtaking optimal *in that class*. The same remark is valid for all other types of optimality.

The next two examples confirm that the notions of strong- and strong*-overtaking optimality are indeed different.

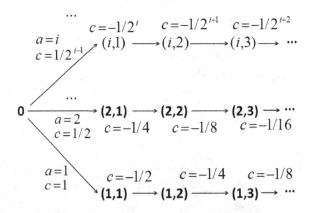

Fig. 4.33 Example 4.2.29: a strategy that is strong-overtaking optimal, but not strong*-overtaking optimal.

Let $\mathbf{X} = \{0, (1,1), (1,2), \ldots, (2,1), (2,2), \ldots\}$, $\mathbf{A} = \{1, 2, \ldots\}$; $p((a,1)|0,a) \equiv 1$, for all $i, j \geq 1$ $p((i, j+1)|(i,j), a) \equiv 1$, with all other transition probabilities zero; $c(0,a) = \left(\frac{1}{2}\right)^{a-1}$, $c((i,j),a) \equiv -\left(\frac{1}{2}\right)^{i+j-1}$ (see Fig. 4.33).

For any T, $V_0^T = \inf_\pi \left\{ E_0^\pi \left[\sum_{t=1}^T c(X_{t-1}, A_t) \right] \right\} = 0$ and, for any strategy π, $\lim_{T \to \infty} E_0^\pi \left[\sum_{t=1}^T c(X_{t-1}, A_t) \right] = 0$, meaning that all strategies are equally strong-overtaking optimal. On the other hand, for any strategy π^*, one can find a selector φ such that $E_0^\varphi [c(0, A_1)] < E_0^{\pi^*} [c(0, A_1)]$ and $E_0^{\pi^*} \left[\sum_{t=1}^T c(X_{t-1}, A_t) \right] > E_0^\varphi \left[\sum_{t=1}^T c(X_{t-1}, A_t) \right]$ for all $T \geq 1$, meaning that no one strategy is strong*-overtaking optimal.

Consider now the following model: $\mathbf{X} = \{0, 1, 2, \ldots\}$, $\mathbf{A} = \{1, 2\}$, $p(0|0, a) \equiv 1$, for all $i \geq 1$ $p(i+1|i, 1) \equiv 1$, $p(0|i, 2) \equiv 1$, with all other transition probabilities zero; $c(0, a) \equiv 1$, for all $i \geq 1$ $c(i, 1) = 0$, $c(i, 2) = -1$ (see Fig. 4.34).

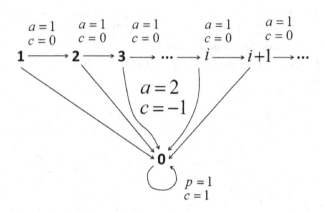

Fig. 4.34 Example 4.2.29: a strategy that is strong*-overtaking optimal, but not strong-overtaking optimal.

For any $x \geq 1$, $T \geq 1$, $V_x^T = -1$: one applies action $a = 2$ only at the last step: $A_T = 2$. But for any strategy π, for each state $x \geq 1$, we have

$$\lim_{T \to \infty} E_x^\pi \left[\sum_{t=1}^T c(X_{t-1}, A_t) \right] = \begin{cases} 0, & \text{if } P_x^\pi(A_t = 1 \text{ for all } t \geq 1) = 1; \\ \infty & \text{otherwise,} \end{cases}$$

meaning that $\lim_{T \to \infty} \left\{ E_x^\pi \left[\sum_{t=1}^T c(X_{t-1}, A_t) \right] - V_x^T \right\} > 0$, so that no one strategy is strong-overtaking optimal. At the same time, the stationary selector $\varphi(x) \equiv 1$ is strong*-overtaking optimal because, for any other strategy π, for each $x \geq 1$, either $E_x^\pi \left[\sum_{t=1}^T c(X_{t-1}, A_t) \right] = E_x^\varphi \left[\sum_{t=1}^T c(X_{t-1}, A_t) \right] = 0$ for all $T \geq 1$, or $\lim_{T \to \infty} E_x^\pi \left[\sum_{t=1}^T c(X_{t-1}, A_t) \right] = \infty$.

There was an attempt to prove that, under appropriate conditions, there exists a strong*-overtaking stationary selector (in the class of AC-optimal stationary selectors): see Lemma 6.2 and Theorems 6.1 and 6.2 in [Fernandez-Gaucherand et al.(1994)]. In fact, the stationary selector described in Theorem 6.2 is indeed strong*-overtaking optimal if it is unique. But the following example, published in [Nowak and Vega-Amaya(1999)], shows that it is possible for *two* stationary selectors to be equal candidates, but neither of them overtakes the other. As a result, a strong*-overtaking optimal stationary selector does not exist. Note that the controlled process under consideration is irreducible and aperiodic under any stationary strategy, and the model is finite.

Let $\mathbf{X} = \{1, 2, 3\}$, $\mathbf{A} = \{1, 2\}$, $p(1|1,1) = p(3|1,2) = 0.7$, $p(3|1,1) = p(1|1,2) = 0.1$, $p(2|1,a) \equiv 0.2$, $p(1|2,a) = p(3|2,a) = p(1|3,a) = p(2|3,a) \equiv 0.5$, with all other transition probabilities zero. We put $c(1,1) = 1.4$, $c(1,2) = 0.2$, $c(2,a) \equiv -9$, $c(3,a) \equiv 6$ (see Fig. 4.35).

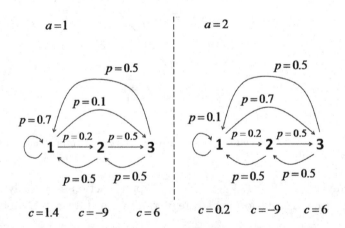

Fig. 4.35 Example 4.2.29: no strong*-overtaking optimal stationary selectors.

The canonical equations (4.2) have the solution $\rho^* = 0$, $h(1) = 8$, $h(2) = 0$, $h(3) = 10$, and both the stationary selectors $\varphi^1(x) \equiv 1$ and $\varphi^2(x) \equiv 2$ provide an infimum, meaning that both $\langle \rho^*, h, \varphi^1 \rangle$ and $\langle \rho^*, h, \varphi^2 \rangle$ are canonical triplets, and both the selectors φ^1 and φ^2 are AC-optimal.

The stationary distributions of the controlled process under the stationary selectors φ^1 and φ^2 are as follows:

$$\hat{\eta}^{\varphi^1}(1) = 15/24, \quad \hat{\eta}^{\varphi^1}(2) = 5/24, \quad \hat{\eta}^{\varphi^1}(3) = 4/24;$$

$$\hat{\eta}^{\varphi^2}(1) = 15/42, \quad \hat{\eta}^{\varphi^2}(2) = 11/42, \quad \hat{\eta}^{\varphi^2}(3) = 16/42,$$

and we see that

$$\sum_{x \in \mathbf{X}} \hat{\eta}^{\varphi^1}(x) h(x) = \sum_{x \in \mathbf{X}} \hat{\eta}^{\varphi^2}(x) h(x) = \frac{20}{3}.$$

The conditions of Theorem 6.2 in [Fernandez-Gaucherand et al.(1994)] hold for the selectors φ^1 and φ^2, but we now show that neither of them is strong*-overtaking optimal. Direct calculations based on the induction argument show that

$$E_x^{\varphi^1}\left[\sum_{t=1}^{T} c(X_{t-1}, A_t)\right] = \begin{cases} \frac{4}{3} + \frac{1}{21}\left[-10\left(-\frac{1}{2}\right)^T - 18\left(\frac{1}{5}\right)^T\right], & \text{if } x = 1; \\ -\frac{20}{3} + \frac{1}{21}\left[110\left(-\frac{1}{2}\right)^T + 30\left(\frac{1}{5}\right)^T\right], & \text{if } x = 2; \\ \frac{10}{3} + \frac{1}{21}\left[-100\left(-\frac{1}{2}\right)^T + 30\left(\frac{1}{5}\right)^T\right], & \text{if } x = 3, \end{cases}$$

$$E_x^{\varphi^2}\left[\sum_{t=1}^{T} c(X_{t-1}, A_t)\right] = \begin{cases} \frac{4}{3} + \frac{1}{3}\left[50\left(-\frac{1}{2}\right)^T - 54\left(\frac{-2}{5}\right)^T\right], & \text{if } x = 1; \\ -\frac{20}{3} + \frac{1}{3}\left[-10\left(-\frac{1}{2}\right)^T + 30\left(\frac{-2}{5}\right)^T\right], & \text{if } x = 2; \\ \frac{10}{3} + \frac{1}{3}\left[-40\left(-\frac{1}{2}\right)^T + 30\left(\frac{-2}{5}\right)^T\right], & \text{if } x = 3. \end{cases}$$

Therefore,

$$21 \cdot 2^T \left[E_x^{\varphi^1}\left[\sum_{t=1}^{T} c(X_{t-1}, A_t)\right] - E_x^{\varphi^2}\left[\sum_{t=1}^{T} c(X_{t-1}, A_t)\right]\right]$$

$$= \begin{cases} -360(-1)^T + o(1), & \text{if } x = 1; \\ 180(-1)^T + o(1), & \text{if } x = 2; \\ 180(-1)^T + o(1), & \text{if } x = 3, \end{cases}$$

where $\lim_{T \to \infty} o(1) = 0$. Inequality (4.28) holds neither for selector φ^1 nor for φ^2. In what follows, $F(x) \triangleq \begin{cases} \frac{4}{3}, & \text{if } x = 1; \\ -\frac{20}{3}, & \text{if } x = 2; \\ \frac{10}{3}, & \text{if } x = 3. \end{cases}$

The selectors φ^1 and φ^2 are not overtaking optimal, because they do not overtake the selector $\varphi_t(x) = \begin{cases} 2, & \text{if } t = \tau, 2\tau, \ldots; \\ 1 & \text{otherwise} \end{cases}$ under large enough τ. The values $E_x^{\varphi^{1,2}}\left[\sum_{t=1}^{n\tau-1} c(X_{t-1}, A_t)\right]$ and $E_x^{\varphi}\left[\sum_{t=1}^{n\tau-1} c(X_{t-1}, A_t)\right]$ are almost equal to $F(x)$, the distribution of $X_{n\tau-1}$ under φ almost coincides with $\hat{\eta}^{\varphi^1}$, the stationary distribution under control strategy φ^1, and $E_x^{\varphi^{1,2}}[c(X_{n\tau-1}, A_{n\tau})] \approx 0$; $E_x^{\varphi}[c(X_{n\tau-1}, A_{n\tau})] \approx -\frac{22}{24}$.

We investigate the discounted version of this MDP when the discount factor β is close to 1. The optimality equation (3.2) takes the form

$$v(1) = \min\{1.4 + \beta[0.7\ v(1) + 0.2\ v(2) + 0.1\ v(3)],$$
$$0.2 + \beta[0.1\ v(1) + 0.2\ v(2) + 0.7\ v(3)]\}$$
$$v(2) = -9 + 0.5\beta[v(1) + v(3)],$$
$$v(3) = 6 + 0.5\beta[v(1) + v(2)].$$

From the second and third equations we obtain:

$$v(3) = \frac{6 - 4.5\beta + (0.5\beta + 0.25\beta^2)v(1)}{1 - 0.25\beta^2}.$$

After we substitute this expression into the two other equations, we obtain the following:

$$[1 - 0.25\beta^2]v(1) = \min\{1.4 - 1.2\beta - 0.2\beta^2 + v(1)\beta[0.7 + 0.15\beta - 0.1\beta^2];$$
$$0.2 + 2.4\beta - 2.6\beta^2 + v(1)\beta[0.1 + 0.45\beta + 0.2\beta^2]\}.$$

From the second line we obtain

$$v(1) = \frac{0.2 + 2.4\beta - 2.6\beta^2}{1 - 0.1\beta - 0.7\beta^2 - 0.2\beta^3} = \frac{4}{3} - \frac{26}{63}(1-\beta) + o(1-\beta),$$

where $\lim_{\varepsilon \to 0} \frac{o(\varepsilon)}{\varepsilon} = 0$. Now, the difference between the first and the second lines equals

$$1.2 - 3.6\beta + 2.4\beta^2 + v(1)\beta[0.6 - 0.3\beta - 0.3\beta^2] = 2(1-\beta)^2 + o((1-\beta)^2) > 0,$$

meaning that, indeed, $v_1^{*,\beta} = \frac{0.2+2.4\beta-2.6\beta^2}{1-0.1\beta-0.7\beta^2-0.2\beta^3}$ for all β close enough to 1, and the stationary selector φ^2 is Blackwell optimal. The values for $v_2^{*,\beta}$ and $v_3^{*,\beta}$ can easily be calculated using the formulae provided.

Note that both the selectors φ^1 and φ^2 are bias optimal, because

$$\lim_{\beta \to 1-} v_1^{\varphi^1,\beta} = \lim_{\beta \to 1-} v_1^{\varphi^2,\beta} = \lim_{\beta \to 1-} v_1^{*,\beta} = F(1) = \frac{4}{3};$$

$$\lim_{\beta \to 1-} v_2^{\varphi^1,\beta} = \lim_{\beta \to 1-} v_2^{\varphi^2,\beta} = \lim_{\beta \to 1-} v_2^{*,\beta} = F(2) = -\frac{20}{3};$$

$$\lim_{\beta \to 1-} v_3^{\varphi^1,\beta} = \lim_{\beta \to 1-} v_3^{\varphi^2,\beta} = \lim_{\beta \to 1-} v_3^{*,\beta} = F(3) = \frac{10}{3}.$$

It is also interesting to emphasize that the introduced function F solves the optimality equation (2.2), which has no other solutions except for

$F(x) + r$ ($r \in \mathbb{R}$ is an arbitrary constant). Both the selectors φ^1 and φ^2 provide the minimum (and also the maximum!) in that equation, and $\sum_{x \in \mathbf{X}} \hat{\eta}^{\varphi^1}(x) F(x) = \sum_{x \in \mathbf{X}} \hat{\eta}^{\varphi^2}(x) F(x) = 0$, meaning that

$$\lim_{t \to \infty} E_x^{\varphi^1}[F(X_t)] = \lim_{t \to \infty} E_x^{\varphi^2}[F(X_t)] = 0,$$

because the controlled process is ergodic. Note also that for each control strategy π, for any initial distribution P_0,

$$E_{P_0}^{\pi}\left[\sum_{t=1}^{\infty} c^+(X_{t-1}, A_t)\right] = +\infty, \quad E_{P_0}^{\pi}\left[\sum_{t=1}^{\infty} c^-(X_{t-1}, A_t)\right] = -\infty,$$

so that Condition 2.1 is violated. Thus, if one wants to investigate the version with the expected total loss, then formula (2.1) needs to be explained. For example, one can define $E_{P_0}^{\pi}[\sum_{t=1}^{\infty} c(X_{t-1}, A_t)]$ as either

$$\limsup_{T \to \infty} E_{P_0}^{\pi}\left[\sum_{t=1}^{T} c(X_{t-1}, A_t)\right], \text{ or } \limsup_{\beta \to 1-} E_{P_0}^{\pi}\left[\sum_{t=1}^{\infty} \beta^{t-1} c(X_{t-1}, A_t)\right].$$

4.2.30 Parrondo's paradox

This paradox can be described as follows [Parrondo and Dinis(2004)]: "Two losing gambling games, when alternated in a periodic or random fashion, can produce a winning game." There exist many examples to illustrate this; we present the simplest one.

Let $\mathbf{X} = \{1, 2, 3\}$, $\mathbf{A} = \{1, 2\}$, $p(2|1, 1) = p(3|2, 1) = p(1|3, 1) = 0.49$, $p(3|1, 1) = p(1|2, 1) = p(2|3, 1) = 0.51$, $p(2|1, 2) = 1 - p(3|1, 2) = 0.09$, $p(3|2, 2) = 1 - p(1|2, 2) = p(1|3, 2) = 1 - p(2|3, 2) = 0.74$, with all other transition probabilities zero. We put $c(1, a) = p(3|1, a) - p(2|1, a)$, $c(2, a) = p(1|2, a) - p(3|2, a)$, $c(3, a) = p(2|3, a) - p(1|3, a)$ (see Fig. 4.36).

One can say that the process moves clockwise or anticlockwise, with the probabilities depending on the actions. The gambler gains one pound for each clockwise step of the walk, and loses one pound for each anticlockwise step. The objective is to minimize the expected average loss per time unit.

After we put $h(1) = 0$, as usual, the canonical equations (4.2) take the form:

$$\rho = \min\{0.02 + 0.49\, h(2) + 0.51\, h(3);$$
$$0.82 + 0.09\, h(2) + 0.91\, h(3)\};$$
$$\rho + h(2) = \min\{0.02 + 0.49\, h(3); \quad -0.48 + 0.74\, h(3)\};$$
$$\rho + h(3) = \min\{0.02 + 0.51\, h(2); \quad -0.48 + 0.26\, h(2)\}.$$

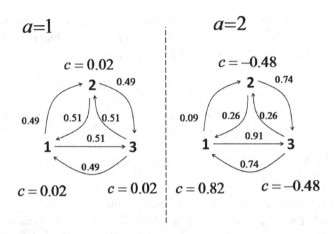

Fig. 4.36 Example 4.2.30: Parrondo's paradox. The arrows are marked with their corresponding transition probabilities.

One can check that the solution is given by $h(2) = -\frac{14500}{38388}$; $h(3) = -\frac{125}{457}$; $\rho = 0.02 + 0.49\ h(2) + 0.51\ h(3) \approx -0.305$; the stationary selector
$$\varphi^*(x) = \begin{cases} 1, & \text{if } x = 1; \\ 2, & \text{if } x = 2 \text{ or } 3 \end{cases}$$
is AC-optimal according to Theorems 4.1 and 4.2.

Consider the stationary selector $\varphi^1(x) \equiv 1$. Analysis of the canonical equations (4.2) gives the following values: $h^1(x) \equiv 0$, $\rho^1 = 0.02$. Similarly, for the stationary selector $\varphi^2(x) \equiv 2$ we obtain: $h^2(1) = 0$, $h^2(2) = -\frac{13195}{12313}$, $h^2(3) = -\frac{1365}{1759}$, $\rho^2 = 0.82 + 0.09\ h^2(2) + 0.91\ h^2(3) \approx 0.017$. This means that the outcomes for both pure games, where action 1 or action 2 is always chosen, are unfavourable: the expected average loss per time unit is positive. The optimal strategy φ^* results in a winning game, but more excitingly, a random choice of actions 1 and 2 at each step also results in a winning game. To be more precise, we analyze the stationary randomized strategy $\pi(1|x) = \pi(2|x) \equiv 0.5$. The canonical equations are as follows:

$$\tilde{\rho} + \tilde{h}(1) = 0.42 + 0.29\ \tilde{h}(2) + 0.71\ \tilde{h}(3),$$
$$\tilde{\rho} + \tilde{h}(2) = -0.23 + 0.615\ \tilde{h}(3) + 0.385\ \tilde{h}(1),$$
$$\tilde{\rho} + \tilde{h}(3) = -0.23 + 0.615\ \tilde{h}(1) + 0.385\ \tilde{h}(2),$$

and the solution is given by $\tilde{h}(1) = 0$, $\tilde{h}(2) = -\frac{11{,}631{,}230}{24{,}541{,}369}$, $\tilde{h}(3) = -\frac{36010}{88597}$, $\tilde{\rho} = 0.42 + 0.29\ \tilde{h}(2) + 0.71\ \tilde{h}(4) \approx -0.006$. Since $\tilde{\rho} < 0$, a random choice of losing games (i.e. actions 1 and 2) results in a winning game.

4.2.31 An optimal service strategy in a queueing system

Consider a two-server queueing system in which the service time distributions of the two servers are different, namely the service time for server 1 is stochastically less than that for server 2. There is no space for waiting, so that customers confronted by an occupied system are lost. If the system is empty, the arriving customer can be served by either of the servers. Intuitively, the best strategy is to send that customer to server 1, in order to minimize the average number of lost customers. The following example, based on [Seth(1977), Section 2], shows that this decision is not necessarily optimal.

We assume that only one customer can arrive during one time slot, with probability λ. Server 1 has a mixed service-time distribution: $T_1 = 0$ with probability $1/2$ and $T_1 \sim geometric(\mu/2)$ with probability $1/2$. Server 2 also has a mixed service-time distribution: $T_2 \sim geometric(\mu)$ with probability $1/2$ and $T_2 \sim geometric(\mu/2)$ with probability $1/2$. The numbers $\lambda, \mu \in (0, 2/3)$ are fixed. It is easy to see that T_1 is stochastically less than T_2: $P(T_1 > z) \leq P(T_2 > z)$ for all $z \geq 0$. The state of the process is encoded as (i, j), where $i = 0$ if server 1 is free, and $i = 2$ if server 1 is performing the $geometric(\mu/2)$ service; $j = 0$ if server 2 is free, $j = 1$ (or 2) if server 2 is performing the $geometric(\mu)$ (or $geometric(\mu/2)$) service. Thus, $\mathbf{X} = \{(0,0), (0,1), (0,2), (2,0), (2,1), (2,2)\}$. Action $a \in \mathbf{A} = \{1, 2\}$ means that a new customer arriving at the free system is sent to server a. Note that we ignore the probability of two (or more) events occurring during one time slot. In fact, we consider the discrete-time approximation of the usual continuous-time queueing system: the time slot is very small, as are the probabilities λ and μ. According to the verbal description of the model,

$$p((0,0)|(0,0),1) = 1 - \lambda/2, \quad p((2,0)|(0,0),1) = \lambda/2,$$

$$p((0,0)|(0,0),2) = 1 - \lambda, \quad p((0,1)|(0,0),2) = \lambda/2,$$

$$p((0,2)|(0,0),2) = \lambda/2,$$

$$p(y|(0,1),a) \equiv \begin{cases} 1 - \lambda/2 - \mu, & \text{if } y = (0,1); \\ \lambda/2, & \text{if } y = (2,1); \\ \mu, & \text{if } y = (0,0), \end{cases}$$

$$p(y|(0,2),a) \equiv \begin{cases} 1 - \lambda/2 - \mu/2, & \text{if } y = (0,2); \\ \lambda/2, & \text{if } y = (2,2); \\ \mu/2, & \text{if } y = (0,0), \end{cases}$$

$$p(y|(2,0),a) \equiv \begin{cases} 1-\lambda-\mu/2, & \text{if } y=(2,0); \\ \lambda/2, & \text{if } y=(2,1); \\ \lambda/2, & \text{if } y=(2,2); \\ \mu/2, & \text{if } y=(0,0), \end{cases}$$

$$p(y|(2,1),a) \equiv \begin{cases} 1-3\mu/2, & \text{if } y=(2,1); \\ \mu, & \text{if } y=(2,0); \\ \mu/2, & \text{if } y=(0,1), \end{cases}$$

$$p(y|(2,2),a) \equiv \begin{cases} 1-\mu, & \text{if } y=(2,2); \\ \mu/2, & \text{if } y=(0,2); \\ \mu/2, & \text{if } y=(2,0), \end{cases}$$

and all the other transition probabilities are zero.

$$c(x,a) \equiv \begin{cases} \lambda, & \text{if } x=(2,1); \text{ or } (2,2), \\ 0 & \text{otherwise} \end{cases}$$

(see Fig. 4.37).

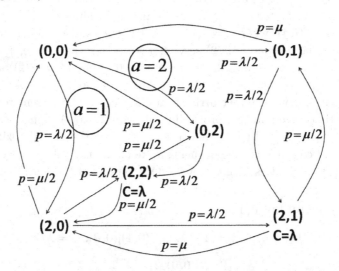

Fig. 4.37 Example 4.2.31: queueing system. The probabilities for the loops $p(x|x,a)$ are not shown.

The canonical equations (4.2) for states $(0,0)$, $(0,1)$, $(0,2)$, $(2,0)$, $(2,1)$ and $(2,2)$, respectively have the form

$$\rho = \min\left\{-\frac{\lambda}{2}h(0,0) + \frac{\lambda}{2}h(2,0); \quad -\lambda h(0,0) + \frac{\lambda}{2}h(0,1) + \frac{\lambda}{2}h(0,2)\right\};$$

$$\rho = -(\mu + \lambda/2)h(0,1) + \mu h(0,0) + \frac{\lambda}{2}h(2,1);$$

$$\rho = -(\mu/2 + \lambda/2)h(0,2) + \frac{\mu}{2}h(0,0) + \frac{\lambda}{2}h(2,2);$$

$$\rho = -(\mu/2 + \lambda)h(2,0) + \frac{\mu}{2}h(0,0) + \frac{\lambda}{2}h(2,2) + \frac{\lambda}{2}h(2,1);$$

$$\rho = \lambda - \frac{3\mu}{2}h(2,1) + \mu h(2,0) + \frac{\mu}{2}h(0,1);$$

$$\rho = \lambda - \mu h(2,2) + \frac{\mu}{2}h(2,0) + \frac{\mu}{2}h(0,2).$$

If we put $h(0,0) \triangleq 0$, as usual, and write $g \triangleq \lambda/\mu$ then, after some trivial algebra, we obtain

$$\rho = \lambda g^2 \frac{3g^2 + 9g + 4}{3g^4 + 14g^3 + 23g^2 + 19g + 6};$$

$$h(2,0) = g^2 \frac{12g^3 + 68g^2 + 115g + 50}{6g^5 + 43g^4 + 116g^3 + 153g^2 + 107g + 30}.$$

Other values of the function h are of no importance, but they can certainly be calculated as well; for example:

$$h(0,1) = g^2 \frac{4g^3 + 20g^2 + 29g + 10}{6g^5 + 43g^4 + 116g^3 + 153g^2 + 107g + 30},$$

and so on. Note that all these formulae come from the equation

$$\rho = -\lambda h(0,0) + \frac{\lambda}{2}h(0,1) + \frac{\lambda}{2}h(0,2);$$

i.e. we accepted $\varphi^*(x) \equiv 2$. To prove the optimality of the selector φ^*, it only remains to compare ρ with $-\frac{\lambda}{2}h(0,0) + \frac{\lambda}{2}h(2,0) = \frac{\lambda}{2}h(2,0)$:

$$\frac{\lambda}{2}h(2,0) - \rho = \lambda g^2 \left[\frac{12g^3 + 68g^2 + 115g + 50}{2(3g^4 + 14g^3 + 23g^2 + 19g + 6)(2g + 5)} \right.$$
$$\left. - \frac{3g^2 + 9g + 4}{3g^4 + 14g^3 + 23g^2 + 19g + 6} \right]$$
$$= \lambda g^2 \frac{2g^2 + 9g + 10}{2(3g^4 + 14g^3 + 23g^2 + 19g + 6)(2g + 5)} > 0$$

for any values of λ and μ. Thus, $\langle \rho, h, \varphi^* \rangle$ is a canonical triplet, and the stationary selector φ^* is AC-optimal. If the system is free then it is better to send the arriving customer to server 2 with a stochastically *longer* service time. To understand this better, we present comments similar to those given in [Seth(1977), Table 1]. When three customers arrive, four different situations can exist with equal probability 1/4:

	Server 1	Server 2	Optimal decision about the first customer
Situation 1	0	$geomteric(\mu)$	both equally good
Situation 2	0	$geomteric(\mu/2)$	both equally good
Situation 3	$geometric(\mu/2)$	$geomteric(\mu/2)$	both equally good
Situation 4	$geometric(\mu/2)$	$geomteric(\mu)$	send to server 2.

If one server has service time zero, then all three customers are served, no matter which strategy is used.

Afterword

This book contains about 100 examples, mainly illustrating that the conditions imposed in the known theorems are important. Several meaningful examples, leading to unexpected and sometimes surprising answers, are also given, such as voting, optimal search, queues, and so on. Real-life applications of Markov Decision Processes are beyond the scope of this book; however, we briefly mention several of them here.

1. **Control of a moving object.** Here, the state is just the position of the object subject to random disturbances, and the action corresponds to the power of an engine. The objective can be, for example, reaching the goal with the minimal expected energy. Such models have been studied in [Dynkin and Yushkevich(1979), Chapter 2, Section 11] and [Piunovskiy(1997), Section 5.4].
2. **Control of water resources.** Here the state is the amount of water in a reservoir, depending on rainfall and on decisions about using the water. The performance to be maximized corresponds to the expected utility of the water consumed. Such models have been studied in [Dynkin and Yushkevich(1979), Chapter 2, Section 8] and [Sniedovich(1980)].
3. **Consumption–investment problems.** Here one has to split the current capital (the state of the process) into two parts; for example, in order to minimize the total expected consumption over the planning interval. Detailed examples can be found in [Bäuerle and Rieder(2011), Sections 4.3, 9.1], [Dynkin and Yushkevich(1979), Chapter 2, Section 7], and [Puterman(1994), Section 3.5.3].
4. **Inventory control.** The state is the amount of product in a warehouse, subject to random demand. Actions are the ordering of new portions. The goal is to maximize the total expected profit from

selling the product. Such models have been considered in [Bertsekas(2005), Section 4.2], [Bertsekas(2001), Section 3.3], [Borkar and Ghosh(1995)], and [Puterman(1994), Section 3.2].
5. **Reliability.** The state of a deteriorating device is subject to random disturbances, and one has to make decisions about preventive maintenance or about replacing the device with a new one. The goal is to minimize the total expected loss resulting from failures and from the maintenance cost. Detailed examples can be found in [Hu and Yue(2008), Chapter 9], [Ross(1970), Section 6.3], and [Ross(1983), Ex. 3.1].
6. **Financial mathematics.** The state is the current wealth along with the vector of stock prices in a random market. The action represents the restructuring of the self-financing portfolio. The goal might be the maximization of the expected utility associated with the final wealth. Such problems have been investigated in [Bäuerle and Rieder(2011), Chapter 4], [Bertsekas(2005), Section 4.3], and [Dokuchaev(2007), Section 3.12].
7. **Selling an asset.** The state is the current random market price of the asset (e.g. a house), and one must decide whether to accept or reject the offer. There is a running maintenance cost, and the objective is to maximize the total expected profit. Such models have been considered in [Bäuerle and Rieder(2011), Section 10.3.1] and [Ross(1970), Section 6.3].
8. **Gambling.** Such models have already appeared in the earlier sections. We mention also [Bertsekas(2001), Section 3.5], [Dubins and Savage(1965)], [Dynkin and Yushkevich(1979), Chapter 2, Section 9], and [Ross(1983), Chapter I, Section 2 and Chapter IV, Section 2].

Finally, many other meaningful examples have been considered in articles and books. Some examples are:

- quality control in a production line [Yao and Zheng(1998)];
- forest management [Forsell *et al.*(2011)];
- controlled populations [Piunovskiy(1997), Section 5.2];
- participating in a quiz show [Bäuerle and Rieder(2011), Section 10.3.2];
- organizing of teaching and examinations [Bertsekas(1987), Section 3.4];

- optimization of publicity efforts [Piunovskiy(1997), Section 5.6];
- insurance [Schmidli(2008)].

It is nearly impossible to name an area where MDPs cannot be applied.

Appendix A

Borel Spaces and Other Theoretical Issues

In this Appendix, familiar definitions and assertions are collected together for convenience. More information can be found in [Bertsekas and Shreve(1978); Goffman and Pedrick(1983); Hernandez-Lerma and Lasserre(1996a); Parthasarathy(2005)].

A.1 Main Concepts

Definition A.1. If $(\mathbf{X}, \tau)$ is a topological space and $\mathbf{Y} \subset \mathbf{X}$, then we understand $\mathbf{Y}$ to be a topological space with open sets $\mathbf{Y} \cap \Gamma$, where Γ ranges over τ. This is called the *relative topology*.

Definition A.2. Let $(\mathbf{X}, \tau)$ be a topological space. A metric ρ in $\mathbf{X}$ is *consistent* with τ if every set of the form $\{y \in \mathbf{X} : \rho(x,y) < c\}$, $x \in \mathbf{X}$, $c > 0$ is in τ, and every non-empty set in τ is the union of sets of this form. The space $(\mathbf{X}, \tau)$ is *metrizable* if such a metric exists.

Definition A.3. Let $(\mathbf{X}_1, \tau_1)$ and $(\mathbf{X}_2, \tau_2)$ be two topological spaces. Suppose that $\varphi : \mathbf{X}_1 \longrightarrow \mathbf{X}_2$ is a one-to-one and continuous mapping, and φ^{-1} is continuous on $\varphi(\mathbf{X}_1)$ with the relative topology. Then we say that φ is a *homeomorphism* and $\mathbf{X}_1$ is *homeomorphic* to $\varphi(\mathbf{X}_1)$.

Definition A.4. Let $\mathbf{X}$ be a metrizable topological space. (In what follows, the topology sign τ is omitted.) The space $\mathbf{X}$ is *separable* if it contains a denumerable dense set.

Definition A.5. A collection of subsets of a topological space $\mathbf{X}$ is called a *base* of the topology if any open set can be represented as the union of subsets from that collection. If a base can be constructed as a collection of finite intersections of subsets from another collection, then the latter is called a *sub-base*.

A metrizable topological space is separable if and only if the topology has a denumerable base.

Definition A.6. Let $\mathbf{X}_\alpha$, $\alpha \in \mathcal{A}$, be an arbitrary collection of topological spaces, and let $\mathbf{X} = \prod_{\alpha \in \mathcal{A}} \mathbf{X}_\alpha$ be their direct product. We take an arbitrary finite number of indices $\alpha_1, \alpha_2, \ldots, \alpha_M$ and fix an open set u_{α_m} in $\mathbf{X}_{\alpha_m}$ for every $m = 1, 2, \ldots, M$. The set of all $x \in \mathbf{X}$ for which $x_{\alpha_1} \in u_{\alpha_1}, x_{\alpha_2} \in u_{\alpha_2}, \ldots, x_{\alpha_M} \in u_{\alpha_M}$ is called the elementary open set $O(u_{\alpha_1}, u_{\alpha_2}, \ldots, u_{\alpha_M})$ (other x_α are arbitrary). The elementary sets form the base of topology in $\mathbf{X}$; this topology turns $\mathbf{X}$ into the topological space known as the *topological (Tychonoff) product*.

Let $\mathbf{X}_1, \mathbf{X}_2, \ldots$ be a sequence of topological spaces, and let $\mathbf{X}$ be their Tychonoff product. Then $x^n \xrightarrow[n \to \infty]{} x$ in the space $\mathbf{X}$ if and only if, $\forall m = 1, 2, \ldots,$ $x_m^n \xrightarrow[n \to \infty]{} x_m$ in the space $\mathbf{X}_m$. (Here $x_m \in \mathbf{X}_m$ is the mth component of the point $x \in \mathbf{X}$.)

Let $\mathbf{X}_1, \mathbf{X}_2, \ldots$ be a sequence of separable metrizable topological spaces. Consider the component-wise convergence in $\mathbf{X} = \mathbf{X}_1 \times \mathbf{X}_2 \times \cdots$ and the corresponding topology with the help of the closure operation. Then we obtain the Tychonoff topology in $\mathbf{X}$. In this case, $\mathbf{X}$ is the separable metrizable space.

Theorem A.1. *(Tychonoff) Let $\mathbf{X}_1, \mathbf{X}_2, \ldots$ be a sequence of metrizable compact spaces, and let $\mathbf{X}$ be their Tychonoff product. Then $\mathbf{X}$ is compact.*

This theorem also holds for an arbitrary (not denumerable) Tychonoff product of compact spaces, which may not be metrizable.

If $\mathbf{X}$ is discrete (i.e. finite or countable) with the *discrete* topology containing all singletons, then all subsets of $\mathbf{X}$ are simultaneously open and closed, and the Borel σ-algebra coincides with the collection of all subsets of $\mathbf{X}$.

Definition A.7. The *Hilbert cube* $\mathcal{H}$ is the topological product of denumerably many copies of the unit interval. Clearly, $\mathcal{H}$ is a separable metrizable space.

Definition A.8. The *Bair null space* $\mathcal{N}$ is the topological product of denumerably many copies of the set $\mathbf{N}$, natural numbers with discrete topology.

Theorem A.2. *(Urysohn) Every separable metrizable space is homeomorphic to a subset of the Hilbert cube $\mathcal{H}$.*

Definition A.9. A metric space $(\mathbf{X}, \rho)$ is *totally bounded* if, for every $\varepsilon > 0$, there exists a finite subset $\Gamma_\varepsilon \subseteq \mathbf{X}$ for which

$$\mathbf{X} = \bigcup_{x \in \Gamma_\varepsilon} \{y \in \mathbf{X} : \ \rho(x,y) < \varepsilon\}.$$

Theorem A.3. *The Hilbert cube is totally bounded under any metric consistent with its topology, and every separable metrizable space has a totally bounded metrization.*

If $\mathbf{X}$ is a metrizable space, the set of all bounded continuous real-valued functions on $\mathbf{X}$ is denoted $\mathbf{C}(\mathbf{X})$. As is well known, $\mathbf{C}(\mathbf{X})$ is a complete (i.e. Banach) space under the norm $\|f(\cdot)\| = \sup_{x \in \mathbf{X}} |f(x)|$.

Definition A.10. If $(\mathbf{X}, \tau)$ is a topological space, the smallest σ-algebra of subsets of $\mathbf{X}$ which contains all open subsets of $\mathbf{X}$ is called the *Borel σ-algebra* and is denoted by $\mathcal{B}(\mathbf{X}) \triangleq \sigma\{\tau\}$.

Theorem A.4. *Let $(\mathbf{X}, \tau)$ be a metrizable space. Then τ is the weakest topology with respect to which every function in $\mathbf{C}(\mathbf{X})$ is continuous; $\mathcal{B}(\mathbf{X})$ is the smallest σ-algebra with respect to which every function in $\mathbf{C}(\mathbf{X})$ is measurable.*

Definition A.11. Let $\mathbf{X}$ be a topological space. If there exists a complete separable metric space $\mathbf{Y}$ and a Borel subset $B \in \mathcal{B}(\mathbf{Y})$ such that $\mathbf{X}$ is homeomorphic to B, then $\mathbf{X}$ is said to be a *Borel space*.

If $\mathbf{X}$ is a Borel space and $B \in \mathcal{B}(\mathbf{X})$, then B is also a Borel space.

Theorem A.5. *Let $\mathbf{X}_1, \mathbf{X}_2, \ldots$ be a sequence of Borel spaces and $\mathbf{Y}_n = \mathbf{X}_1 \times \mathbf{X}_2 \times \cdots \times \mathbf{X}_n$; $\mathbf{Y} = \mathbf{X}_1 \times \mathbf{X}_2 \times \cdots$. Then $\mathbf{Y}$ and each $\mathbf{Y}_n$ with the product topology (i.e. Tychonoff topology) is a Borel space, and their σ-algebras coincide with the product σ-algebras, i.e. $\mathcal{B}(\mathbf{Y}_n) = \mathcal{B}(\mathbf{X}_1) \times \mathcal{B}(\mathbf{X}_2) \times \cdots \times \mathcal{B}(\mathbf{X}_n)$ and $\mathcal{B}(\mathbf{Y}) = \mathcal{B}(\mathbf{X}_1) \times \mathcal{B}(\mathbf{X}_2) \times \cdots$.*

Definition A.12. Let $\mathbf{X}$ and $\mathbf{Y}$ be Borel spaces and let $\varphi(\cdot) : \mathbf{X} \longrightarrow \mathbf{Y}$ be a Borel-measurable, one-to-one function (other types of measurability almost never occur in the present book). Assume that $\varphi^{-1}(\cdot)$ is Borel-measurable on $\varphi(\mathbf{X})$. Then $\varphi(\cdot)$ is called a *Borel isomorphism*, and we say that $\mathbf{X}$ and $\varphi(\mathbf{X})$ are *isomorphic*.

Theorem A.6. *Two Borel spaces are isomorphic if and only if they have the same cardinality. Every uncountable Borel space has cardinality c (the continuum) and is isomorphic to the segment* $[0,1]$ *and to the Bair null space* $\mathcal{N}$.

If **X** is an uncountable Borel space then there exist many different natural enough σ-algebras containing $\mathcal{B}(\mathbf{X})$. We discuss the analytical and universal σ-algebras.

Definition A.13. A subset $\Gamma \subseteq \mathbf{X}$ is called *analytical* if there exist an uncountable Borel space **Y**, a measurable map $\varphi : \mathbf{Y} \to \mathbf{X}$, and a set $B \in \mathcal{B}(\mathbf{Y})$ such that $\Gamma = \varphi(B)$.

Every Borel subset of **X** is also analytical. On the other hand, it is known that any uncountable Borel space contains an analytical subset which is not Borel-measurable. Basically, any analytical subset coincides with the projection on **X** of some Borel (even closed) subset of $\mathbf{X} \times \mathcal{N}$.

Definition A.14. The minimal σ-algebra $\mathcal{A}(\mathbf{X})$ in **X** containing all analytical subsets is called an *analytical* σ-algebra; it contains $\mathcal{B}(\mathbf{X})$, and its elements are called *analytically measurable* subsets.

Definition A.15. Let **X** be a Borel space. The function $f : \mathbf{X} \to \mathbb{R}^*$ is called *lower semi-analytical* if the set $\{x \in \mathbf{X} : f(x) \leq c\}$ is analytical at any $c \in \mathbb{R}^*$.

Every such function is analytically measurable, i.e. $\{x \in \mathbf{X} : f(x) \leq c\} \in \mathcal{A}(\mathbf{X})$, but not *vice versa*.

A.2 Probability Measures on Borel Spaces

Recall that the support **Supp** μ of a measure μ on $(\mathbf{X}, \mathcal{B}(\mathbf{X}))$, where **X** is a topological space, is the set of all points $x \in \mathbf{X}$ for which every open neighbourhood of x has a positive μ measure [Hernandez-Lerma and Lasserre(1999), Section 7.3].

Theorem A.7. *Let* **X** *be a metrizable space. Every probability measure* $p(dx)$ *on* $(\mathbf{X}, \mathcal{B}(\mathbf{X}))$ *is regular, i.e.* $\forall \Gamma \in \mathcal{B}(\mathbf{X})$,

$$p(\Gamma) = \sup\{p(F) : \quad F \subseteq \Gamma, \ F \text{ is closed }\}$$

$$= \inf\{p(G) : \quad G \supseteq \Gamma, \ G \text{ is open }\}.$$

Definition A.16. The real random variable ξ defined on the probability space $(\Omega, \mathcal{F}, P)$ (that is, the measurable function $\Omega \longrightarrow \mathbb{R}^*$) is said to be integrable if $\int_\Omega \xi^+(\omega) P(d\omega) < +\infty$ and $\int_\Omega \xi^-(\omega) P(d\omega) > -\infty$. If only the first (second) integral is finite, then the real random variable is called *quasi-integrable above (below)*. If

$$\int_\Omega \xi^+(\omega) P(d\omega) = +\infty \text{ and } \int_\Omega \xi^-(\omega) P(d\omega) = -\infty$$

then we put

$$\int_\Omega \xi(\omega) P(d\omega) \stackrel{\triangle}{=} +\infty.$$

Definition A.17. Let $\mathbf{X}$ be a metrizable space. The set of all probability measures on $(\mathbf{X}, \mathcal{B}(\mathbf{X}))$ will be denoted by $\mathbf{P}(\mathbf{X})$. The *weak topology* in $\mathbf{P}(\mathbf{X})$ is the weakest topology with respect to which every mapping $\theta_c(\cdot) : \mathbf{P}(\mathbf{X}) \longrightarrow \mathbb{R}^1$ of the type

$$\theta_c(p) \stackrel{\triangle}{=} \int_\mathbf{X} c(x) p(dx) \tag{A.1}$$

is continuous. Here $c(\cdot) \in \mathbf{C}(\mathbf{X})$ is a bounded continuous function.

We always assume that the space $\mathbf{P}(\mathbf{X})$ is equipped with weak topology.

Theorem A.8. Let $\mathbf{X}$ be a separable metrizable space and $p, p_n \in \mathbf{P}(\mathbf{X})$, $n = 1, 2, \ldots$ Then $p_n \xrightarrow[n\to\infty]{} p$ if and only if $\int_\mathbf{X} c(x) p_n(dx) \xrightarrow[n\to\infty]{} \int_\mathbf{X} c(x) p(dx)$ for every bounded continuous function $c(\cdot) \in \mathbf{C}(\mathbf{X})$.

Theorem A.9. *If $\mathbf{X}$ is a Borel space, then $\mathbf{P}(\mathbf{X})$ is a Borel space. If $\mathbf{X}$ is a compact metrizable space, then $\mathbf{P}(\mathbf{X})$ is also a compact metrizable space.*

Definition A.18. Let $\mathbf{X}$ and $\mathbf{Y}$ be separable metrizable spaces. A *stochastic kernel* $q(dy|x)$ on $\mathbf{Y}$ given $\mathbf{X}$ (or the *transition probability* from $\mathbf{X}$ to $\mathbf{Y}$) is a collection of probability measures in $\mathbf{P}(\mathbf{Y})$ parameterized by $x \in \mathbf{X}$. If the mapping $\gamma : \mathbf{X} \longrightarrow \mathbf{P}(\mathbf{Y})$: $\gamma(x) \stackrel{\triangle}{=} q(\cdot|x)$ is measurable or (weakly) continuous, then the stochastic kernel q is said to be *measurable* or *weakly continuous*, respectively. ($\mathbf{P}(\mathbf{Y})$ is equipped with weak topology and the corresponding Borel σ-algebra $\mathcal{B}(\mathbf{P}(\mathbf{Y}))$.)

Theorem A.10. Let $\mathbf{X}$, $\mathbf{Y}$ and $\mathbf{Z}$ be Borel spaces, and let $q(d(y,z)|x)$ be a measurable stochastic kernel on $\mathbf{Y} \times \mathbf{Z}$ given $\mathbf{X}$. Then there exist measurable stochastic kernels $r(dz|x,y)$ and $s(dy|x)$ on $\mathbf{Z}$ given $\mathbf{X} \times \mathbf{Y}$ and on $\mathbf{Y}$ given $\mathbf{X}$, respectively, such that $\forall \Gamma^{\mathbf{Y}} \in \mathcal{B}(\mathbf{Y})$ $\forall \Gamma^{\mathbf{Z}} \in \mathcal{B}(\mathbf{Z})$

$$q(\Gamma^{\mathbf{Y}} \times \Gamma^{\mathbf{Z}}|x) = \int_{\Gamma^{\mathbf{Y}}} r(\Gamma^{\mathbf{Z}}|x,y) s(dy|x).$$

If there is no dependence on the parameter x, then every probability measure $q \in \mathbf{P}(\mathbf{Y} \times \mathbf{Z})$ can be expressed in the form

$$q(d(y,z)) = r(dz|y) s(dy).$$

Here s is the projection of the q measure on $\mathbf{Y}$ (the marginal), and r is a measurable stochastic kernel on $\mathbf{Z}$ given $\mathbf{Y}$.

Definition A.19. Let E be a family of probability measures on a metric space $\mathbf{X}$.

(a) E is *tight* if, for every $\varepsilon > 0$, there is a compact set $K \subset \mathbf{X}$ such that $\forall p \in E$ $p(K) > 1 - \varepsilon$.
(b) E is *relatively compact* if every sequence in E contains a weakly convergent sub-sequence.

Definition A.20. A stochastic kernel $q(dy|x)$ on $\mathbf{X}$ given $\mathbf{X}$ (or the homogeneous Markov chain with transition probability q) is called *λ-irreducible* if there is a σ-finite measure λ on $(\mathbf{X}, \mathcal{B}(\mathbf{X}))$ such that $q(B|x) > 0$ for all $x \in \mathbf{X}$ whenever $\lambda(B) > 0$.

Definition A.21. A stochastic kernel $q(dy|x)$ on $\mathbf{X}$ given $\mathbf{X}$ (or the homogeneous Markov chain with transition probability q) is called *geometric ergodic* if there is a probability measure μ on $(\mathbf{X}, \mathcal{B}(\mathbf{X}))$ such that

$$\left| \int_{\mathbf{X}} u(y) Q^t(dy|x) - \int_{\mathbf{X}} u(x) d\mu(x) \right| \leq \|u\| R \rho^t, \quad t = 0,1,2,\ldots,$$

where $R > 0$ and $0 < \rho < 1$ are constants, and

$$Q^0(dy|x) \stackrel{\triangle}{=} \delta_x(dy), \quad Q^t(dy|x) \stackrel{\triangle}{=} \int_{\mathbf{X}} q(dy|z) Q^{t-1}(dz|x)$$

is the t-step transition probability, $u(\cdot)$ is an arbitrary measurable bounded function, $\|u(x)\| \stackrel{\triangle}{=} \sup_{x \in \mathbf{X}} |u(x)|$.

According to the Prohorov Theorem, if E is tight then it is relatively compact. The converse statement is also correct if $\mathbf{X}$ is separable and complete.

Definition A.22. If $\mathbf{X}$ is a Borel space and $p \in \mathbf{P}(\mathbf{X})$, then $\forall \Gamma \subseteq \mathbf{X}$ we can write
$$p^*(\Gamma) \stackrel{\Delta}{=} \inf\{p(B)|\Gamma \subseteq B,\ B \in \mathcal{B}(\mathbf{X})\};$$
the function p^* is called the *outer measure*. The collection of subsets
$$B_{\mathbf{X}}(p) \stackrel{\Delta}{=} \{\Gamma \subseteq \mathbf{X}:\ p^*(\Gamma) + p^*(\Gamma^c) = 1\}$$
is a σ-algebra called the *completion* of $\mathcal{B}(\mathbf{X})$ w.r.t. p.

Incidentally, the Lebesgue measurable subsets of $\mathbf{X} = [0,1]$ form $B_{\mathbf{X}}(p)$ w.r.t. the probability measure $p(dx)$ defined by its values on intervals $p([a,b]) = b - a$.

Definition A.23. The σ-algebra $\mathcal{U}(\mathbf{X}) \stackrel{\Delta}{=} \bigcap_{p \in \mathbf{P}(\mathbf{X})} B_{\mathbf{X}}(p)$ is called a *universal σ-algebra*; its elements are called *universally measurable* subsets.

It is known that $\mathcal{B}(\mathbf{X}) \subseteq \mathcal{A}(\mathbf{X}) \subseteq \mathcal{U}(\mathbf{X})$, and the inclusions are strict if $\mathbf{X}$ is uncountable.

If $p \in \mathbf{P}(\mathbf{X})$ then the integrals $\int_{\mathbf{X}} f(x) p(dx)$ are also well defined for universally measurable functions $f(\cdot):\ \mathbf{X} \to \mathbb{R}^*$ (see Definition A.16).

A.3 Semi-continuous Functions and Measurable Selection

Recall the definitions of the upper and lower limits. Let $\mathbf{X}$ be a metric space with the distance function ρ, and let $f(\cdot):\ \mathbf{X} \longrightarrow \mathbb{R}^*$ be a real-valued function.

Definition A.24. The *lower limit* of the $f(\cdot)$ function in the point x is the number $\varliminf_{y \to x} f(y) = \lim_{\delta \downarrow 0} \inf_{\rho(x,y) < \delta} f(y) \in \mathbb{R}^*$; the *upper limit* is defined by the formula $\varlimsup_{y \to x} f(y) = \lim_{\delta \downarrow 0} \sup_{\rho(x,y) < \delta} f(y) \in \mathbb{R}^*$.

One can introduce similar definitions for real-valued sequences:
$$\varliminf_{n \to \infty} f_n \stackrel{\Delta}{=} \lim_{n \to \infty} \inf_{i \geq n} f_i;\quad \varlimsup_{n \to \infty} f_n \stackrel{\Delta}{=} \lim_{n \to \infty} \sup_{i \geq n} f_i.$$

We present the obvious enough properties of the lower and upper limits. Let a_n and b_n, $n = 1, 2, \ldots$ be two numerical sequences in $\mathbb{R}^*$. Then

(a) $\varliminf_{n\to\infty}(a_n+b_n) \geq \varliminf_{n\to\infty} a_n + \varliminf_{n\to\infty} b_n$;

(b) $\varlimsup_{n\to\infty}(a_n+b_n) \leq \varlimsup_{n\to\infty} a_n + \varlimsup_{n\to\infty} b_n$;

(c) if $y \geq 0$, then $\varliminf_{n\to\infty} y a_n = y \varliminf_{n\to\infty} a_n$; $\varlimsup_{n\to\infty} y a_n = y \varlimsup_{n\to\infty} a_n$.

Definition A.25. Let $\mathbf{X}$ be a metric space with the distance function ρ. The function $f(\cdot): \mathbf{X} \longrightarrow \mathbb{R}^*$ is called *lower semi-continuous* at the point $x \in \mathbf{X}$, if $\forall \varepsilon > 0 \; \exists \delta > 0 \; \forall y \in \mathbf{X} \; \rho(x,y) < \delta \Longrightarrow f(y) \geq f(x) - \varepsilon$.

The equivalent definition is $\varliminf_{y\to x} f(y) \geq f(x)$.

Definition A.26. Let $\mathbf{X}$ be a metrizable space. If the function $f(\cdot): \mathbf{X} \longrightarrow \mathbb{R}^*$ is lower semi-continuous at every point, then it is called *lower semi-continuous*.

Theorem A.11. *The function $f(\cdot): \mathbf{X} \longrightarrow \mathbb{R}^*$ is lower semi-continuous on the metrizable space $\mathbf{X}$ if and only if the set $\{x \in \mathbf{X}: f(x) \leq c\}$ is closed for every real c.*

Definition A.27. The function $f(\cdot): \mathbf{X} \longrightarrow \mathbb{R}^*$ is called *upper semi-continuous* (everywhere or at point x) if $-f(\cdot)$ is lower semi-continuous (everywhere or at point x).

Obviously, the function $f(\cdot)$ is continuous (everywhere or at point x) if and only if it is simultaneously lower and upper semi-continuous (everywhere or at point x).

If the metrizable space $\mathbf{X}$ is compact, then any lower (upper) semi-continuous function is necessarily bounded below (above).

Note that all the assertions concerning upper semi-continuous functions can be obtained from the corresponding assertions concerning lower semi-continuous functions with the help of Definition A.27.

Theorem A.12. *Let $\mathbf{X}$ be a metrizable space and $f(\cdot): \mathbf{X} \longrightarrow \mathbb{R}^*$.*

(a) The function $f(\cdot)$ is lower (upper) semi-continuous if and only if there exists a sequence of continuous functions $f_n(\cdot)$ such that $\forall x \in \mathbf{X} \; f_n(x) \uparrow f(x) \; (f_n(x) \downarrow f(x))$.

(b) The function $f(\cdot)$ is lower (upper) semi-continuous and bounded below (above) if and only if there exists a sequence of bounded continuous functions $f_n(\cdot)$ such that $\forall x \in \mathbf{X} \; f_n(x) \uparrow f(x) \; (f_n(x) \downarrow f(x))$.

Theorem A.13. *Let $\mathbf{X}$ and $\mathbf{Y}$ be separable metrizable spaces, let $q(dy|x)$ be a continuous stochastic kernel on $\mathbf{Y}$ given $\mathbf{X}$, and let $f(\cdot): \mathbf{X} \times \mathbf{Y} \longrightarrow \mathbb{R}^*$*

be a measurable function. Define
$$g(x) \stackrel{\triangle}{=} \int_Y f(x,y) q(dy|x).$$

(a) If $f(\cdot)$ is lower semi-continuous and bounded below, then $g(\cdot)$ is lower semi-continuous and bounded below.
(b) If $f(\cdot)$ is upper semi-continuous and bounded above, then $g(\cdot)$ is upper semi-continuous and bounded above.

Theorem A.14. *Let $\mathbf{X}$ and $\mathbf{Y}$ be metrizable spaces, and let $f(\cdot) : \mathbf{X} \times \mathbf{Y} \longrightarrow \mathbb{R}^*$ be given. Define*
$$g(x) \stackrel{\triangle}{=} \inf_{y \in \mathbf{Y}} f(x,y).$$

(a) If $f(\cdot)$ is lower semi-continuous and $\mathbf{Y}$ is compact, then $g(\cdot)$ is lower semi-continuous and for every $x \in \mathbf{X}$ the infimum is attained by some $y \in \mathbf{Y}$. Furthermore, there exists a (Borel)-measurable function $\varphi : \mathbf{X} \longrightarrow \mathbf{Y}$ such that $f(x, \varphi(x)) = g(x)$ for all $x \in \mathbf{X}$.
(b) If $f(\cdot)$ is upper semi-continuous, then $g(\cdot)$ is also upper semi-continuous.

Let $\mathbf{X}$ be a metrizable space. When considering the set L of all bounded lower (upper) semi-continuous functions $f(x)$, one can introduce the metric $r(f_1, f_2) \stackrel{\triangle}{=} \sup_{x \in \mathbf{X}} |f_1(x) - f_2(x)|.$

Theorem A.15. *The constructed metric space L is complete.*

A.4 Abelian (Tauberian) Theorem

If $\{z_i\}_{i=1}^\infty$ is a sequence of non-negative numbers, then
$$\liminf_{n \to \infty} \frac{1}{n} \sum_{i=1}^n z_i \leq \liminf_{\beta \to 1-} (1-\beta) \sum_{i=1}^\infty \beta^{i-1} z_i$$
$$\leq \limsup_{\beta \to 1-} (1-\beta) \sum_{i=1}^\infty \beta^{i-1} z_i \leq \limsup_{n \to \infty} \frac{1}{n} \sum_{i=1}^n z_i$$

(see [Hernandez-Lerma and Lasserre(1999), p. 139] and [Puterman(1994), Lemma 8.10.6]). The same inequalities also hold for non-positive z_i. Since the first several values of z_i do not affect these inequalities, it is sufficient to require that all z_i be non-negative (or non-positive) for all $i \geq I \geq$

1. Moreover, the inequalities presented also hold for the case where the sequence $\{z_i\}_{i=1}^{\infty}$ is bounded (below or above), as one can always add or subtract a constant from $\{z_i\}_{i=1}^{\infty}$.

The presented inequalities can be strict. For instance, in [Liggett and Lippman(1969)], a sequence $\{z_i\}_{i=1}^{\infty}$ of the form $(1, 1, \ldots, 1, 0, 0, \ldots, 0, 1, 1, \ldots)$ is built, such that

$$\liminf_{n\to\infty} \frac{1}{n} \sum_{i=1}^{n} z_i < \liminf_{\beta\to 1-}(1-\beta) \sum_{i=1}^{\infty} \beta^{i-1} z_i.$$

The sub-sequences of ones and zeros become longer and longer.

For $C_i \stackrel{\triangle}{=} 1 - z_i \geq 0$ we have

$$\limsup_{n\to\infty} \frac{1}{n} \sum_{i=1}^{n} C_i > \limsup_{\beta\to 1-}(1-\beta) \sum_{i=1}^{\infty} \beta^{i-1} C_i.$$

In Lemma 4 of [Flynn(1976)], a bounded sequence $\{u_i\}_{i=1}^{\infty}$ is built such that

$$\lim_{n\to\infty} \frac{1}{n} \sum_{i=1}^{n} \sum_{j=1}^{i} u_j = \infty \text{ and } \liminf_{n\to\infty} \frac{1}{n} \sum_{j=1}^{n} u_j < 0.$$

Additionally, in Lemma 5 of [Flynn(1976)], a bounded sequence $\{z_i\}_{i=1}^{\infty}$ is constructed (actually using only values 0 and ± 1), such that

$$\liminf_{n\to\infty} \frac{1}{n} \sum_{i=1}^{n} \sum_{j=1}^{i} z_j = -1$$

and

$$\lim_{\beta\to 1-} \sum_{j=1}^{\infty} \beta^{j-1} z_j = \limsup_{n\to\infty} \frac{1}{n} \sum_{i=1}^{n} \sum_{j=1}^{i} z_j = 0.$$

Appendix B

Proofs of Auxiliary Statements

Lemma B.1. *Let* **A** *and* **X** *be two Borel spaces, $P_0(dx)$ be a non-atomic probability distribution on* **X**, *and $f(x)$ be a real-valued measurable function on* **X**. *Then, for any measurable stochastic kernel $\pi(da|x)$ on* **A** *given* **X**, *there exists a measurable mapping (selector) $\varphi :$* **X** $\to$ **A** *such that, for any real measurable bounded function $\rho(a)$,*

$$\int_{\mathbf{X}} \int_{\mathbf{A}} \rho(a) f(x) \pi(da|x) P_0(dx) = \int_{\mathbf{X}} \rho(\varphi(x)) f(x) P_0(dx) \qquad (B.1)$$

(we call π and φ strongly equivalent w.r.t. $f(\cdot)$).

Proof. Without loss of generality, we assume that **A** = **X** = $[0, 1]$, so that we deal with random variables and their distributions. (The case of discrete **A** is obviously a simplified version.)

Firstly, suppose that $f(x) \geq 0$ and $\int_{[0,1]} f(x) P_0(dx) < \infty$. Then

$$F_{\mathbf{X}}(x) \stackrel{\Delta}{=} \int_0^x f(x) P_0(dx) \bigg/ \int_0^1 f(x) P_0(dx)$$

is the cumulative distribution function (CDF) of a non-atomic probability measure on **X**. (If $\int_0^1 f(x) P_0(dx) = 0$ then the statement of the Lemma is trivial.) Let

$$F_{\mathbf{A}}(a) \stackrel{\Delta}{=} \int_0^1 \left[\int_{[0,a]} \pi(da|x) \right] f(x) P_0(dx) \bigg/ \int_0^1 f(x) P_0(dx)$$

be the cumulative distribution function of a probability measure on **A**. Now put

$$\varphi(x) \stackrel{\Delta}{=} \inf\{a : F_{\mathbf{A}}(a) \geq F_{\mathbf{X}}(x)\}.$$

267

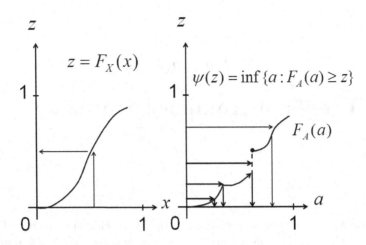

Fig. B.1 Construction of the selector φ.

We know that the image of measure $f(x)P_0(dx)/\int_0^1 f(x)P_0(dx)$ w.r.t. the map $z = F_{\mathbf{X}}(x)$ is uniform on $[0,1] \ni z$; we also know that the image of the uniform measure on $[0,1]$ w.r.t. the map $\psi(z) = \inf\{a : F_{\mathbf{A}}(a) \geq z\}$ coincides with the distribution defined by the CDF $F_{\mathbf{A}}(\cdot)$ (see Fig. B.1). Therefore the image of the measure $f(x)P_0(dx)/\int_0^1 f(x)P_0(dx)$ w.r.t. $\varphi: \mathbf{X} \to \mathbf{A}$ coincides with the distribution defined by the CDF $F_{\mathbf{A}}(\cdot)$. Hence, $\forall \rho(a)$,

$$\frac{\int_{\mathbf{X}} \int_{\mathbf{A}} \rho(a) f(x) \pi(da|x) P_0(dx)}{\int_0^1 f(x) P_0(dx)} = \frac{\int_{\mathbf{X}} \rho(\varphi(x)) f(x) P_0(dx)}{\int_0^1 f(x) P_0(dx)}.$$

Now, if $f(x) \geq 0$ and $\int_{[0,1]} f(x) P_0(dx) = \infty$, one should consider the subsets

$$\mathbf{X}_j \triangleq \{x \in \mathbf{X} : j - 1 \leq f(x) < j\}, \quad j = 1, 2, \ldots,$$

and build the selector φ strongly equivalent to π w.r.t. $f(\cdot)$ separately on each subset $\mathbf{X}_j$.

Finally, if the function $f(\cdot)$ is not non-negative, one should consider the subsets $\mathbf{X}_+ \triangleq \{x \in \mathbf{X} : f(x) \geq 0\}$ and $\mathbf{X}_- \triangleq \{x \in \mathbf{X} : f(x) < 0\}$ and build the selectors $\varphi_+(x)$ and $\varphi_-(x)$ strongly equivalent to π w.r.t. $f(\cdot)$ and $-f(\cdot)$ correspondingly. The combined selector $\varphi(x) \stackrel{\triangle}{=} \varphi_+(x)I\{x \in \mathbf{X}_+\} + \varphi_-(x)I\{x \in \mathbf{X}_-\}$ will satisfy equality (B.1). □

Remark B.1. If function $f(\cdot)$ is non-negative or non-positive, then (B.1) holds for any function $\rho(\cdot)$ (not necessarily bounded).

Theorem B.1. *Let Ω be the collection of all ordinals up to (and excluding) the first uncountable one, or, in other words, let Ω be the first uncountable ordinal. Let $h(\alpha)$ be a real-valued non-increasing function on Ω taking non-negative values and such that, in the case where $\inf_{\gamma<\alpha} h(\gamma) > 0$, the strict inequality $h(\alpha) < \inf_{\gamma<\alpha} h(\gamma)$ holds.*
Then $h(\alpha) = 0$ for some $\alpha \in \Omega$.

Proof. Suppose $h(\alpha) > 0$ for all $\alpha \in \Omega$. For each α, consider the open interval $(h(\alpha), \inf_{\gamma<\alpha} h(\gamma))$. Such intervals are non-empty and disjoint for different α. The total collection of such intervals is not more than countable, because each interval contains a rational number. However, Ω is not countable, so that $h(\alpha) = 0$ for some $\alpha \in \Omega$ (and for all $\gamma > \alpha$ as well). □

Lemma B.2. *Suppose positive numbers λ_i, $i = 1, 2, \ldots$, and μ_i, $i = 2, 3, \ldots$, are such that $\lambda_1 \leq 1$, $\lambda_i + \mu_i \leq 1$ for $i \geq 2$, and $\sum_{j=2}^{\infty} \left(\frac{\mu_2 \mu_3 \cdots \mu_j}{\lambda_2 \lambda_3 \cdots \lambda_j} \right) < \infty$. Then the equations*
$$\eta(1) = 1 + \mu_2 \eta(2);$$
$$\eta(i) = \lambda_{i-1} \eta(i-1) + \mu_{i+1} \eta(i+1), \quad i = 2, 3, 4, \ldots$$
have a (minimal non-negative) solution satisfying the inequalities
$$\eta(1) \leq \sum_{j=1}^{\infty} \left(\frac{\mu_2 \mu_3 \cdots \mu_j}{\lambda_2 \lambda_3 \cdots \lambda_j} \right) = 1 + \sum_{j=2}^{\infty} \left(\frac{\mu_2 \mu_3 \cdots \mu_j}{\lambda_2 \lambda_3 \cdots \lambda_j} \right);$$
$$\eta(i) \leq \sum_{j=i}^{\infty} \left(\frac{\mu_2 \mu_3 \cdots \mu_j}{\lambda_2 \lambda_3 \cdots \lambda_j} \right) \bigg/ \left(\frac{\mu_2 \mu_3 \cdots \mu_i}{\lambda_2 \lambda_3 \cdots \lambda_{i-1}} \right), \quad i = 2, 3, \ldots.$$

Proof. The minimal non-negative solution can be built by successive approximations:
$$\eta_0(i) \equiv 0;$$
$$\eta_{n+1}(1) = 1 + \mu_2 \eta_n(2);$$
$$\eta_{n+1}(i) = \lambda_{i-1} \eta_n(i-1) + \mu_{i+1} \eta_n(i+1), \quad i = 2, 3, 4, \ldots;$$
$$n = 0, 1, 2, \ldots.$$

For each $i \geq 1$, the value $\eta_n(i)$ increases with n, and we can prove the inequalities

$$\eta_n(1) \leq \sum_{j=1}^{\infty} \left(\frac{\mu_2 \mu_3 \cdots \mu_j}{\lambda_2 \lambda_3 \cdots \lambda_j} \right) = 1 + \sum_{j=2}^{\infty} \left(\frac{\mu_2 \mu_3 \cdots \mu_j}{\lambda_2 \lambda_3 \cdots \lambda_j} \right);$$

$$\eta_n(i) \leq \sum_{j=i}^{\infty} \left(\frac{\mu_2 \mu_3 \cdots \mu_j}{\lambda_2 \lambda_3 \cdots \lambda_j} \right) \bigg/ \left(\frac{\mu_2 \mu_3 \cdots \mu_i}{\lambda_2 \lambda_3 \cdots \lambda_{i-1}} \right), \quad i = 2, 3, \cdots$$

by induction w.r.t. n. These inequalities hold for $n = 0$. Suppose they are satisfied for some n. Then

$$\eta_{n+1}(1) \leq 1 + \mu_2 \sum_{j=2}^{\infty} \left(\frac{\mu_2 \mu_3 \cdots \mu_j}{\lambda_2 \lambda_3 \cdots \lambda_j} \right) \bigg/ \mu_2;$$

for $i \geq 2$,

$$\eta_{n+1}(i) \leq \lambda_{i-1} \sum_{j=i-1}^{\infty} \left(\frac{\mu_2 \mu_3 \cdots \mu_j}{\lambda_2 \lambda_3 \cdots \lambda_j} \right) \bigg/ \left(\frac{\mu_2 \mu_3 \cdots \mu_{i-1}}{\lambda_2 \lambda_3 \cdots \lambda_{i-2}} \right)$$

$$+ \mu_{i+1} \sum_{j=i+1}^{\infty} \left(\frac{\mu_2 \mu_3 \cdots \mu_j}{\lambda_2 \lambda_3 \cdots \lambda_j} \right) \bigg/ \left(\frac{\mu_2 \mu_3 \cdots \mu_{i+1}}{\lambda_2 \lambda_3 \cdots \lambda_i} \right)$$

$$= \left\{ \sum_{j=i}^{\infty} \left(\frac{\mu_2 \mu_3 \cdots \mu_j}{\lambda_2 \lambda_3 \cdots \lambda_j} \right) \bigg/ \left(\frac{\mu_2 \mu_3 \cdots \mu_i}{\lambda_2 \lambda_3 \cdots \lambda_{i-1}} \right) \right\} \{\mu_i + \lambda_i\},$$

and, for $i = 2$, similar calculations lead to the inequality

$$\eta_{n+1}(2) \leq \lambda_1 + \left(\lambda_1 + \frac{\lambda_2}{\mu_2} \right) \sum_{j=2}^{\infty} \left(\frac{\mu_2 \mu_3 \cdots \mu_j}{\lambda_2 \lambda_3 \cdots \lambda_j} \right) - 1$$

$$\leq \left\{ \sum_{j=2}^{\infty} \left(\frac{\mu_2 \mu_3 \cdots \mu_j}{\lambda_2 \lambda_3 \cdots \lambda_j} \right) \bigg/ \mu_2 \right\} (\lambda_2 + \lambda_1 \mu_2). \qquad \square$$

Remark B.2. The proof also remains correct if some values of μ_i are zero.

Proof of Lemma 2.1.

(a) It is sufficient to compare the actions $\hat{b}$ and $\hat{c}$: their difference equals

$$\delta = \alpha \bar{\gamma} \, v(AC) - \alpha \bar{\beta} \, v(AB) = \alpha^2 \left[\frac{\bar{\beta}}{\alpha + \bar{\alpha}\beta} - \frac{\bar{\gamma}}{\alpha + \bar{\alpha}\gamma} \right].$$

But the function $\frac{1-\beta}{\alpha + \bar{\alpha}\beta}$ decreases in β, so that $\delta > 0$ because $\gamma > \beta$.

(b) The given formula for $v(ABC)$ comes from the equation

$$v(ABC) = \bar{\beta}\bar{\gamma}\, v(ABC) + \beta\, v(AB) + \bar{\beta}\gamma\, v(AC),$$

and it is sufficient to prove that $\delta \triangleq \alpha\bar{\beta}\bar{\gamma}\, v(ABC) + (\alpha\beta - \alpha\bar{\beta})v(AB) + \alpha\bar{\beta}\gamma\, v(AC) \leq 0$, as this expression equals the difference between the first and third formulae in the optimality equation for the state ABC. Now

$$\delta[(\beta+\alpha\bar{\beta})(\alpha+\bar{\alpha}\gamma)(\beta+\bar{\beta}\gamma)] = \alpha^2\{\alpha[-\beta^2\bar{\gamma}-(\bar{\beta})^2\gamma^2]+(\bar{\beta})^2\gamma^2 - \beta\gamma\},$$

and $\delta \leq 0$ if and only if $\alpha \geq \dfrac{(\bar{\beta})^2\gamma^2 - \beta\gamma}{\beta^2\bar{\gamma} + (\bar{\beta})^2\gamma^2}$.

(c) The given formula comes from the equation

$$v(ABC) = \bar{\alpha}\bar{\beta}\bar{\gamma}\, v(ABC) + (\alpha\bar{\beta} + \bar{\alpha}\beta)v(AB) + \bar{\alpha}\bar{\beta}\gamma\, v(AC),$$

and it is sufficient to prove that $\delta \triangleq v(ABC) - \bar{\beta}\bar{\gamma}\, v(ABC) - \beta\, v(AB) - \bar{\beta}\gamma\, v(AC) \leq 0$, as this expression equals the difference between the third and first formulae in the optimality equation for the state ABC. Now

$$\delta[(\beta + \alpha\bar{\beta})(\alpha + \bar{\alpha}\gamma)(1 - \bar{\alpha}\bar{\beta}\bar{\gamma})]/\alpha$$

$$= -(1 - \bar{\beta}\bar{\gamma})[\alpha^2\bar{\beta} + \alpha\bar{\alpha}\beta + \alpha\bar{\alpha}\bar{\beta}\gamma + (\bar{\alpha})^2\beta\gamma + \bar{\alpha}\beta\bar{\beta}\gamma + \alpha\bar{\alpha}(\bar{\beta})^2\gamma]$$

$$- \beta(\alpha + \bar{\alpha}\gamma)(1 - \bar{\alpha}\bar{\beta}\bar{\gamma}) - \bar{\beta}\gamma(\beta + \alpha\bar{\beta})(1 - \bar{\alpha}\bar{\beta}\bar{\gamma})$$

$$= -\alpha[(\bar{\beta})^2\gamma^2 - \beta\gamma] + \alpha^2[(\bar{\beta})^2\gamma^2 + \beta^2\bar{\gamma}],$$

and $\delta \leq 0$ if and only if $\alpha \leq \dfrac{(\bar{\beta})^2\gamma^2 - \beta\gamma}{\beta^2\bar{\gamma} + (\bar{\beta})^2\gamma^2}$. □

Proof of Lemma 3.2.

(a) Let $i \leq n - 1$. Then

$$\delta \triangleq \frac{1 - \beta^{n-i+1} + 2\beta^{2n-i+1}}{1 - \beta} - \left[1 + \frac{2\beta^{i+1}}{1 - \beta}\right]$$

$$= \frac{\beta}{1-\beta}[2\beta^{2n-i} - \beta^{n-i} + 1 - 2\beta^i] = \frac{\beta^{1-i}}{1-\beta}[2\beta^{2n} - \beta^n + \beta^i - 2\beta^{2i}].$$

Since

$$\beta^{i+1} - 2\beta^{2(i+1)} - \beta^i + 2\beta^{2i} = \beta^i(\beta - 1)[1 - 2\beta^i(1 + \beta)] > 0,$$

the function $\beta^i - 2\beta^{2i}$ increases with $i \in \{1, 2, \ldots, n-1\}$, and the inequality

$$2\beta^{2n} - \beta^n + \beta^{n-1} - 2\beta^{2(n-1)} = \beta^{n-1}[2\beta^{n+1} - \beta + 1 - 2\beta^{n-1}]$$

$$= \beta^{n-1}(1-\beta)[1 - 2\beta^{n-1}(1+\beta)] < 0$$

implies that $\delta < 0$.

For all $i < n-1$, the equality $v^\beta((i,0)) = 1 + \beta v^\beta((i+1,0))$ is obvious; it holds also for $i = n-1$:

$$1 + \beta v^\beta((n,0)) = 1 + \beta \left[1 + \frac{2\beta^{n+1}}{1-\beta}\right] = \frac{1 - \beta^2 + 2\beta^{n+2}}{1-\beta}.$$

(b) Let $i \geq n$. Then

$$\delta \stackrel{\triangle}{=} 1 + \frac{2\beta^{i+1}}{1-\beta} - [1 + \beta v^\beta((i+1,0))]$$

$$= \frac{2\beta^{i+1} - \beta(1-\beta) - 2\beta^{i+3}}{1-\beta} = 2\beta^{i+1}(1+\beta) - \beta \leq 0.$$

□

Proof of Proposition 4.1.

(a) Suppose the canonical equations (4.2) have a solution $\langle \rho, h, \varphi \rangle$. Then, for any $x \geq 1$, $\rho(x) = \rho(x-1)$, so that $\rho(x) \equiv \rho$. From the second equation (4.2) we obtain

$$\rho + h(x) = 1 + h(x-1), \quad x \geq 1,$$

so that

$$h(x) = h(0) + (1-\rho)x$$

and, for $x = 0$, we have

$$\rho + h(0) = \min\{0 + h(0) + (1-\rho)\sum_{x \geq 1} xq_x; \quad 1 + h(0) + 1 - \rho\}.$$

ρ cannot be greater than 1, because otherwise $\rho + h(0) = -\infty$. If $\rho < 1$ then

$$\rho + h(0) = 2 - \rho + h(0),$$

and hence $\rho = 1$. Therefore, $\rho = 1$ and $\rho + h(0) = 0 + h(0)$, which is a contradiction.

(b) Condition 4.2(ii) is certainly not satisfied, since otherwise there would have existed a canonical triplet [Hernandez-Lerma and Lasserre(1996a), Th. 5.5.4]. A more straightforward argument can be written as follows. If, for $x = 0$, $a = 1$, the function $b(y)$ is summable w.r.t. the distribution $p(y|0,1) = q_y$, then, by the Lebesgue Dominated Convergence Theorem [Goffman and Pedrick(1983), Section 3.7],

$$\lim_{\beta \to 1-} \sum_{y \geq 1} h^\beta(y) q_y = \sum_{y \geq 1} \lim_{\beta \to 1-} h^\beta(y) q_y.$$

But on the left-hand side we have

$$\lim_{\beta \to 1-} \frac{1 - \sum_{y \geq 1} \beta^y q_y}{1 - \beta \sum_{y \geq 1} \beta^y q_y} = \lim_{\beta \to 1-} \frac{\sum_{y \geq 1} y \beta^{y-1} q_y}{\sum_{y \geq 1} \beta^y q_y + \sum_{y \geq 1} y \beta^y q_y} = 1,$$

and on the right we have zero because

$$h(x) = \lim_{\beta \to 1-} h^\beta(x) = \frac{x}{1 + \sum_{y \geq 1} y q_y} = 0.$$

(c) If a stationary distribution $\eta(x)$ on $\mathbf{X}$ exists, then it satisfies the equations

$$\eta(0) = \eta(1); \qquad \eta(x) = \eta(0) q_x + \eta(x+1).$$

After we write $\gamma \triangleq \eta(0)$, the value of γ comes from the normalization condition:

$$\gamma[(1) + (1) + (1 - q_1) + (1 - q_1 - q_2) + (1 - q_1 - q_2 - q_3) + \cdots]$$
$$= \gamma \left[1 + \lim_{n \to \infty} \{n - (n-1)q_1 - (n-2)q_2 - \cdots - q_{n-1}\} \right] \qquad \text{(B.2)}$$
$$= \gamma \left[1 + \lim_{n \to \infty} \{n - n \sum_{i=1}^{n-1} q_i + \sum_{i=1}^{n-1} i q_i\} \right] = 1.$$

But $n - n \sum_{i=1}^{n-1} q_i \geq 0$ and $\lim_{n \to \infty} \sum_{i=1}^{n-1} i q_i = +\infty$, so that $\gamma \cdot \infty = 1$, which is a contradiction.

Let $\sum_{y \geq 1} y q_y < \infty$. Then equation (B.2) implies that $\gamma = \eta(0) = \frac{1}{1 + \sum_{y \geq 1} y q_y}$, because

$$\lim_{n \to \infty} n \left[1 - \sum_{i=1}^{n-1} q_i \right] = \lim_{n \to \infty} n \sum_{n=n}^{\infty} q_i \leq \lim_{n \to \infty} \sum_{i=n}^{\infty} i q_i = 0,$$

and the assertion follows:

$$\eta(x) = \gamma \left[1 - \sum_{i=1}^{x-1} q_i \right], \qquad x \geq 1.$$

□

Proof of Proposition 4.2.

(a) For $i \geq 1$, the mean recurrence time from state i' to state 0 equals $M_{i'0}(\pi) = 2^i$ for any control strategy π. Similarly, $M_{i0}(\varphi^n) = 2^i$ if $i \geq n$. In what follows, φ^n is the stationary selector defined in (4.8). For $0 < i < n < \infty$ we have

$$M_{i0}(\varphi^n) = 1 + \frac{1}{2}M_{i+1,0} = 1 + \frac{1}{2} + \frac{1}{4} + \cdots + \left(\frac{1}{2}\right)^{n-i-1}$$
$$+ \left(\frac{1}{2}\right)^{n-i} M_{n0} = 2 + 2^i - \left(\frac{1}{2}\right)^{n-i-1},$$

and $M_{i0}(\varphi^\infty) = 2$. Therefore,

$$M_{00}(\varphi^n) = 1 + \sum_{i=1}^{\infty} \frac{3}{2}\left(\frac{1}{4}\right)^i \left[M_{i0}(\varphi^n) + 2^i\right] =$$
$$= 5 - \left(\frac{3}{2}\right)\sum_{i=1}^{n-1}\left(\frac{1}{2}\right)^{n+i-1} - 3\sum_{i=n}^{\infty}\left(\frac{1}{4}\right)^i$$
$$= 5 - 3\left(\frac{1}{2}\right)^n + 2\left(\frac{1}{4}\right)^n < 5,$$

$\lim_{n\to\infty} M_{00}(\varphi^n) = 5$, and the convergence is monotonic. Note that $M_{00}(\varphi^\infty) = \frac{7}{2}$.

Now let π be an arbitrary control strategy, suppose $X_{T-1} = 0$, and estimate the mean recurrence time to state 0:

$$E_{P_0}^\pi [\tau : \tau = \min\{t \geq T : X_t = 0\} - (T-1)|X_{T-1} = 0].$$

Note that the actions in states 0 and i' ($i \geq 1$) play no role, state $i \geq 1$ can increase only if action 2 is applied, and, after action 1 is used in state $i \geq 1$, the process will reach state 0, possibly after several loops in state i'. Therefore, assuming that $X_T = i$ or $X_T = i'$, only the following three types of trajectories can be realized:

$$(x_{T-1} = 0, a_T, x_T = i, a_{T+1} = 2, x_{T+1} = i+1, \ldots, a_{T+k-i} = 2,$$

$$x_{T+k-i} = k, a_{T+k-i+1} = 2 \text{ or } 1, x_{T+k-i+1=\tau} = 0), \quad k \geq i;$$

$$(x_{T-1} = 0, a_T, x_T = i, a_{T+1} = 2, x_{T+1} = i+1, \ldots, a_{T+n-i} = 2,$$

$$x_{T+n-i} = n, a_{T+n-i+1} = 1, x_{T+n-i+1} = n', \ldots, x_\tau = 0), \quad n \geq i;$$

$$(x_{T-1} = 0, \ a_T, \ x_T = i', \ldots, x_\tau = 0).$$

In the third case, which is realized with probability $\frac{3}{2}\left(\frac{1}{4}\right)^i$, the expected value of τ equals 2^i. In the first two cases (when $X_T = i$) one can say that the stationary selector φ^n ($n \geq i$) is used, and the probability of this event (given that we observed the trajectory h_{T-1} up to time $T-1$, when $X_{T-1} = 0$) equals

$$P_{P_0}^\pi(A_{T+1} = 2, \ldots, A_{T+n-i} = 2, A_{T+n-i+1} = 1 | h_{T-1})$$

if $n < \infty$, and equals

$$P_{P_0}^\pi(A_{T+1} = 2, A_{T+2} = 2, \ldots | h_{T-1})$$

if $n = \infty$. All these probabilities for $n = i, i+1, \ldots, \infty$ sum to one. Now, assuming that $X_T = i$, the conditional expectation (given h_{T-1}) of the recurrence time from state i to state 0 equals

$$M_{i0}(\pi, h_{T-1})$$

$$= \sum_{n=i}^\infty P_{P_0}^\pi(A_{T+1} = 2, \ldots, A_{T+n-i} = 2, A_{T+n-i+1} = 1 | h_{T-1})$$
$$\times M_{i0}(\varphi^n) + P_{P_0}^\pi(A_{T+1} = 2, A_{T+2} = 2, \ldots | h_{T-1}) M_{i0}(\varphi^\infty)$$
$$< \left\{ \sum_{n=i}^\infty P_{P_0}^\pi(A_{T+1} = 2, \ldots, A_{T+n-i} = 2, A_{T+n-i+1} = 1 | h_{T-1}) \right.$$
$$\left. + P_{P_0}^\pi(A_{T+1} = 2, A_{T+2} = 2, \ldots | h_{T-1}) \right\} [2 + 2^i] = 2 + 2^i.$$

In fact, we have proved that, for any strategy, if $X_T = i$, then the expected recurrence time to state 0 is smaller than $2 + 2^i$. This also holds for $T = 0$.

Finally,

$$E_{P_0}^\pi [\tau : \tau = \min\{t \geq T : X_t = 0\} - (T-1)|X_{T-1} = 0]$$

$$= 1 + \sum_{i=1}^\infty \frac{3}{2}\left(\frac{1}{4}\right)^i M_{i0}(\pi, h_{T-1}) + \sum_{i=1}^\infty \frac{3}{2}\left(\frac{1}{4}\right)^i 2^i$$

$$< 1 + \sum_{i=1}^\infty \frac{3}{2}\left(\frac{1}{4}\right)^i [2 + 2 \cdot 2^i] = 5.$$

For any stationary strategy π^{ms}, the controlled process X_t is positive recurrent; it was shown previously that the mean recurrence time $M_{00}(\pi^{ms})$ from 0 to 0 is strictly smaller than 5. Therefore, for any initial distribution P_0, the stationary probability of state 0, $\eta^{\pi^{ms}}(0) = \frac{1}{M_{00}(\pi^{ms})}$ [Kemeny et al.(1976), Prop. 6.25] is strictly greater than $\frac{1}{5}$. But, in the case under consideration, $v^{\pi^{ms}} = \eta^{\pi^{ms}}(0) > \frac{1}{5}$. Note also that $v^{\varphi^n} = \frac{1}{M_{00}(\varphi^n)} \downarrow \frac{1}{5}$ when $n \to \infty$.

Consider now the discounted functional (3.1) with $\beta \in (0,1)$. Since only cost $c(0,a) = 1$ is non-zero, it is obvious that $v_x^{*,\beta} < v_0^{*,\beta}$ for all $x \in \mathbf{X} \setminus \{0\}$. This inequality also follows from the Bellman equation. Indeed,

$$v_{i'}^{*,\beta} = \frac{\beta \left(\frac{1}{2}\right)^i v_0^{*,\beta}}{1 - \beta[1 - \left(\frac{1}{2}\right)^i]}$$

and

$$v_i^{*,\beta} \leq \frac{\beta \left(\frac{1}{2}\right)^i v_0^{*,\beta}}{1 - \beta[1 - \left(\frac{1}{2}\right)^i]}.$$

But the function $\frac{\beta\left(\frac{1}{2}\right)^i}{1-\beta[1-\left(\frac{1}{2}\right)^i]}$ decreases with i, so that, for all $x \in \mathbf{X} \setminus \{0\}$,

$$v_x^{*,\beta} \leq \frac{\frac{\beta}{2} v_0^{*,\beta}}{1 - \beta + \frac{\beta}{2}} < v_0^{*,\beta}.$$

Fix an arbitrary control strategy π and an initial state $X_0 = 0$. Let $T_0 = 0, T_1, \ldots$ be the sequence of time moments when $X_{T_n} = 0$. Then

$$v_0^{\pi,\beta} = E_0^{\pi}[1 + \beta^{T_1} + \beta^{T_2} + \cdots].$$

We have proved that $E_0^{\pi}[T_1] < 5$, $E_0^{\pi}[T_2] < 10$, ..., $E_0^{\pi}[T_n] < 5n$. Thus, from the Jensen inequality, we have

$$v_0^{\pi,\beta} \geq 1 + \sum_{n=1}^{\infty} \beta^{E_0^{\pi}[T_n]} > \sum_{n=0}^{\infty} \beta^{5n} = \frac{1}{1 - \beta^5}.$$

According to the Tauberian/Abelian Theorem (Section A.4, see also [Hernandez-Lerma and Lasserre(1996a), Lemma 5.3.1]),

$$v_0^{\pi} = \limsup_{T \to \infty} \frac{1}{T} E_0^{\pi}\left[\sum_{t=1}^{T} c(X_{t-1}, A_t)\right]$$

$$\geq \limsup_{\beta \to 1-} (1-\beta) E_0^{\pi}\left[\sum_{t=1}^{\infty} \beta^{t-1} c(X_{t-1}, A_t)\right]$$

$$= \limsup_{\beta \to 1-} (1-\beta) v_0^{\pi,\beta} \geq \lim_{\beta \to 1-} (1-\beta)\frac{1}{1-\beta^5} = \frac{1}{5}.$$

If the initial state x is $i > 0$ or i' then the mean first-recurrence time to state 0 is smaller than $2 + 2^i$ and

$$v_x^{\pi,\beta} \geq \beta^{2+2^i} \frac{1}{1-\beta^5},$$

so that $v_x^\pi \geq \frac{1}{5}$ as well. The same is true for the initial distribution P_0 satisfying the requirement $\sum_{i=1}^\infty [P_0(i) + P_0(i')]2^i < \infty$.

(b) We know that, for the stationary selector φ^n, $M_{00}(\varphi^n) < 5$ and $M_{00}(\varphi^n) \uparrow 5$ as $n \to \infty$. Therefore, as mentioned above, for an arbitrary initial distribution P_0, $v^{\varphi^n} > 5$ and $v^{\varphi^n} \downarrow \frac{1}{5}$ as $n \to \infty$. Let φ^{n_k} be a selector (4.8) such that $v^{\varphi^{n_k}} \leq \frac{1}{5} + \frac{1}{2k}$, $1 \leq n_1 < n_2 < \cdots$, and fix $N_1 \geq 1$ such that

$$\frac{1}{T} E_{P_0}^{\varphi^{n_1}} \left[\sum_{t=1}^T c(X_{t-1}, A_t) \right] \leq v^{\varphi^{n_1}} + \frac{1}{2}$$

for all $T \geq N_1$. Similarly, fix $N_k \geq 1$ such that

$$\frac{1}{T} E_0^{\varphi^{n_k}} \left[\sum_{t=1}^T c(X_{t-1}, A_t) \right] \leq v^{\varphi^{n_k}} + \frac{1}{2k}$$

for all $T \geq N_k$, $k = 2, 3, \ldots$. Let $\bar{N}_1 > N_1$ be such that

$$\frac{1}{\bar{N}_1} E_{P_0}^{\varphi^{n_1}} \left[\sum_{t=1}^{\bar{N}_1} c(X_{t-1}, A_t) + N_2 \right] \leq v^{\varphi^{n_1}} + \frac{1}{2}$$

and define $\bar{N}_k > N_k$, $n = 2, 3, \ldots$, recursively by letting $\bar{N}_k$ be such that

$$\frac{1}{\bar{N}_k} E_0^{\varphi^{n_k}} \left[\sum_{t=1}^{\bar{N}_k} c(X_{t-1}, A_t) + \sum_{j=1}^{k-1} \bar{N}_j + N_{k+1} \right] \leq v^{\varphi^{n_k}} + \frac{1}{2k}.$$

We put $T_0 = 0$, $T_k = \min\{t \geq T_{k-1} + \bar{N}_k : X_t = 0\}$, and we define the selector

$$\varphi_t^*(x) = \varphi^{n_k}(x) \cdot I\{T_{k-1} < t \leq T_k\}, \quad k = 1, 2, \ldots$$

(see Figure B.2).

Fix an arbitrary T such that $\sum_{i=1}^k \bar{N}_i < T \leq \sum_{i=1}^{k+1} \bar{N}_i$, where $k > 1$, and prove that

$$\frac{1}{T} E_{P_0}^{\varphi^*} \left[\sum_{t=1}^T c(X_{t-1}, A_t) \right] \leq v^{\varphi^{n_k}} + \frac{1}{2k} \leq \frac{1}{5} + \frac{1}{k}. \tag{B.3}$$

Obviously, $T \leq T_k + \bar{N}_{k+1}$ because $\sum_{i=1}^k \bar{N}_i \leq T_k$. We shall consider two cases:

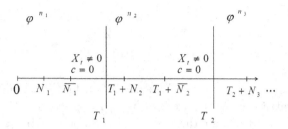

Fig. B.2 Construction of the selector φ^*.

(i) $T \leq T_k + N_{k+1}$. Now $\sum_{t=1}^{T} c(X_{t-1}, A_t) \leq \sum_{j=1}^{k-1} \bar{N}_j + \sum_{t=T_{k-1}+1}^{T_{k-1}+\bar{N}_k} c(X_{t-1}, A_t) + N_{k+1}$ (recall that, $c(X_{t-1}, A_t) = 0$ for all $t = T_0 + \bar{N}_1 + 1, \ldots, T_1$, for all $t = T_1 + \bar{N}_2 + 1, \ldots, T_2$ and so on, for all $t = T_{k-1} + \bar{N}_k + 1, \ldots, T_k$). Therefore,

$$\frac{1}{T} E_{P_0}^{\varphi^*} \left[\sum_{t=1}^{T} c(X_{t-1}, A_t) \cdot I\{T \leq T_k + N_{k+1}\} \right]$$

$$\leq \frac{1}{\bar{N}_k} \left\{ E_0^{\varphi^{n_k}} \left[\sum_{t=1}^{\bar{N}_k} c(X_{t-1}, A_t) \right] + \sum_{j=1}^{k-1} \bar{N}_j + N_{k+1} \right\} \quad (B.4)$$

$$\leq v^{\varphi^{n_k}} + \frac{1}{2k}.$$

(ii) $T_k + N_{k+1} < T \leq T_k + \bar{N}_{k+1}$. Below, we write the event $T_k + N_{k+1} < T \leq T_k + \bar{N}_{k+1}$ as D for brevity. Now

$$\frac{1}{T} E_{P_0}^{\varphi^*} \left[\sum_{t=1}^{T} c(X_{t-1}, A_t) \cdot I\{D\} \right]$$

$$= \frac{1}{T} E_{P_0}^{\varphi^*} \left[\left(\sum_{t=1}^{T_k} c(X_{t-1}, A_t) + \sum_{t=T_k+1}^{T} c(X_{t-1}, A_t) \right) I\{D\} \right]$$

$$\leq \frac{\bar{N}_k}{T} \cdot \frac{1}{\bar{N}_k} E_0^{\varphi^{n_k}} \left[\sum_{i=1}^{k-1} \bar{N}_i + \sum_{t=0}^{\bar{N}_k} c(X_{t-1}, A_t) \right]$$

$$+ \frac{T - \bar{N}_k}{T} \sum_{i \geq 1} E_0^{\varphi^{n_{k+1}}} \left[\frac{1}{T - \bar{N}_k} \sum_{t=1}^{T-i} c(X_{t-1}, A_t) \right]$$

$$\times E_{P_0}^{\varphi^*}\left[I\{T_k = i\} \cdot I\{D\}\right].$$

Since, under assumption D, $T - \bar{N}_k \geq T - T_k > N_{k+1}$ ($P_{P_0}^{\varphi^*}$-a.s.), we conclude that only the terms

$$E_0^{\varphi^{n_{k+1}}}\left[\frac{1}{T - \bar{N}_k} \sum_{t=1}^{T-i} c(X_{t-1}, A_t)\right]$$

$$\leq v^{\varphi^{n_{k+1}}} + \frac{1}{2(k+1)} \leq v^{\varphi^{n_k}} + \frac{1}{2k}$$

appear in the last sum with positive probabilities $E_{P_0}^{\varphi^*}\left[I\{T_k = i\} \cdot I\{D\}\right]$. The inequality

$$\frac{1}{\bar{N}_k} E_0^{\varphi^{n_k}}\left[\sum_{i=1}^{k-1} \bar{N}_i + \sum_{t=0}^{\bar{N}_k} c(X_{t-1}, A_t)\right] \leq v^{\varphi^{n_k}} + \frac{1}{2k}$$

follows from the definition of $\bar{N}_k$. Therefore,

$$\frac{1}{T} E_{P_0}^{\varphi^*}\left[\sum_{t=1}^{T} c(X_{t-1}, A_t) \cdot I\{D\}\right] \leq v^{\varphi^{n_k}} + \frac{1}{2k}.$$

This inequality and inequality (B.4) complete the proof of (B.3).

Now statement (b) of Proposition 4.2 is obvious. □

Proof of Proposition 4.4.

(a) We introduce events $B_n = \{\exists n \geq 1 : X_l = (n, 1) \text{ for some } l > 0\}$. Event $B_0 \triangleq \Omega \setminus \{\cup_{n=1}^{\infty} B_n\}$ means that the controlled process x_l takes values $k, k+1, k+2, \ldots$, so that

$$E_k^{\pi}\left[\sum_{\tau=1}^{t} c(X_{\tau-1}, A_{\tau}) | B_0\right] = 0.$$

For $n \geq 1$ we have

$$E_k^{\pi}\left[\sum_{\tau=1}^{t} c(X_{\tau-1}, A_{\tau}) | B_n\right] = n$$

for sufficiently large t meaning that

$$\liminf_{T \to \infty} \frac{1}{T} \sum_{t=1}^{T} E_k^{\pi}\left[\sum_{\tau=1}^{t} c(X_{\tau-1}, A_{\tau}) | B_n\right] = n > 0.$$

Finally,
$$\liminf_{T\to\infty} \frac{1}{T} \sum_{t=1}^{T} E_k^{\pi} \left[\sum_{\tau=1}^{t} c(X_{\tau-1}, A_{\tau}) \right]$$
$$\geq \sum_{n=0}^{\infty} P_k^{\pi}(B_n) \liminf_{T\to\infty} \frac{1}{T} \sum_{t=1}^{T} E_k^{\pi} \left[\sum_{\tau=1}^{t} c(X_{\tau-1}, A_{\tau}) | B_n \right] \geq 0.$$

(b) Consider the stationary selectors $\varphi^N(x) = I\{x = N\}$, $N \geq 1$. It is not hard to calculate $v_1^{\varphi^N,\beta} = \frac{3\beta^{2N} - 2\beta^{3N} - \beta^N}{1-\beta}$. The function $g(y) \triangleq 3y^2 - 2y^3 - y$ decreases in the interval $\left[0, y^* \triangleq \frac{3-\sqrt{3}}{6}\right]$, and has the minimal value $\min_{y \in [0,1]} g(y) = g(y^*) < 0$. Since the function g is continuous, there exist $\varepsilon < 0$ and $\underline{\beta} \in (0,1)$ such that, for all $\beta \in [\underline{\beta}, 1)$, the inequality $g(\beta y^*) \leq -\varepsilon$ holds. Now, for each β from the above interval, we can find a unique $N(\beta)$ such that $\beta^{N(\beta)} \in [\beta y^*, y^*)$ and
$$v_1^{*,\beta} \leq v_1^{\varphi^{N(\beta)},\beta} = \frac{g(\beta^{N(\beta)})}{1-\beta} < g(\beta y^*) \leq -\varepsilon.$$

□

Notation

A	action space	1,3
A_t	action (as a random element)	1
a, a_t	action (as a variable, argument of a function, etc.)	1,2
$\mathcal{B}(\mathbf{X})$	Borel σ-algebra	257
$c_t(x,a), c(x,a)$	loss function	1,3
$C(x)$	terminal loss	2,3
$\mathcal{D}, \mathcal{D}^N$ etc	spaces of strategic measures	4,17,54
$E^\pi_{P_0}$	mathematical expectation w.r.t. $P^\pi_{P_0}$	2
$\tilde{f}^T_{\pi,x}$	expected frequency	220
g_t	history	4
$\mathcal{H}$	Hilbert cube	256
$\mathbf{H}, \mathbf{H}_t$	spaces of trajectories (histories)	3
h_t	history	4
$h(x)$	element of a canonical triplet	177
$\mathcal{N}$	Bair null space	256
$P_0(dx)$	initial distribution of X_0	2,4
$P^\pi_{P_0}, P^\pi_{x_0}, P^\pi_{h_\tau}$	strategic measure	2,4,6
$p_t(dy|x,a),$ $p(dy|x,a)$	transition probability	2,3
sp	span	212
Supp μ	support of measure μ	258
t	(discrete) time	
T	time horizon	1,3
$v^\pi, v^\pi_{h_\tau}, v^\pi_x, v^{\pi,\beta}$	performance functional	2,7,51,127,177
$v_t(x), v(x)$	Bellman function (solution to the Bellman or optimality equation)	5,52,127

$v_x^*, v_x^{*,\beta}$	minimal possible loss starting from $X_0 = x$ (Bellman function)	7,51,127,177
V_x^T	minimal possible loss in the finite-horizon case	7
$v^n(x)$	Bellman function approximated using value iteration	63,128,211
$v^\infty(x)$	limit of the approximated Bellman function	64,128
$\mathcal{V}, \mathcal{V}^N$	performance spaces	17
$W, w(h)$	total realized loss	2,4
$\mathbf{X}$	state space	1,3
X_t	state of the controlled process (as a random element)	1
x, x_t, y, y_t	state of the controlled process (as a variable, argument of a function, etc.)	1
Y_t^π	estimating process	32
$y(\tau)$	fluid approximation to a random walk	94
β	discount factor	127
Δ (or 0)	absorbing state (cemetery)	53,71
Δ^{All} (Δ^M)	collection of all (Markov) strategies	3
Δ^{MN}	collection of all Markov selectors	3
Δ^S (Δ^{SN})	collection of all stationary strategies (selectors)	4
η^π	occupation measure	101,149
$\hat{\eta}^\pi$	marginal of an occupation measure	102,151
$\eta, \tilde{\eta}$	admissible solution to a linear program (state–action frequency)	215,219
$\mu(x)$	Lyapunov function	83,103
$\nu(x, a)$	weight function	83
π	control strategy	2,3
π^*	(uniformly) optimal control strategy	2,7
π^m	Markov strategy	3
$\pi^{\text{ms}}, \pi^{\text{s}}$	(Markov) stationary strategy	3
$\langle \rho, h, \varphi^* \rangle$	canonical triplet	177
$\rho, \rho(x)$	element of a canonical triplet (minimal average loss)	177
$\varphi, \varphi(x), \varphi_t(x)$	selector (non-randomized strategy)	3

List of the Main Statements

Condition 2.1	51		
Condition 2.2	53	Proposition 3.1	153
Condition 2.3	85	Proposition 4.1	190
Condition 3.1	171	Proposition 4.2	195
Condition 4.1	188	Proposition 4.3	227
Condition 4.2	188	Proposition 4.4	238
Condition 4.3	212		
Condition 4.4	219		
Condition 4.5	220		

Corollary 1.1	8	Lemma 1.1	7
Corollary 1.2	8	Lemma 2.1	124
		Lemma 3.1	151
		Lemma 3.2	174

Definition 1.1	20				
Definition 2.1	72	Remark 1.1	16	Theorem 1.1	21
Definition 2.2	101	Remark 1.2	38	Theorem 2.1	53
Definition 2.3	103	Remark 1.3	39	Theorem 2.2	56
Definition 2.4	104	Remark 2.1	52	Theorem 2.3	61
Definition 3.1	141	Remark 2.2	58	Theorem 2.4	62
Definition 3.2	143	Remark 2.3	80	Theorem 2.5	83
Definition 3.3	149	Remark 2.4	83	Theorem 2.6	85
Definition 3.4	158	Remark 2.5	92	Theorem 2.7	92
Definition 3.5	160	Remark 2.6	111	Theorem 2.8	92
Definition 3.6	163	Remark 3.1	128	Theorem 2.9	95
Definition 3.7	163	Remark 3.2	165	Theorem 3.1	132
Definition 3.8	165	Remark 4.1	178	Theorem 3.2	160
Definition 3.9	168	Remark 4.2	182	Theorem 3.3	171
Definition 3.10	175	Remark 4.3	183	Theorem 3.4	173
Definition 4.1	177	Remark 4.4	197	Theorem 4.1	178
Definition 4.2	181	Remark 4.5	201	Theorem 4.2	178
Definition 4.3	182	Remark 4.6	226	Theorem 4.3	188
Definition 4.4	186	Remark 4.7	236	Theorem 4.4	192
Definition 4.5	226	Remark 4.8	242	Theorem 4.5	194
Definition 4.6	229			Theorem 4.6	219
Definition 4.7	230			Theorem 4.7	220
Definition 4.8	233			Theorem 4.8	223
Definition 4.9	239				
Definition 4.10	239				

Bibliography

Altman, E. and Shwartz, A. (1991a). Adaptive control of constrained Markov chains: criteria and policies, *Ann. Oper. Res.*, **28**, pp. 101–134.
Altman, E. and Shwartz, A. (1991b). Markov decision problems and state–action frequences, *SIAM J. Control and Optim.*, **29**, pp. 786–809.
Altman, E. and Shwartz, A. (1993). Time-sharing policies for controlled Markov chains, *Operations Research*, **41**, pp. 1116–1124.
Altman, E. (1999). *Constrained Markov Decision Processes* (Chapman and Hall/CRC, Boca Raton, FL, USA).
Altman, E., Avrachenkov, K.E. and Filar, J.A. (2002). An asymptotic simplex method and Markov decision processes, in Petrosjan, L.A and Zenkevich, N.A. (eds.), *Proc. of the 10th Intern. Symp. on Dynamic Games, Vol.I*, (St. Petersburg State University, Institute of Chemistry, St. Petersburg, Russia), pp. 45–55.
Arapostathis, A., Borkar, V.S., Fernandez-Gaucherand, E., *et al.* (1993). Discrete-time controlled Markov processes with average cost criterion: a survey, *SIAM J. Control and Optim.*, **31**, pp. 282–344.
Avrachenkov, K.E., Filar, J. and Haviv, M. (2002). Singular perturbations of Markov chains and decision processes, in Feinberg, E. and Shwartz, A. (eds.), *Handbook of Markov Decision Processes*, (Kluwer, Boston, USA), pp. 113–150.
Ball, K. (2004). An elementary introduction to monotone transportation, *Geometric Aspects of Functional Analysis, Lecture Notes in Math.*, Vol. 1850, pp. 41–52.
Bäuerle, N. and Rieder, U. (2011). *Markov Decision Processes with Applications to Finance* (Springer-Verlag, Berlin, Germany).
Bellman, R. (1957). *Dynamic Programming* (Princeton University Press, Princeton, NJ, USA).
Bertsekas, D. and Shreve, S. (1978). *Stochastic Optimal Control* (Academic Press, New York, USA).
Bertsekas, D. (1987). *Dynamic Programming: Deterministic and Stochastic Models* (Prentice-Hall, Englewood Cliffs, NJ, USA).

Bertsekas, D. (2001). *Dynamic Programming and Optimal Control, V.II* (Athena Scientific, Belmont, MA, USA).

Bertsekas, D. (2005). *Dynamic Programming and Optimal Control, V.I* (Athena Scientific, Belmont, MA, USA).

Blackwell, D. (1962). Discrete dynamic programming, *Ann. Math. Stat.*, **33**, pp. 719–726.

Blackwell, D. (1965). Discounted dynamic programming, *Ann. Math. Stat.*, **36**, pp. 226–235.

Boel, R. (1977). Martingales and dynamic programming, in *Markov Decision Theory, Proc. Adv. Sem., Netherlands, 1976*, (Math. Centre Tracts, No. 93, Math. Centr. Amsterdam, Netherlands), pp. 77–84.

Borkar, V.S. and Ghosh, M.K. (1995). Recent trends in Markov decision processes, *J. Indian Inst. Sci.*, **75**, pp. 5–24.

Carmon, Y. and Shwartz, A. (2009). Markov decision processes with exponentially representable discounting, *Oper. Res. Letters*, **37**, pp. 51–55.

Cavazos-Cadena, R. (1991). A counterexample on the optimality equation in Markov decision chains with the average cost criterion, *Systems and Control Letters*, **16**, pp. 387–392.

Cavazos-Cadena, R., Feinberg, E. and Montes-de-Oca, R. (2000). A note on the existence of optimal policies in total reward dynamic programs with compact action sets, *Math. Oper. Res.*, **25**, pp. 657–666.

Chen, R.W., Shepp, L.A. and Zame, A. (2004). A bold strategy is not always optimal in the presence of inflation, *J. Appl. Prob.*, **41**, pp. 587–592.

Dekker, R. (1987). Counter examples for compact action Markov decision chains with average reward criteria, *Commun. Statist. Stochastic Models*, **3**, pp. 357–368.

Denardo, E.V. and Miller, B.L. (1968). An optimality condition for discrete dynamic programming with no discounting, *Ann. Math. Stat.*, **39**, pp. 1220–1227.

Denardo, E.V. and Rothblum, U.G. (1979). Optimality for Markov decision chains, *Math. Oper. Res.*, **4**, pp. 144–152.

Derman, C. (1964). On sequential control processes, *Ann. Math. Stat.*, **35**, pp. 341–349.

Dokuchaev, N. (2007). *Mathematical Finance* (Routledge, London, UK).

Dubins, L.E. and Savage, L.J. (1965). *How to Gamble if You Must* (McGraw-Hill, New York, USA).

Dufour, F. and Piunovskiy, A.B. (2010). Multiobjective stopping problem for discrete-time Markov processes: convex analytic approach, *J. Appl. Probab.*, **47**, pp. 947–996.

Dufour, F. and Piunovskiy, A.B. (submitted). The expected total cost criterion for Markov Decision Processes under constraints, *J. Appl. Probab.*

Dynkin, E.B. and Yushkevich, A.A. (1979). *Controlled Markov Processes and their Applications* (Springer-Verlag, Berlin, Germany).

Fainberg, E.A. (1977). Finite controllable Markov chains, *Uspehi Mat. Nauk*, **32**, pp. 181–182, (in Russian).

Fainberg, E.A. (1980). An ε-optimal control of a finite Markov chain with an average reward criterion, *Theory Probab. Appl.*, **25**, pp. 70–81.

Feinberg, E.A. (1982). Controlled Markov processes with arbitrary numerical criteria, *Theory Probab, Appl.*, **27**, pp. 486–503.

Feinberg, E.A. (1987). Sufficient classes of strategies in discrete dynamic programming. I. Decomposition of randomized strategies and embedded models, *Theory Probab. Appl.*, **31**, pp. 658–668.

Feinberg, E.A. and Shwartz, A. (1994). Markov decision models with weighted discounted criteria, *Math. Oper. Res.*, **19**, pp. 152–168.

Feinberg, E.A. (1996). On measurability and representation of strategic measures in Markov decision processes, in Ferguson, T. (ed.), *Statistics, Probability and Game Theory: Papers in Honor of David Blackwell, IMS Lecture Notes Monographs Ser.*, **30**, pp. 29–43.

Feinberg, E.A. and Sonin, I.M. (1996). Notes on equivalent stationary policies in Markov decision processes with total rewards, *Math. Meth. Oper. Res.*, **44**, pp. 205–221.

Feinberg, E.A. (2002). Total reward criteria, in Feinberg, E. and Shwartz, A. (eds.), *Handbook of Markov Decision Processes*, (Kluwer, Boston, USA), pp. 173–207.

Feinberg, E.A. and Piunovskiy, A.B. (2002). Nonatomic total rewards Markov decision processes with multiple criteria, *J.Math. Anal. Appl.*, **273**, pp. 93–111.

Feinberg, E.A. and Piunovskiy, A.B. (2010). On strongly equivalent nonrandomized transition probabilities, *Theory Probab. Appl.*, **54**, pp. 300–307.

Fernandez-Gaucherand, E., Ghosh, M.K. and Marcus, S.I. (1994). Controlled Markov processes on the infinite planning horizon: weighted and overtaking cost criteria, *ZOR – Methods and Models of Oper. Res.*, **39**, pp. 131–155.

Fisher, L. and Ross, S.M. (1968). An example in denumerable decision processes, *Ann. Math. Statistics*, **39**, pp. 674–675.

Flynn, J. (1974). Averaging vs. discounting in dynamic programming: a counterexample, *The Annals of Statistics*, **2**, pp. 411–413.

Flynn, J. (1976). Conditions for the equivalence of optimality criteria in dynamic programming, *The Annals of Statistics*, **4**, pp.936–953.

Flynn, J. (1980). On optimality criteria for dynamic programs with long finite horizons, *J.Math. Anal. Appl.*, **76**, pp. 202–208.

Forsell, N., Wilkström, P., Garcia, F., et al. (2011). Management of the risk of wind damage in forestry: a graph-based Markov decision process approach, *Ann. Oper. Res.*, **190**, pp.57–74.

Frid, E.B. (1972). On optimal strategies in control problems with constraints, *Theory Probab. Appl.*, **17**, pp. 188–192.

Gairat, A. and Hordijk, A. (2000). Fluid approximation of a controlled multiclass tandem network, *Queueing Systems*, **35**, pp. 349-380.

Gelbaum, B.R. and Olmsted, J.M.H. (1964). *Counterexamples in Analysis* (Holden-Day, San Francisco, USA).

Goffman, C. and Pedrick, G. (1983). *First Course in Functional Analysis* (Chelsea, New York, USA).

Golubin, A.Y. (2003). A note on the convergence of policy iteration in Markov decision processes with compact action spaces, *Math. Oper. Res.*, **28**, pp. 194–200.

Haviv, M. (1996). On constrained Markov decision processes, *Oper. Res. Letters*, **19**, pp. 25–28.

Heath, D.C., Pruitt, W.E. and Sudderth, W.D. (1972). Subfair red-and-black with a limit, *Proc. of the AMS*, **35**, pp. 555–560.

Hernandez-Lerma, O. and Lasserre, J.B. (1996a). *Discrete-Time Markov Control Processes. Basic Optimality Criteria* (Springer-Verlag, New York, USA).

Hernandez-Lerma, O. and Lasserre, J.B. (1996b). Average optimality in Markov control processes via discounted-cost problems and linear programming, *SIAM J. Control and Optimization*, **34**, pp. 295–310.

Hernandez-Lerma, O. and Vega-Amaya, O. (1998). Infinite-horizon Markov control processes with undiscounted cost criteria: from average to overtaking optimality, *Applicationes Mathematicae*, **25**, pp. 153–178.

Hernandez-Lerma, O. and Lasserre, J.B. (1999). *Further Topics on Discrete-Time Markov Control Processes* (Springer-Verlag, New York, USA).

Hordijk, A. and Tijms, H.C. (1972). A counterexample in discounted dynamic programming, *J. Math. Anal. Appl.*, **39**, pp. 455–457.

Hordijk, A. and Puterman, M.L. (1987). On the convergence of policy iteration in finite state undiscounted Markov decision processes: the unichain case, *Math. Oper. Res.*, **12**, pp. 163–176.

Hordijk, A. and Yushkevich, A.A. (2002). Blackwell optimality, in Feinberg, E. and Shwartz, A. (eds.), *Handbook of Markov Decision Processes*, (Kluwer, Boston, USA), pp. 231–267.

Hu, Q. and Yue, W. (2008). *Markov Decision Processes with their Applications* (Springer Science, New York, USA).

Kallenberg, L.C.M. (2010). *Markov Decision Processes, Lecture Notes* (University of Leiden, The Netherlands).

Kemeny, J.G., Snell, J.L. and Knapp, A.W. (1976). *Denumerable Markov Chains* (Springer-Verlag, New York, USA).

Kertz, R.P. and Nachman, D.C. (1979). Persistently optimal plans for nonstationary dynamic programming: the topology of weak convergence case, *The Annals of Probability*, **1**, pp. 811–826.

Kilgour, D.M. (1975). The sequential truel, *Intern. J. Game Theory*, **4**, pp. 151–174.

Langford, E., Schwertman, N., and Owens M. (2001). Is the property of being positively correlated transitive? *The American Statistician*, **55**, pp. 322–325.

Liggett, T.M. and Lippman, S.A. (1969). Stochastic games with perfect information and time average payoff, *SIAM Review*, **11**, pp. 604–607.

Lippman, S.A. (1969). Criterion equivalence in discrete dynamic programming, *Oper. Res.*, **17**, pp. 920–923.

Loeb, P. and Sun, Y. (2006). Purification of measure-valued maps, *Illinois J. of Mathematics*, **50**, pp. 747–762.

Luque-Vasquez, F. and Hernandez-Lerma, O. (1995). A counterexample on the semicontinuity minima, *Proc. of the American Mathem. Society*, **123**, pp. 3175–3176.

Magaril-Il'yaev, G.G. and Tikhomirov, V.M. (2003). *Convex Analysis: Theory and Applications* (AMS, Providence, RI, USA).

Maitra, A. (1965). Dynamic programming for countable state systems, *Sankhya, Ser.A*, **27**, pp. 241–248.

Mine, H. and Osaki, S. (1970). *Markovian Decision Processes* (American Elsevier, New York, USA).

Nowak, A.S. and Vega-Amaya, O. (1999). A counterexample on overtaking optimality, *Math. Meth. Oper. Res.*, **49**, pp. 435–439.

Ornstein, D. (1969). On the existence of stationary optimal strategies, *Proc. of the American Mathem. Society*, **20**, pp. 563–569.

Pang, G. and Day, M. (2007). Fluid limits of optimally controlled queueing networks, *J. Appl. Math. Stoch. Anal.*, vol.2007, 1–20. [Online] Available at: doi:10.1155/2007/68958 [Accessed 26 April 2012].

Parrondo, J.M.R. and Dinis, L. (2004). Brownian motion and gambling: from ratchets to paradoxical games, *Contemporary Physics*, **45**, pp. 147–157.

Parthasarathy, K.R. (2005). *Probability Measures on Metric Spaces* (AMS Chelsea Publishing, Providence, RI, USA).

Piunovskiy, A.B. (1997). *Optimal Control of Random Sequences in Problems with Constraints* (Kluwer, Dordrecht, Netherlands).

Piunovskiy, A. and Mao, X. (2000). Constrained Markovian decision processes: the dynamic programming approach, *Oper. Res. Letters*, **27**, pp. 119–126.

Piunovskiy, A.B. (2006). Dynamic programming in constrained Markov decision processes, *Control and Cybernetics*, **35**, pp. 645–660.

Piunovskiy, A.B. (2009a). When Bellman's principle fails, *The Open Cybernetics and Systemics J.*, **3**, pp. 5–12.

Piunovskiy, A. (2009b). Random walk, birth-and-death process and their fluid approximations: absorbing case, *Math. Meth. Oper. Res.*, **70**, pp. 285–312.

Piunovskiy, A and Zhang, Y. (2011). Accuracy of fluid approximation to controlled birth-and-death processes: absorbing case, *Math. Meth. Oper. Res.*, **73**, pp. 159–187.

Priestley, H.A. (1990). *Introduction to Complex Analysis* (Oxford University Press, Oxford, UK).

Puterman, M.L. (1994). *Markov Decision Processes* (Wiley, New York, USA).

Robinson, D.R. (1976). Markov decision chains with unbounded costs and applications to the control of queues, *Adv. Appl. Prob.*, **8**, pp. 159–176.

Rockafellar, R.T. (1970). *Convex Analysis* (Princeton, NJ, USA).

Rockafellar, R.T. (1987). *Conjugate Duality and Optimization* (SIAM, Philadelphia, PA, USA).

Ross, S.M. (1968). Non-discounted denumerable Markovian decision models, *Ann. Math. Stat.*, **39**, pp. 412–423.

Ross, S.M. (1970). *Applied Probability Models with Optimization Applications* (Dover Publications, New York, USA).

Ross, S.M. (1971). On the nonexistence of ε-optimal randomized stationary policies in average cost Markov decision models, *Ann. Math. Stat.*, **42**, pp. 1767–1768.
Ross, S.M. (1983). *Introduction to Stochastic Dynamic Programming* (Academic Press, San Diego, CA, USA).
Schäl, M. (1975a). On dynamic programming: compactness of the space of policies, *Stoch. Processes and their Appl.*, **3**, pp. 345–364.
Schäl, M. (1975b). Conditions for optimality in dynamic programming and for the limit of n-stage optimal policies to be optimal, *Z. Wahrscheinlichkeitstheorie verw. Gebiete*, **32**, pp. 179–196.
Schmidli, H. (2008). *Stochastic Control in Insurance* (Springer-Verlag, London, UK).
Schweitzer, P.J. (1987). A Brouwer fixed-point mapping approach to communicating Markov decision processes. *J. Math. Anal. Appl.*, **123**, pp. 117–130.
Sennott, L. (1989). Average cost optimal stationary policies in infinite state Markov decision processes with unbounded costs, *Oper. Res.*, **37**, pp. 626–633.
Sennott, L. (1991). Constrained discounted Markov decision chains, *Prob. in the Engin. and Inform. Sciences*, **5**, pp.463–475.
Sennott, L. (2002). Average reward optimization theory for denumerable state spaces, in Feinberg, E. and Shwartz, A. (eds.), *Handbook of Markov Decision Processes*, (Kluwer, Boston, USA), pp. 153–172.
Seth, K. (1977). Optimal service policies, just after idle periods in two-server heterogeneous queuing systems, *Oper. Res.*, **25**, pp. 356–360.
Sniedovich, M. (1980). A variance-constrained reservoir control problem, *Water Resources Res.*, **16**, pp. 271–274.
Stoyanov, J.M. (1997). *Counterexamples in Probability* (Wiley, Chichester, UK).
Strauch, R.E. (1966). Negative dynamic programming, *Ann. Math. Stat.*, **37**, pp. 871–890.
Suhov, Y. and Kelbert, M. (2008). *Probability and Statistics by Example. V.II: Markov Chains* (Cambridge University Press, Cambridge, UK).
Szekely, G.J. (1986). *Paradoxes in Probability Theory and Mathematical Statistics* (Akademiai Kiado, Budapest, Hungary).
Wal, J. van der and Wessels, J. (1984). On the use of information in Markov decision processes, *Statistics and Decisions*, **2**, pp. 1–21.
Whittle, P. (1983). *Optimization over Time* (Wiley, Chichester, UK).
Yao, D.D. and Zheng, S. (1998). Markov decision programming for process control in batch production, *Prob. in the Engin. and Inform. Sci.*, **12**, pp. 351–371.

Index

σ-algebra
    analytical, 260
    Borel, 259
    universal, 263

Abelian theorem, 265
action space, 1
algorithm
    strategy iteration, 61, 204, 208
    value iteration, 63, 128

Bair null space, 258
base of topology, 257
Bellman function, 5, 7, 51, 128
Bellman principle, 5
blackmailer's dilemma, 87
bold strategy, 112

canonical equations, 178
canonical triplet, 177
completion of σ-algebra, 263
controller, 143
convex analytic approach, 101, 150

decision epoch, 1
discount factor, 127
disturbance, 143
dual functional, 154
Dual Linear Program, 108

expected frequencies, 220

feedback, 143
function
    exponentially representable, 160
    inf-compact, 48, 219
    lower semi-analytical, 260
    lower semi-continuous, 264
    piece-wise continuous, vi
    piece-wise continuously differentiable, vii
    piece-wise Lipschitz, vii
    upper semi-continuous, 264

gambling, 80, 112, 115

Hilbert cube, 258
histories, 3
homeomorphism, 257

initial distribution, 2, 4
isomorphism, 259

Lagrange function, 154
limit
    lower, 263
    upper, 263
loss
    final (terminal), 2
    one-step loss (or simply loss function), 2
    total expected loss, 2
    total realized loss, 2, 4
Lyapunov function, 83, 103

marginal (projection), 262
Markov Decision Process (MDP)
   constrained, 15, 152, 225
   singularly perturbed, 202
   stable, 64
   with average loss, 88, 171
   with discounted loss, 58, 63, 64, 127
   with expected total loss, 51
   with finite horizon, 3
martingale, 32
measure
   occupation, 101, 149
   outer, 263
   regular, 260
   strategic, 4, 51
measures set
   relatively compact, 262
   tight, 262
metric
   consistent, 257
mixture of strategies, 226
model
   absorbing, 53, 101, 127
   communicating, 186
   discrete, 53
   finite, 62
   fluid, 95
      refined, 98
   homogeneous, 51
   multichain, 208
   negative, 53, 61
   positive, 53
   recurrent, 217
   semi-continuous, 46, 85, 182
   transient, 104
   unichain, 85, 181
multifunction
   lower semi-continuous, 48

opportunity loss, 164
optimal stopping, 53, 71
   stable, 72
optimality (Bellman) equation, 5, 52, 127

performance functional, 2, 51, 127, 177
performance space, 16
polytope condition, 88
Primal Linear Program, 104
process
   controlled, 1
      λ-irreducible, 262
      geometric ergodic, 262
   estimating, 32

queueing model, 56

random variable, 261
   integrable, 261
   quasi-integrable, 261

search strategy, 119
secretary problem, 13
selector, 3
   canonical, 178
   conserving (thrifty), 52, 135
   equalizing, 52, 127
   Markov, 3
   (N, ∞)-stationary, 158
   semi-Markov, 3
   stationary, 3
Slater condition, 155, 227
space
   Borel, 259
   metric
      totally bounded, 259
   metrizable, 257
   separable, 257
span, 212
stable
   controller, 143
   system, 143
state, 1
   absorbing, 53
   cemetery, 53
state space, 1
   continuous, 1
   discrete, 1
stochastic basis, 4
stochastic kernel, 261

λ-irreducible, 262
geometric ergodic, 262
measurable, 261
(weakly) continuous, 261
strategy, 3
   AC-ε-optimal, 177
   AC-optimal, 177
   admissible, 15, 152, 225
   average-overtaking optimal, 233
   bias optimal, 230
   Blackwell optimal, 163
   D-optimal, 231
   ε-optimal, 7, 52
   equivalent, 17
   good, 175
   induced, 215
   Maitra optimal, 168
   Markov, 3
   mixed, 54
   myopic, 141
   n-discount optimal, 165
   nearly optimal, 163
   non-randomized, 3
   opportunity-cost optimal, 229
   optimal, 2, 5, 51, 52
      in the class Δ, 242
   overtaking optimal, 229
   persistently ε-optimal, 7
   semi-Markov, 3
   stationary, 3
   strong*-overtaking optimal, 242
   strong-average optimal, 239
   strong-overtaking optimal, 239
   strongly equivalent, 20, 267

time-sharing, 228
transient, 104
transient-optimal, 66
uniformly ε-optimal, 7, 52
uniformly optimal, 7, 52
weakly overtaking optimal, 232
sub-base of a topology, 258
subset
   analytical, 260
   analytically measurable, 260
   universally measurable, 263
sufficient statistic, 119
support of a measure, 260
system equation, 128, 143

Tauberian theorem, 265
time horizon, 3
   infinite, 5
topology
   discrete, 90, 258
   relative, 257
   weak, 261
   ws^∞, 109
trajectories, 3
transition probability, 1, 261
truel, 122
Tychonoff product, 258
Tychonoff theorem, 258

Urysohn theorem, 259

voting problems, 11

weight function, 83